Maritime Electrical

Level One

Trainee Guide

 Pearson

Boston   Columbus   Indianapolis   New York   San Francisco   Amsterdam
Cape Town Dubai London Madrid Milan Munich Paris Montreal Toronto Delhi
Mexico City Sao Paulo Sydney Hong Kong Seoul Singapore Taipei Tokyo

NCCER

President and Chief Executive Officer: Don Whyte
President: Boyd Worsham
Vice President: Steve Greene
Chief Operations Officer: Katrina Kersch
Electrical Project Manager: Chris Wilson
Senior Development Manager: Mark Thomas
Senior Production Manager: Tim Davis

Art Manager: Kelly Sadler
Managing Editor: Natalie Richoux
Desktop Publishing Manager: James McKay
Permissions Specialist: Adrienne Payne
Production Specialist: Gene Page
Editors: Graham Hack, Jordan Hutchinson, Steve Robb, Rebecca Bobb

Writing and development services provided by Topaz Publications, Liverpool, NY

Lead Writer/Project Manager: Veronica Westfall
Desktop Publisher: Joanne Hart
Art Director: Alison Richmond

Permissions Editors: Andrea LaBarge, Briana Morgan
Writer: Veronica Westfall

Pearson

Director of Alliance/Partnership Management: Kelly Trakalo
Program Manager: Alexandrina B. Wolf
Assistant Content Producer: Alma Dabral
Digital Content Producer: Jose Carchi
Senior Marketing Manager: Brian Hoehl

Composition: NCCER
Printer/Binder: LSC Communications
Cover Printer: LSC Communications
Text Fonts: Palatino and Univers

Credits and acknowledgments for content borrowed from other sources and reproduced, with permission, in this textbook appear at the end of each module.

ScoutAutomatedPrintCode

Perfect Bound   ISBN-13:   978-0-13-569834-1
                    ISBN-10:   0-13-569834-0

Preface

To the Trainee

Electricity powers the applications that make our daily lives more productive and efficient. The demand for electricity has led to vast job opportunities in the electrical field. Electricians constitute one of the largest construction occupations in the United States. The shipbuilding industry also requires electricians to install, maintain, and repair shipboard electrical systems.

Electricians install electrical systems in structures such as homes, office buildings, and vessels. These systems include wiring and other components, such as circuit breaker panels, switches, and light fixtures. They use specialized tools and testing equipment, such as ammeters, ohmmeters, and voltmeters.

We wish you success as you embark on your first year of training in the electrical craft, and hope that you will continue your training beyond this textbook.

New with *Maritime Electrical*

NCCER and Pearson are pleased to present *Maritime Electrical*. The modules in this four-level program were hand-picked by some of the shipbuilding industry's foremost subject matter experts from NCCER's *Electrical* curriculum for construction work. While this program does not cover marine-specific construction or standards, it does cover electrical competencies required of maritime electricians as well as an understanding of electrical safety, theory, tools, and equipment.

We wish you success as you progress through this training program. If you have any comments on how NCCER might improve upon this textbook, please complete the User Update form located at the back of each module and send it to us. We will always consider and respond to input from our customers.

We invite you to visit the NCCER website at **www.nccer.org** for information on the latest product releases and training, as well as online versions of the *Cornerstone* magazine and Pearson's NCCER product catalog.

Your feedback is welcome. You may email your comments to **curriculum@nccer.org** or send general comments and inquiries to **info@nccer.org**.

NMEC

The National Maritime Education Council (NMEC) is a multi-regional, industry-centered organization working to address critical workforce issues facing its member yards and the maritime industry. Established in 2012 to provide a centralized voice for the industry as it relates to workforce development, NMEC's primary mission is to lead the maritime industry in the development, promotion, and implementation of a national workforce development system that includes standardized craft training, portable credentials, and journey-level assessments.

Through its partnership with NCCER, NMEC is working in collaboration with its member yards to:

- Establish a system of career awareness and recruiting that will allow the industry to speak with one voice and deliver a consistent message to students, student influencers, the underemployed, women, and returning military.
- Establish and implement an industry-wide, comprehensive, standardized training and credentialing system for maritime craft professionals.
- Grow a pipeline of skilled craft professionals trained and credentialed through an industry-driven, uniform process.
- Improve industry outcomes by developing a better skilled, safer, more productive workforce.

NCCER Standardized Curricula

NCCER is a not-for-profit 501(c)(3) education foundation established in 1996 by the world's largest and most progressive construction companies and national construction associations. It was founded to address the severe workforce shortage facing the industry and to develop a standardized training process and curricula. Today, NCCER is supported by hundreds of leading construction and maintenance companies, manufacturers, and national associations. The NCCER Standardized Curricula was developed by NCCER in partnership with Pearson, the world's largest educational publisher.

Some features of the NCCER Standardized Curricula are as follows:

- An industry-proven record of success
- Curricula developed by the industry, for the industry
- National standardization providing portability of learned job skills and educational credits
- Compliance with the Office of Apprenticeship requirements for related classroom training (*CFR 29:29*)
- Well-illustrated, up-to-date, and practical information

NCCER also maintains the NCCER Registry, which provides transcripts, certificates, and wallet cards to individuals who have successfully completed a level of training within a craft in NCCER's Curricula. Training programs must be delivered by an NCCER Accredited Training Sponsor in order to receive these credentials.

Special Features

In an effort to provide a comprehensive and user-friendly training resource, this curriculum showcases several informative features. Whether you are a visual or hands-on learner, these features are intended to enhance your knowledge of the construction industry as you progress in your training. Some of the features you may find in the curriculum are explained below.

Introduction

This introductory page, found at the beginning of each module, lists the module Objectives, Performance Tasks and Trade Terms. The Objectives list the knowledge you will acquire after successfully completing the module. The Performance Tasks give you an opportunity to apply your knowledge to real-world tasks. The Trade Terms are industry specific vocabulary that you will learn as you study this module.

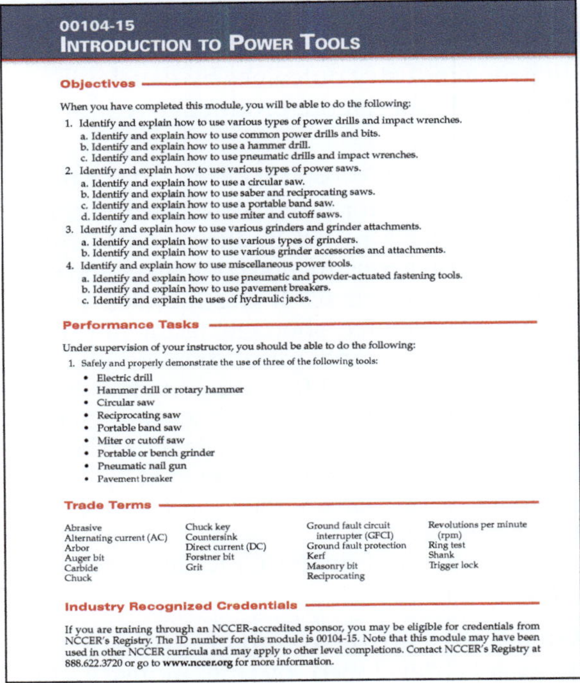

Trade Features

Trade features present technical tips and professional practices based on real-life scenarios similar to those you might encounter on the job site.

Bowline Trivia

Some people use this saying to help them remember how to tie a bowline: "The rabbit comes out of his hole, around a tree, and back into the hole."

Figures and Tables

Photographs, drawings, diagrams, and tables are used throughout each module to illustrate important concepts and provide clarity for complex instructions. Text references to figures and tables are emphasized with *italic* type.

Notes, Cautions, and Warnings

Safety features are set off from the main text in highlighted boxes and categorized according to the potential danger involved. Notes simply provide additional information. Cautions flag a hazardous issue that could cause damage to materials or equipment. Warnings stress a potentially dangerous situation that could result in injury or death to workers.

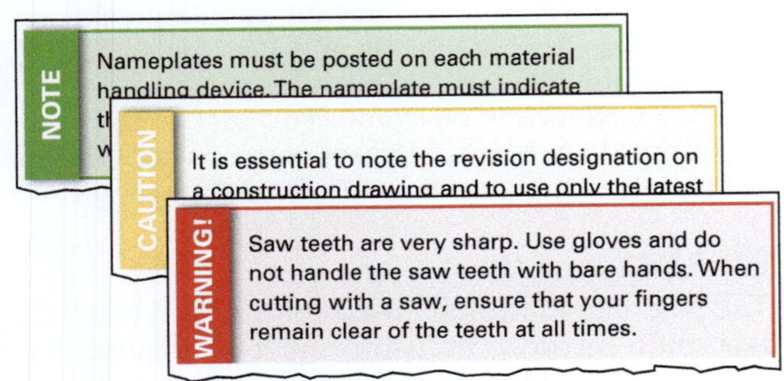

Case History

Case History features emphasize the importance of safety by citing examples of the costly (and often devastating) consequences of ignoring best practices or OSHA regulations.

Case History

Requesting an Outage

An electrical contractor requested an outage when asked to install two bolt-in, 240V breakers in panels in a data processing room. It was denied due to the 24/7 worldwide information processing hosted by the facility. The contractor agreed to proceed only if the client would sign a letter agreeing not to hold them responsible if an event occurred that damaged computers or resulted in loss of data. No member of upper management would accept liability for this possibility, and the outage was scheduled.

The Bottom Line: If you can communicate the liability associated with an electrical event, you can influence management's decision to work energized.

Going Green

Going Green features present steps being taken within the construction industry to protect the environment and save energy, emphasizing choices that can be made on the job to preserve the health of the planet.

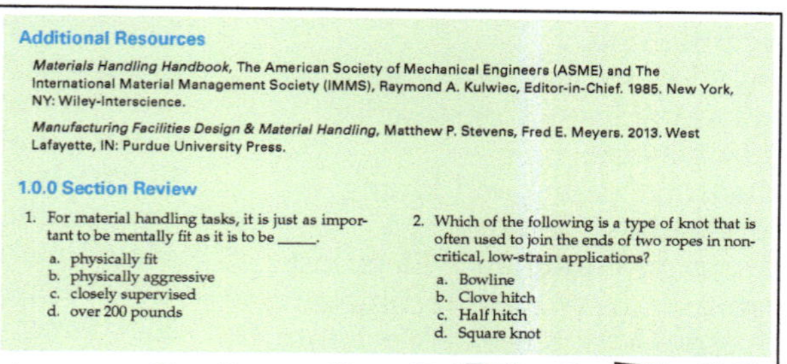

Did You Know

Did You Know features introduce historical tidbits or interesting and sometimes surprising facts about the trade.

Did You Know?

Safety First

Safety training is required for all activities. Never operate tools, machinery, or equipment without prior training. Always refer to the manufacturer's instructions.

Step-by-Step Instructions

Step-by-step instructions are used throughout to guide you through technical procedures and tasks from start to finish. These steps show you how to perform a task safely and efficiently.

Perform the following steps to erect this system area scaffold:

Step 1 Gather and inspect all scaffold equipment for the scaffold arrangement.

Step 2 Place appropriate mudsills in their approximate locations.

Step 3 Attach the screw jacks to the mudsills.

Trade Terms

Each module presents a list of Trade Terms that are discussed within the text and defined in the Glossary at the end of the module. These terms are presented in the text with **bold, blue** type upon their first occurrence. To make searches for key information easier, a comprehensive Glossary of Trade Terms from all modules is located at the back of this book.

During a rigging operation, the **load** being lifted or moved must be connected to the apparatus, such as a crane, that will provide the power for movement. The connector—the link between the load and the apparatus—is often a sling made of synthetic, chain, or **wire rope** materials. This section focuses on three types of slings:

Section Review

Each section of the module wraps up with a list of Additional Resources for further study and Section Review questions designed to test your knowledge of the Objectives for that section.

Additional Resources

Materials Handling Handbook, The American Society of Mechanical Engineers (ASME) and The International Material Management Society (IMMS), Raymond A. Kulwiec, Editor-in-Chief. 1985. New York, NY: Wiley-Interscience.

Manufacturing Facilities Design & Material Handling, Matthew P. Stevens, Fred E. Meyers. 2013. West Lafayette, IN: Purdue University Press.

1.0.0 Section Review

1. For material handling tasks, it is just as important to be mentally fit as it is to be _____.
   a. physically fit
   b. physically aggressive
   c. closely supervised
   d. over 200 pounds

2. Which of the following is a type of knot that is often used to join the ends of two ropes in non-critical, low-strain applications?
   a. Bowline
   b. Clove hitch
   c. Half hitch
   d. Square knot

Review Questions

The end-of-module Review Questions can be used to measure and reinforce your knowledge of the module's content.

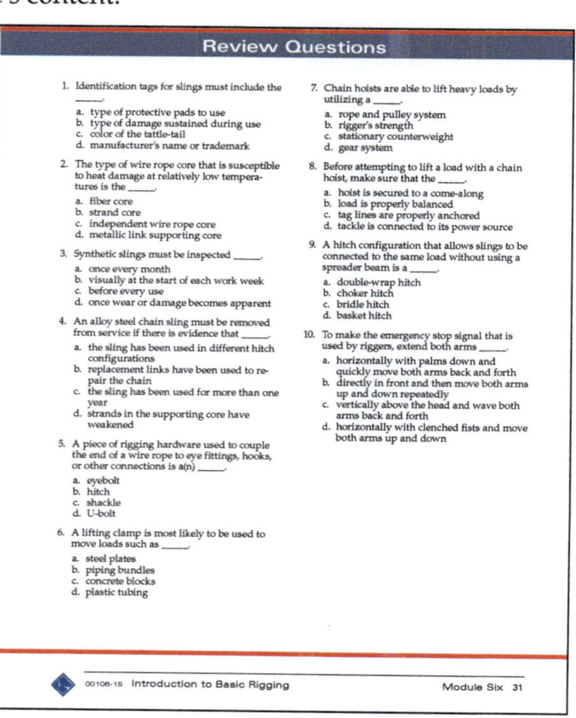

Review Questions

1. Identification tags for slings must include the _____.
   a. type of protective pads to use
   b. type of damage sustained during use
   c. color of the tattle-tail
   d. manufacturer's name or trademark

2. The type of wire rope core that is susceptible to heat damage at relatively low temperatures is the _____.
   a. fiber core
   b. strand core
   c. independent wire rope core
   d. metallic link supporting core

3. Synthetic slings must be inspected _____.
   a. once every month
   b. visually at the start of each work week
   c. before every use
   d. once wear or damage becomes apparent

4. An alloy steel chain sling must be removed from service if there is evidence that _____.
   a. the sling has been used in different hitch configurations
   b. replacement links have been used to repair the chain
   c. the sling has been used for more than one year
   d. strands in the supporting core have weakened

5. A piece of rigging hardware used to couple the end of a wire rope to eye fittings, hooks, or other connections is a(n) _____.
   a. eyebolt
   b. hitch
   c. shackle
   d. U-bolt

6. A lifting clamp is most likely to be used to move loads such as _____.
   a. steel plates
   b. piping bundles
   c. concrete blocks
   d. plastic tubing

7. Chain hoists are able to lift heavy loads by utilizing a _____.
   a. rope and pulley system
   b. rigger's strength
   c. stationary counterweight
   d. gear system

8. Before attempting to lift a load with a chain hoist, make sure that the _____.
   a. hoist is secured to a come-along
   b. load is properly balanced
   c. tag lines are properly anchored
   d. tackle is connected to its power source

9. A hitch configuration that allows slings to be connected to the same load without using a spreader beam is a _____.
   a. double-wrap hitch
   b. choker hitch
   c. bridle hitch
   d. basket hitch

10. To make the emergency stop signal that is used by riggers, extend both arms _____.
    a. horizontally with palms down and quickly move both arms back and forth
    b. directly in front and then move both arms up and down repeatedly
    c. vertically above the head and wave both arms back and forth
    d. horizontally with clenched fists and move both arms up and down

00106-15 Introduction to Basic Rigging   Module Six 31

NCCER Standardized Curricula

NCCER's training programs comprise more than 80 construction, maintenance, pipeline, and utility areas and include skills assessments, safety training, and management education.

Boilermaking
Cabinetmaking
Carpentry
Concrete Finishing
Construction Craft Laborer
Construction Technology
Core Curriculum: Introductory
 Craft Skills
Drywall
Electrical
Electronic Systems Technician
Heating, Ventilating, and Air
 Conditioning
Heavy Equipment Operations
Highway/Heavy Construction
Hydroblasting
Industrial Coating and Lining
 Application Specialist
Industrial Maintenance Electrical
 and Instrumentation Technician
Industrial Maintenance Mechanic
Instrumentation
Insulating
Ironworking
Masonry
Millwright
Mobile Crane Operations
Painting
Painting, Industrial
Pipefitting
Pipelayer
Plumbing
Reinforcing Ironwork
Rigging
Scaffolding
Sheet Metal
Signal Person
Site Layout
Sprinkler Fitting
Tower Crane Operator
Welding

Maritime

Maritime Industry Fundamentals
Maritime Pipefitting
Maritime Structural Fitter

Green/Sustainable Construction

Building Auditor
Fundamentals of Weatherization
Introduction to Weatherization
Sustainable Construction
 Supervisor
Weatherization Crew Chief
Weatherization Technician
Your Role in the Green
 Environment

Energy

Alternative Energy
Introduction to the Power Industry
Introduction to Solar Photovoltaics
Introduction to Wind Energy
Power Industry Fundamentals
Power Generation Maintenance
 Electrician
Power Generation I&C
 Maintenance Technician
Power Generation Maintenance
 Mechanic
Power Line Worker
Power Line Worker: Distribution
Power Line Worker: Substation
Power Line Worker: Transmission
Solar Photovoltaic Systems Installer
Wind Turbine Maintenance
 Technician

Pipeline

Control Center Operations, Liquid
Corrosion Control
Electrical and Instrumentation
Field Operations, Liquid
Field Operations, Gas
Maintenance
Mechanical

Safety

Field Safety
Safety Orientation
Safety Technology

Supplemental Titles

Applied Construction Math
Tools for Success

Management

Construction Workforce
 Development Professional
Fundamentals of Crew Leadership
Mentoring for Craft Professionals
Project Management
Project Supervision

Spanish Titles

Acabado de concreto: nivel uno
 (*Concrete Finishing Level One*)
Aislamiento: nivel uno
 (*Insulating Level One*)
Albañilería: nivel uno
 (*Masonry Level One*)
Andamios (*Scaffolding*)
Carpintería: Formas para
 carpintería, nivel tres
 (*Carpentry: Carpentry Forms, Level
 Three*)
Currículo básico: habilidades
 introductorias del oficio
 (*Core Curriculum: Introductory Craft
 Skills*)
Electricidad: nivel uno
 (*Electrical Level One*)
Herrería: nivel uno
 (*Ironworking Level One*)
Herrería de refuerzo: nivel uno
 (*Reinforcing Ironwork Level One*)
Instalación de rociadores: nivel uno
 (*Sprinkler Fitting Level One*)
Instalación de tuberías: nivel uno
 (*Pipefitting Level One*)
Instrumentación: nivel uno, nivel
 dos, nivel tres, nivel cuatro
 (*Instrumentation Levels One through
 Four*)
Orientación de seguridad
 (*Safety Orientation*)
Paneles de yeso: nivel uno
 (*Drywall Level One*)
Seguridad de campo
(*Field Safety*)

Acknowledgments

This curriculum was revised as a result of the farsightedness and leadership of the following sponsors:

ABC of Iowa
ABC Northern California Chapter
ABC Southern California Chapter
Austal USA
Beacon Electrical Contractors
Bollinger Shipyards
Cianbro Corporation
Faith Technologies, Inc.
Gaylor Electric, Inc.
Harbor Energy Solutions
Ingalls Shipbuilding
Industrial Management and Training Institute, Inc.

ISC Constructors, LLC
Lamphear Electric
Lee College
Madison Comprehensive High School
Pro Circuit, Inc.
Putnam Career and Technical Center
Specialized Services
Técnico Corporation
Tri-City Electrical Contractors
Vigor
Zenith Education Group

This curriculum would not exist were it not for the dedication and unselfish energy of those volunteers who served on the Authoring Team. A sincere thanks is extended to the following:

Chuck Ackland
Harold Black
Christopher Comstock
Tim Dean
Tim Ely
Ronnie Gulino
Justin Johnson
Robert Kolb
Dan Lamphear

L.J. LeBlanc
David Lewis
John Lupacchino
Scott Mitchell
Todd Moody
John Mueller
Mike Powers
Louis Rivera
Raymond Saldivar

Brandon Sampey
Greg Schuman
Joshua Simpson
Wayne Stratton
Eddie Thompson
Jay Taylor
Marcel Veronneau

NCCER Partners

American Council for Construction Education
American Fire Sprinkler Association
Associated Builders and Contractors, Inc.
Associated General Contractors of America
Association for Career and Technical Education
Association for Skilled and Technical Sciences
Construction Industry Institute
Construction Users Roundtable
Design Build Institute of America
GSSC – Gulf States Shipbuilders Consortium
ISN
Manufacturing Institute
Mason Contractors Association of America
Merit Contractors Association of Canada
NACE International
National Association of Women in Construction
National Insulation Association
National Technical Honor Society
National Utility Contractors Association
NAWIC Education Foundation
North American Crane Bureau
North American Technician Excellence
Pearson

Prov
SkillsUSA®
Steel Erectors Association of America
U.S. Army Corps of Engineers
University of Florida, M. E. Rinker Sr., School of Construction Management
Women Construction Owners & Executives, USA

Contents

Module Eight

Device Boxes

Covers the hardware and systems used by an electrician to mount and support boxes, receptacles, and other electrical components. Covers *NEC®* fill and pull requirements for device, pull, and junction boxes under 100 cubic inches. (Module ID 26106-17; 10 Hours)

Glossary

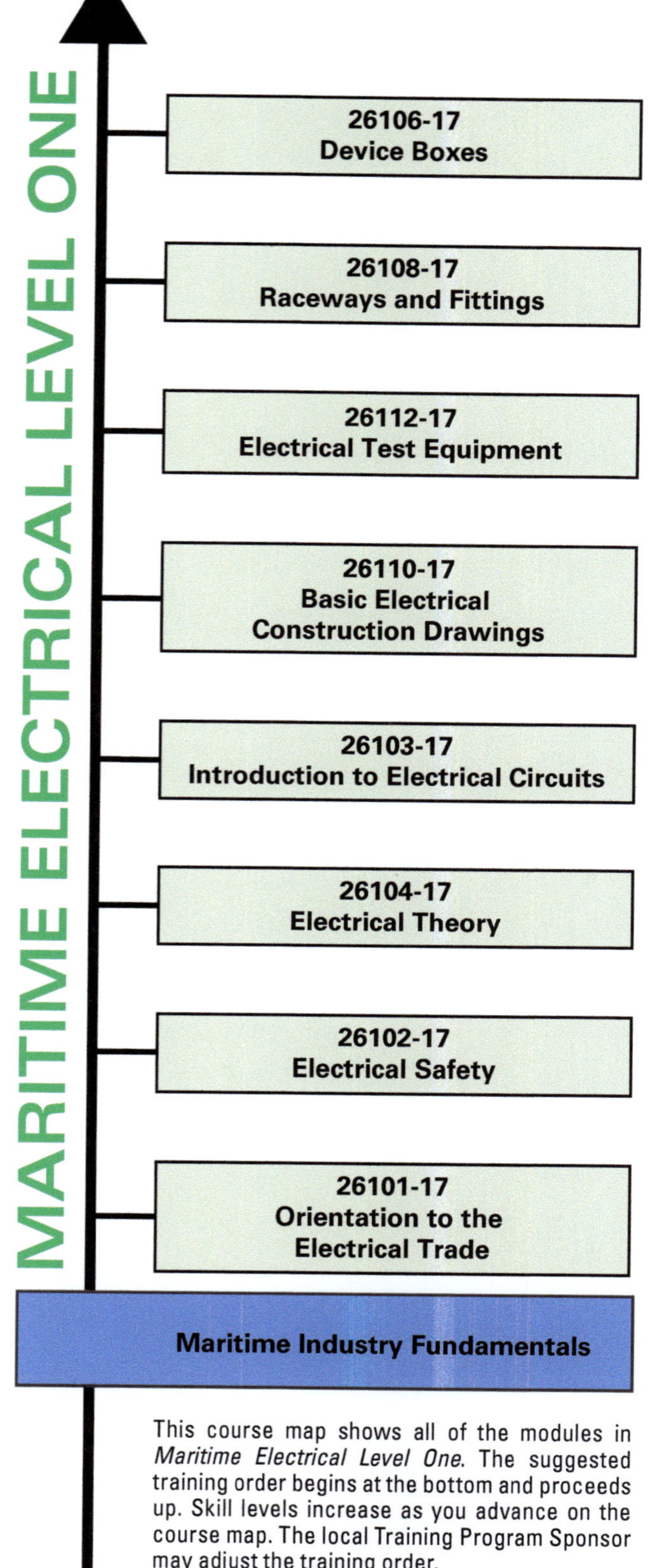

This course map shows all of the modules in *Maritime Electrical Level One*. The suggested training order begins at the bottom and proceeds up. Skill levels increase as you advance on the course map. The local Training Program Sponsor may adjust the training order.

Orientation to the Electrical Trade

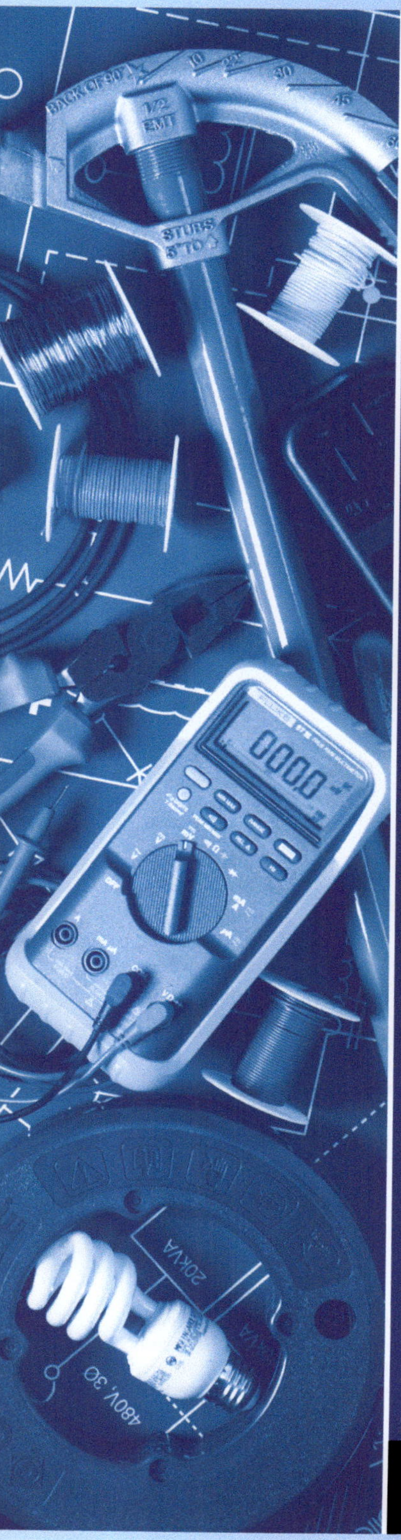

OVERVIEW

Skilled people in the electrical field are essential to maintain electrical systems and equipment in residential, commercial, and industrial settings. This module describes the various career paths in the electrical industry. It also covers the apprenticeship requirements for electricians and discusses employer/employee responsibilities.

Module 26101-17

Objectives

When you have completed this module, you will be able to do the following:

1. Identify the various sectors and trade options in the electrical industry.
   a. Describe the typical components in a residential wiring system.
   b. Describe the typical components in a commercial wiring system.
   c. Describe the typical components in an industrial wiring system.
   d. List various career paths and opportunities in the electrical trade.
2. Understand the apprenticeship/training process for electricians.
   a. List Department of Labor (DOL) requirements for apprenticeship.
   b. Describe various types of training in the electrical field.
3. Understand the responsibilities of the employee and employer.
   a. Identify employee responsibilities.
   b. Identify employer responsibilities.

Performance Tasks

This is a knowledge-based module; there are no performance tasks.

Trade Terms

Electrical service
Occupational Safety and Health Administration
  (OSHA)
On-the-job learning (OJL)

Raceway system
Rough-in
Trim-out

Industry Recognized Credentials

If you are training through an NCCER-accredited sponsor, you may be eligible for credentials from NCCER's Registry. The ID number for this module is 26101-17. Note that this module may have been used in other NCCER curricula and may apply to other level completions. Contact NCCER's Registry at 888.622.3720 or go to **www.nccer.org** for more information.

Note

NFPA 70®, *National Electrical Code*® and *NEC*® are registered trademarks of the National Fire Protection Association, Quincy, MA.

Contents

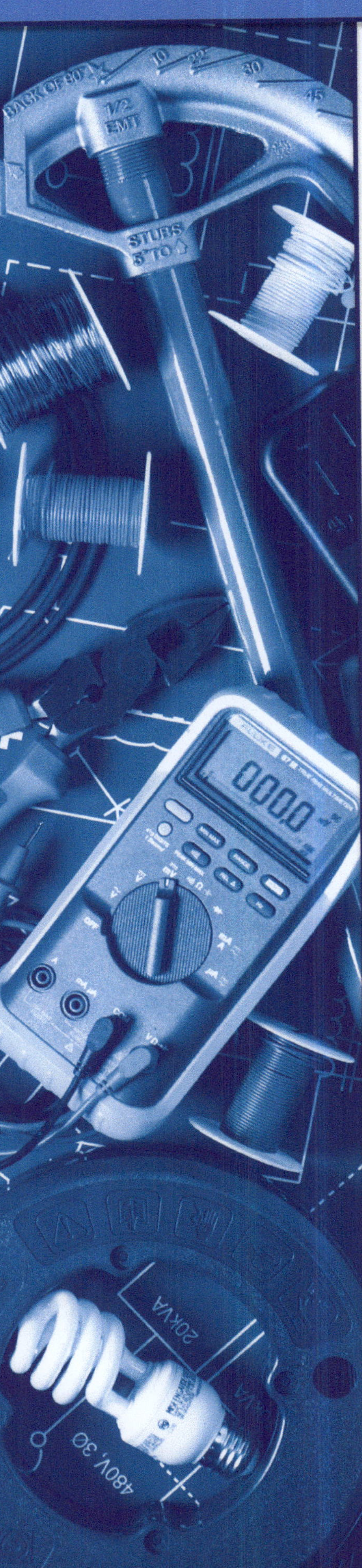

1.0.0 IDENTIFYING SECTORS AND TRADE OPTIONS IN THE ELECTRICAL INDUSTRY

Objective

Identify the various sectors and trade options in the electrical industry.

a. Describe the typical components in a residential wiring system.
b. Describe the typical components in a commercial wiring system.
c. Describe the typical components in an industrial wiring system.
d. List various career paths and opportunities in the electrical trade.

Trade Terms

Electrical service: The electrical components that are used to connect the serving utility to the premises wiring system.

Raceway system: An enclosure that houses the conductors in an electrical system (such as fittings, boxes, and conduit.)

Rough-in: The installation of the raceway system, wiring, or cable.

Trim-out: The installation and termination of devices and fixtures after rough-in.

We live in a world of electrical dependency, where most people take its availability for granted until we experience an unexpected power failure or power outage. It takes an army of electrically skilled individuals to generate, transmit, distribute, and maintain electrical systems and equipment in order to provide the convenience of continuous and quality electrical energy.

The electrical field can be divided into three broad categories: residential, commercial, and industrial. As you study to become an electrician, you will reach a point at which you must decide which area of electrical work you want to pursue. Many skilled electricians become comfortable in residential or commercial wiring, whereas others feel at home in large industrial facilities, such as petrochemical plants, installing or maintaining huge electrical systems including motors and control devices. Later on in your career, you may decide to start your own company or teach others the trade.

1.1.0 Residential Wiring Systems

Components of a residential electrical system include an electrical supply, electrical service, nonmetallic-sheathed cable, nail-on device boxes, panelboards, and fixtures. Phases of residential electrical wiring include rough-in, trim-out, testing, and troubleshooting. The following primary components of residential wiring systems are shown in *Figure 1*:

- Pad-mounted transformer
- Electrical service
- Nail-on device box
- Nonmetallic-sheathed cable
- Interior panel (subpanel)
- Luminaire (lighting fixture)

Interior panel enclosures, such as the one shown in *Figure 1 (E)*, are typically installed and partially terminated during the rough-in stage.

1.2.0 Commercial Wiring Systems

Electrical installations in commercial structures, such as office buildings and stores, contain many of the same elements of residential installations. One exception is that, in commercial and industrial electrical installations, conductors are typically installed in metal raceways, requiring the installing electricians to be skilled in conduit bending. A well-trained electrician can install a metal raceway system with little or no waste in conduit, while an inexperienced beginner will typically go through several pieces of conduit before acquiring the necessary bend.

Some of the elements that make up a commercial electrical system are shown in *Figure 2*. These include a pad-mounted commercial transformer, the commercial electrical service, conduit, a fire alarm system, and office and outdoor lighting.

1.3.0 Industrial Wiring Systems

Because of the hazardous materials that exist in many industrial facilities, the installation and maintenance of electrical systems in these volatile environments must follow rigid requirements governed by the *National Electrical Code®* (*NEC®*). For similar reasons, commercial and residential installations also have strict code requirements that electricians must obey.

(A) PAD-MOUNTED TRANSFORMER

(B) RESIDENTIAL ELECTRICAL SERVICE

(C) NAIL-ON DEVICE BOX

(D) NONMETALLIC-SHEATHED CABLE

(E) INTERIOR PANEL (SUBPANEL)

(F) LUMINAIRE (LIGHTING FIXTURE)

Figure 1 Primary components of residential wiring.

(A) PAD-MOUNTED COMMERCIAL TRANSFORMER

(B) COMMERCIAL ELECTRICAL SERVICE

(C) CONDUIT SYSTEM

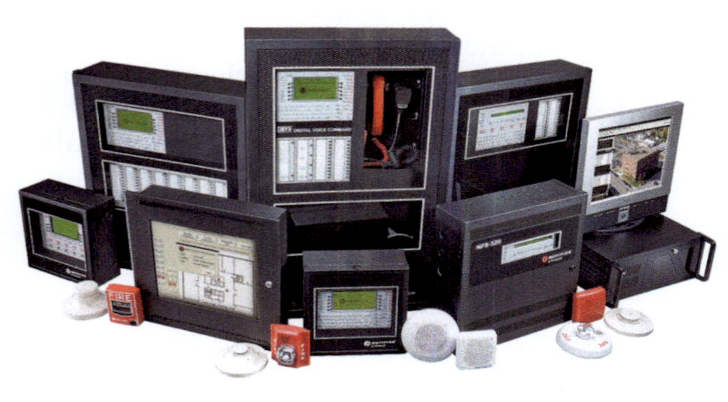

(D) FIRE ALARM SYSTEM

(E) OFFICE LIGHTING

(F) OUTDOOR LIGHTING

Figure 2 Commercial electrical system components.

Conduit systems in volatile environments must be sealed to outside vapors and gases, and any potential sparking or arcing device must be contained within a special enclosure or casing to prevent the ignition of hazardous vapors that might be present.

Industrial electricians are generally split into two groups: installers and maintenance personnel. In large industrial facilities, electrical systems are typically installed by contract electricians who do not work directly for the facility but rather for a contractor hired by the facility. It is the responsibility of these electricians to install conduit systems, conductors, motors, and equipment. The completed system is then turned over to maintenance electricians who work directly for the facility (on-site personnel). These electricians maintain the system once it is energized and operating. In smaller facilities, industrial plant electricians may both install and maintain the electrical equipment.

Figure 3 illustrates some of the electrical equipment that may be found in industrial facilities, including distribution switchgear, rigid metallic conduit (RMC), and a motor control center.

1.4.0 Career Paths and Opportunities in the Electrical Trade

Every electrical system in the United States and all over the world must be installed and maintained by someone qualified to do the work. This creates a great opportunity for work that is rewarding both personally and financially. Growth in the industry, new technology, upgrading and retrofitting of existing equipment, and retirement of current workers create openings for trained and skilled electricians. Examples of electrical occupations include residential electrician, commercial electrician, industrial electrician, and electrical maintenance technician.

1.4.1 Residential Electrician

The primary goal of a residential electrician is to provide a complete electrical system in a residential structure. Elements of a residential wiring installation include installing the electrical service entrance equipment, branch circuit conductors, device boxes, panel enclosures, overcurrent protective devices (circuit breakers), and lighting, smoke detectors, and other fixtures.

Residential electrical contractors are the primary employers of residential electricians. Various methods of employment are available, depending on the policies of the contractor. Residential electricians often work directly for

(A) DISTRIBUTION SWITCHGEAR

(B) RIGID METALLIC CONDUIT (RMC) SYSTEM

(C) MOTOR CONTROL CENTER

Figure 3 Industrial electrical equipment.

the contractor. Under other circumstances, electricians work as individual contractors who are responsible for filing their own taxes and providing their own insurance, as well as supplying their own equipment. The latter type of electricians are generally paid by piece work, which is a fixed price for each house completed. Tract homes (subdivisions), similar to those shown in *Figure 4*, are frequently wired by residential electricians.

1.4.2 Commercial Electrician

Commercial electricians install power, light, and control wiring in various locations including apartment buildings, stores, offices, service stations, and hospitals.

Commercial electricians are generally employed by electrical contractors who work as subcontractors for a general contractor. Many electrical contractors install both residential and commercial wiring, such as in metal frame commercial buildings (shown in *Figure 5*). However, they often rely on electricians who are specialists in one or the other type of wiring.

1.4.3 Industrial Electrician

Electricians who specialize in installing electrical systems in industrial facilities require additional training due to the amount of specialty equipment that must be installed and tested. Electricians working in hazardous locations must understand the special code requirements associated with these locations. These craft workers must differentiate between the hazardous classes and divisions and know the requirements for each class and division. Electricians must be familiar with three-phase power, motors, and motor control systems. Industrial electricians may also be responsible for installing the conduit and wiring for process control instrumentation, such as the installation shown in *Figure 6*. Electricians must be able to troubleshoot any of these systems should they fail during initial testing.

Large corporations and contractors that specialize in building industrial facilities typically employ industrial electricians.

1.4.4 Electrical Maintenance Technician

Electrical maintenance technicians work in commercial and industrial facilities. These technicians typically work directly for the owner or management of the facility. In large facilities usually are members of a maintenance group supervised by a maintenance manager or supervisor.

In industrial facilities, maintenance electricians are frequently responsible for both the electrical and instrumentation systems and equipment and are referred to as *E&I technicians*. Instrumentation expertise is a trade of its own, and it requires additional training over and above the electrical skills training found in this course.

Figure 5 Metal frame commercial building.

Figure 4 Tract homes.

Figure 6 Instrumentation installation.

Electrical maintenance electricians are usually employees of the facility; however, there are contract maintenance groups that provide maintenance personnel who work side-by-side with full-time plant personnel in maintaining electrical and instrumentation systems.

A common component in industrial facilities with which all electrical maintenance personnel must be familiar is the magnetic motor starter, illustrated in *Figure 7*. In indusustrial environments, magnetic motor starters frequently fail, so maintenance electricians must be able to disassemble and reassemble, troubleshoot, and repair these components.

Figure 7 Magnetic motor starter.

Lighting the New York Skyline

The first searchlight on top of the Empire State Building heralded the election of Franklin D. Roosevelt as President in 1932. A series of floodlights were installed in 1964 to illuminate the top 30 floors of the building. Today, the color of the lights is changed for various events. For example, the lights are yellow during the US Open and they are red, white, and blue on Independence Day.

The building is lit from the 72nd floor to the base of the TV antenna by 204 metal halide lamps and 310 fluorescent lamps. In 1984, a color-changing apparatus was added in the uppermost mooring mast. There are 880 vertical and 220 horizontal fluorescent lights. The colors can be changed with the flick of a switch.

Figure Credit: © iStockphoto.com/haveseen

Electricians Connect the New England Patriots to the 21st Century

It took a team of 200 electricians to build Gillette Stadium™ for the New England Patriots in 2000. The main scope of work was a $30 million contract for the stadium's power, lighting, voice and data systems, and the aboveground electrical infrastructure. In 16 months, electricians installed 4,000,000 feet of wire and cable, 900,000 feet of raceway, 310 panel boards, 11,000 interior light fixtures, and 728 sports lighting fixtures. The project also included a 2,000kV generator and five substations for emergency power systems.

Additional Resources

National Electrical Code® Handbook, Latest Edition. Quincy, MA: National Fire Protection Association.

1.0.0 Section Review

1. The termination of devices and fixtures is completed during a construction phase known as ______.

    a. rough-in
    b. trim-out
    c. final finish
    d. cleanup

2. Which of the following is true with regard to commercial electrical work?

    a. Wiring is rarely installed in raceways.
    b. Most wiring is installed in classified (hazardous) locations.
    c. Fire alarm systems are commonly required.
    d. It is exactly the same as residential work.

3. In industrial facilities, the electrical systems are most likely maintained by ______.

    a. on-site maintenance personnel
    b. external electrical contractors
    c. senior management
    d. the local utility

4. Employers of commercial electricians are generally electrical contractors who work as subcontractors for ______.

    a. larger electrical contractors
    b. standards organizations
    c. general contractors
    d. architects

2.0.0 TRAINING AND APPRENTICESHIP PROCESS FOR ELECTRICIANS

Objective

Understand the apprenticeship/training process for electricians.

a. List Department of Labor (DOL) requirements for apprenticeship.
b. Describe various types of training in the electrical field.

Trade Term

On-the-job learning (OJL): Job-related learning an apprentice acquires while working under the supervision of journey-level workers. Also called on-the-job training (OJT).

The demand for skilled electricians is high. New homes, schools, office buildings, malls, airports, industrial plants, and many other types of structures are being constructed every day. According to the US Bureau of Labor Statistics, this new construction creates a very great demand for new workers every year.

While you are in an **on-the-job learning (OJL)** program, you are being paid to learn the trade. This is a huge advantage compared to a typical college student who may be paying a large tuition bill to attend class. Another advantage of learning on the job is the hands-on experience you get, which is invaluable in any job.

Pay in the construction industry is very good, and the pay for electricians is close to the top of the scale for all construction occupations. There are many ways to increase your skills and grow professionally in construction. There are many opportunities to try different types of jobs and earn more money within the electrical trade. You will find that electrical work is demanding but fulfilling. There is a large variety of work to be done. You may be indoors installing boxes or hooking up motor controllers, or you may be outside climbing a ladder or running conduit on a pipe rack high in the air.

Electricity and electrical equipment are needed everywhere. This means jobs for electricians are also needed everywhere.

Apprentice training goes back thousands of years, and its basic principles have not changed. First, it is a means for a person entering the craft to learn from those who have mastered the craft. Second, it focuses on learning by doing—real skills versus theory. Some theory is presented in the classroom. However, it is always presented in a way that helps the trainee understand the purpose behind the skill that is to be learned.

2.1.0 DOL Apprenticeship Standards

The US Department of Labor (DOL) Office of Apprenticeship sets the minimum standards for training programs across the country. These programs rely on mandatory classroom instruction and OJL. They require at least 144 hours of classroom instruction per year and 2,000 hours of OJL per year. In a typical electrical apprenticeship program, trainees spend at least 576 hours in classroom instruction and 8,000 hours in OJL before receiving journeyman certificates issued by registered apprenticeship programs.

To address the training needs of the professional communities, NCCER developed a four-year electrical training program. NCCER uses the minimum DOL standards as a foundation for comprehensive curricula that provide trainees with in-depth classroom and OJL experience.

This NCCER curriculum provides trainees with industry-driven training and education. It adopts a purely competency-based teaching approach. This means that trainees must show the instructor that they possess the knowledge and skills needed to safely perform the hands-on tasks that are covered in each module.

When a certified instructor is satisfied that a trainee has the required knowledge and skills for a given module, that information is sent to NCCER and kept in the Registry system. NCCER's Registry system can then confirm training and skills for workers as they move from state to state, from company to company, or even within a company. See the *Appendix* for examples of the credentials issued by NCCER.

Whether you enroll in an NCCER program or another apprenticeship program, make sure you work for an employer or sponsor who supports a nationally standardized training program that includes credentials to confirm your skill development.

2.2.0 Types of Training

All apprenticeship standards prescribe certain work-related or on-the-job learning (OJL). This training may begin after graduation from high school or before graduation as a part of a youth apprenticeship program. After the training is completed, electricians may sit for one or more licensing exams.

2.2.1 On-the-Job Learning

The OJL is broken down into specific tasks in which the apprentice receives hands-on training. In addition, a specified number of hours is required in each task. The total number of OJL hours for an apprenticeship program is traditionally 8,000, which amounts to four years of training.

In a competency-based program, it may be possible to shorten this time by testing out of specific tasks through a series of performance exams. In a traditional program, the required OJL may be acquired in increments of 2,000 hours per year.

The apprentice must log all work time and turn it in to the apprenticeship committee so that accurate time control can be maintained. After each 1,000 hours of related work, the apprentice typically will receive a pay increase as prescribed by the apprenticeship standards.

For those entering an apprenticeship program, a high school or technical school education is desirable. Courses in shop, mechanical drawing, and general mathematics are helpful. Manual dexterity, good physical conditioning, and quick reflexes are important. The ability to solve problems quickly and accurately and to work closely with others is essential. You must also have high awareness of safety concerns.

The prospective apprentice must submit certain information to the apprenticeship committee. This may include the following:

- Aptitude test (General Aptitude Test Battery or GATB Form Test) results (usually administered by the local Employment Security Commission)
- Proof of educational background (candidate should have school transcripts sent to the committee)
- Letters of reference from past employers and friends
- Proof of age
- If the candidate is a veteran, a copy of Form DD214
- A record of technical training received that relates to the construction industry and/or a record of any pre-apprenticeship training

NOTE

Some companies have physical activity requirements that apprentices must meet. These requirements vary from company to company. Generally, apprentices must do the following:

- Wear proper safety and personal protective equipment (PPE) on the job
- Purchase and maintain tools of the trade as needed and required by the contractor
- Submit a monthly on-the-job training report to the committee
- Report to the committee if a change in employment status occurs
- Attend classroom-related instruction and adhere to all classroom regulations, such as attendance requirements

Note that informal OJL provided by employers is usually less thorough than OJL provided through a formal apprenticeship program. The degree of training and supervision in this type of program often depends on the size of the employer. A small contractor may provide training in only one area, whereas a large company may be able to provide training in several areas.

2.2.2 Youth Apprenticeship Program

Also available is the Youth Apprenticeship Program, which allows students to begin their apprentice training while still in high school. Students entering the program in eleventh grade can complete as much as two years of the NCCER Standardized Craft Training four-year program by high school graduation. In addition, the program, in cooperation with local craft employers, allows students to work in the trade and earn money while still in school. Upon graduation, students can enter the industry at a higher level and with more pay than someone just starting the apprenticeship program.

This training program is similar to the one used by NCCER learning centers, contractors, and colleges across the country. Students are recognized through official transcripts and can enter the next year of the program wherever it is offered. Students may also have the option of applying the credits at a two-year or four-year college that offers degree or certification programs in the construction trades.

2.2.3 Licensing

After completing your training, you will probably want to take your state or local licensing exam. The purpose of licensing is to provide assurance that you are qualified to install and/or maintain electrical systems. Licensing and will allow you to work independently and earn a higher income. As a licensed electrician, you are not only responsible for your work, but you are liable for that work as well. If someone is working for you, then you are also responsible and liable for that person's work.

Licensing requirements vary from state to state and may vary by municipality. Contact your local building department for the requirements in your area. After you receive your license, your state or locality may require continuing education in order to renew your license.

Think About It

Licensing

What are the licensing requirements in your area?

Additional Resources

29 *CFR* 1900–1910, *Standards for General Industry*, Occupational Safety and Health Administration US Department of Labor. **www.ecfr.gov**

29 *CFR* 1926, *Standards for the Construction Industry*, Occupational Safety and Health Administration US Department of Labor. **www.ecrf.gov**

2.0.0 Section Review

1. The NCCER electrical training program applies DOL standards and is a ______.

    a. one-year program
    b. two-year program
    c. three-year program
    d. four-year program

2. Entry into an apprenticeship program is likely to require a(n) ______.

    a. GATB Form Test
    b. license
    c. Scholastic Aptitude Test
    d. OSHA 40-hour course

3.0.0 EMPLOYEE AND EMPLOYER RESPONSIBILITIES

Objective

Understand the responsibilities of the employee and employer.
a. Identify employee responsibilities.
b. Identify employer responsibilities.

Trade Term

Occupational Safety and Health Administration (OSHA): The federal government agency established to ensure a safe and healthy environment in the workplace.

The safe and cost-effective installation and service of electrical systems requires close collaboration between the employer and the employees. All workers must understand the responsibilities of providing a safe and productive workplace.

3.1.0 Employee Responsibilities

To be successful, you must be able to use current trade materials, tools, and equipment to finish the task quickly and efficiently. You must keep up-to-date on technical advancements and continually gain the skills to use them. A professional never takes chances with regard to personal safety or the safety of others.

3.1.1 Professionalism

The term *professionalism* broadly describes the desired overall behavior and attitude expected in the workplace. Professionalism is too often absent from the construction site and various trades. Most persons would argue that professionalism must start at the top in order to be successful. It is true that management support of professionalism is important to its success in the workplace, but it is just as important that individuals recognize personal responsibilities for professionalism.

Professionalism includes honesty, productivity, safety, civility, cooperation, teamwork, clear and concise communication, being on time, and coming prepared to work. It can be demonstrated in a variety of ways every minute you are on the job.

Professionalism is a benefit to both the employer and the employee. It is a personal responsibility. The construction industry is what each individual chooses to make of it—choose professionalism and the industry image will follow.

3.1.2 Honesty

Honesty and personal integrity are important traits of successful professionals. Professionals pride themselves on performing a job well and on being punctual and dependable. Each job is completed in a professional way, never by cutting corners or reducing materials. A valued professional maintains work attitudes and ethics that protect tools, materials, and other property belonging to employers, customers, and other trades from damage or theft at the shop or job site.

Honesty and success go hand-in-hand for both the employer and the professional electrician. It is not simply a choice between good and bad but a choice between success and failure. Dishonesty will always catch up with you. Whether you steal materials, tools, or equipment from the job site or simply lie about your work, it will not take long for your employer to find out.

If you plan to be successful and enjoy continuous employment, consistency of earnings, and being sought after as opposed to seeking employment, then start out with the basic understanding of honesty in the workplace. You will reap the benefits.

Honesty means more, however, than simply not taking things that do not belong to you. It also means giving a fair day's work for a fair day's pay. Employers place a high value on employees who display honesty.

3.1.3 Loyalty

Employees expect employers to look out for their interests, to provide them with steady employment, and to promote them to better jobs as openings occur. Employers feel that they, too, have a right to expect loyalty from their employees— to keep their interests in mind, to speak well of them to others, to keep any minor troubles strictly within the plant or office, and to keep absolutely confidential all matters that pertain to the business. Both employers and employees should keep in mind that loyalty is not something to be demanded; rather, it is something to be earned.

Electricians Are Key Players for the NBA and NHL

Electricians are critical players in building modern sports complexes for professional sports franchises. The Pepsi Center was built in Denver, Colorado for the NBA's Denver Nuggets, the NHL's Colorado Avalanche, and several other teams. During the two-year construction cycle, electricians installed more than 120 miles of conduit, 569 miles of wire, 13,000 fixtures, and 280 panels. The lighting systems are computer controlled and can be preset for basketball games and concerts. In addition to high-quality sound for concerts, the state-of-the-art sound and security system includes CCTV monitoring and card access security controls.

Figure Credit: © iStockphoto.com/AndreyKrav

3.1.4 Willingness to Learn

Every company and job site has its own way of doing things. Employers expect their workers to be willing to learn these ways. You must be willing to adapt to change and learn new methods and procedures as quickly as possible. Sometimes, a change in safety regulations or the purchase of new equipment makes it necessary for even experienced employees to learn new methods and operations. Successful people take every opportunity to learn more about their trade.

3.1.5 Taking Responsibility

Most employers expect their employees to see what needs to be done and do it. After an assignment is received and the procedure and safety guidelines are fully understood, you should assume the responsibility for that task without further reminders.

3.1.6 Cooperation

To cooperate means to work together. In our modern business world, cooperation is the key to getting things done. Learn to work as a member of a team with your employer, supervisor, and fellow workers in a common effort to get the work done efficiently, safely, and on time.

3.1.7 Rules and Regulations

Employees can work well together only if there is some understanding about the nature of work to be done, when and how it will be done, and who will do it. Rules and regulations are a necessity in any work situation and must be followed by all employees.

3.1.8 Tardiness and Absenteeism

Tardiness means being late for work, and absenteeism means being off the job for one reason or another. While occasional absences are unavoidable, consistent tardiness and frequent absences are an indication of poor work habits, unprofessional conduct, and a lack of commitment.

Although workers may not be paid when they are absent or tardy, there is still a cost to the employer. For example, the worker's health care

Ethical Principles for Members of the Construction Trades

Honesty—Be honest and truthful in all dealings. Conduct business according to the highest professional standards. Faithfully fulfill all contracts and commitments. Do not deliberately mislead or deceive others.

Integrity—Demonstrate personal integrity and the courage of your convictions by doing what is right even if there is pressure to do otherwise. Do not sacrifice your principles because it seems easier.

Loyalty—Be worthy of trust. Demonstrate fidelity and loyalty to companies, employers and sponsors, co-workers, trade institutions, and other organizations.

Fairness—Be fair and just in all dealings. Do not take undue advantage of another's mistakes or difficulties. Fair people are open-minded and committed to justice, equal treatment of individuals, and tolerance for and acceptance of diversity.

Respect for others—Be courteous and treat all people with equal respect and dignity.

Obedience—Abide by laws, rules, and regulations relating to all personal and business activities.

Commitment to excellence—Pursue excellence in performing your duties, be well-informed and prepared, and constantly try to increase your proficiency by gaining new skills and knowledge.

Leadership—By your own conduct, seek to be a positive role model for others.

---

insurance must still be paid, even though the worker is not on site. In addition, jobs are bid and scheduled are based on a certain work force size. If you are not there, work is not being done and schedules are not being met. It is important for you to be at work, on time, every day. If you must be absent, call in as soon as possible so that your employer can find a replacement.

3.1.9 Safety

In exchange for the benefits of your employment and your own well-being, you are obligated to work safely. You are also obligated to make sure anyone you supervise or work with is working safely. Your employer is obligated to maintain a safe workplace for all employees. Safety is everyone's responsibility.

You have a responsibility to maintain a safe working environment. This means two things:

- Follow your company's rules for proper working procedures and practices.
- Report any unsafe equipment and conditions directly to your supervisor.

If you see something unsafe while on the job, report it! Do not ignore it. It will not correct itself. In the end, even if you do not think an unsafe condition affects you, it does. Always report unsafe conditions. Do not think your employer will be angry because your productivity suffers while the condition is being reported. On the contrary, your employer will be more likely to criticize you for not reporting a problem.

> **WARNING!**
>
> For the safety of yourself and others, always report unsafe conditions. Ignoring them could cause danger or harm to workers.

Your employer knows that the short time lost in making conditions safe again is nothing compared with shutting down the whole job because of a major disaster. If that happens, you are out of work anyway. In fact, Occupational Safety and Health Administration (OSHA) regulations require you to report hazardous conditions. This applies to every part of the construction industry. Whether you work for a large contractor or a small contractor, you are obligated to report unsafe conditions.

In addition to the OSHA standards, there are specific standards related to electrical systems and devices. The *National Electrical Code®* (*NEC®*) sets the minimum standards for the safe installation of electrical systems. For example, *NEC®* temporary power requirements are more stringent than OSHA standards and can be enforced by the inspector and OSHA. You will become very familiar with the *NEC®* as you progress through your training.

Another standard with which you should become familiar is *NFPA 70E®, Standard for Electrical Safety in the Workplace.* This standard covers safe work practices that must be applied when working on or near exposed energized parts. It describes in detail the steps required for putting a circuit or electrical system into an electrically safe working condition. The standard also covers safe approach distances to exposed energized parts, and it introduces the little understood but very dangerous reality of arc flash hazard. The PPE required to protect against electrical shock and arc flash hazards is also covered. *Figure 8* shows a higher level of PPE (arc flash hood) that will be required when exposed to a potentially high level of arc flash.

3.2.0 Employer Responsibilities

Just as the employee has responsibilities on the job, the employer also has responsibilities. These are set out in the *Occupational Safety and Health Act of 1970*. The job of OSHA is to set occupational safety and health standards for all places of employment, enforce these standards, ensure that employers provide and maintain a safe workplace for all employees, and provide research and educational programs to support safe working practices.

OSHA was adopted with the stated purpose to assure as best as possible every worker in the nation with safe and healthful working conditions and to preserve our human resources.

OSHA requires each employer to provide a safe and hazard-free working environment. OSHA also requires that employees comply with OSHA rules and regulations that relate to workers'

Figure 8 Arc flash hood.

conduct on the job. To gain compliance, OSHA can perform spot inspections of job sites, impose fines for violations, and even stop work from proceeding until the job site is safe.

According to OSHA standards, you are entitled to on-the-job safety training. Your employer must do the following:

- Show you how to do each job safely
- Provide you with the required personal protective equipment
- Warn you about specific hazards
- Supervise you for safety while performing the work

The enforcement for this act of Congress is provided by the federal and state safety inspectors, who have the legal authority to impose fines for safety violations. The law allows states to have their own safety regulations and agencies to enforce them, but the US Secretary of Labor must first approve the states' programs. In states that do not develop such regulations and agencies, federal OSHA standards are mandatory.

OSHA standards are listed in 29 *CFR 1926, OSHA Safety and Health Standards for the Construction Industry* (sometimes called *OSHA Standards 1926*). Other safety standards that apply to construction are published in *29CFR 1900 – 1910,* in *NFPA 70E*®, and in the *NEC*®.

The most important general requirements that OSHA places on employers in the construction industry are as follows:

- The employer must post, in an easily seen area, signs informing employees of their rights and responsibilities.
- The employer must ensure there are no serious hazards on the job site and make sure the workplace complies with OSHA rules and regulations.
- Warning signs, posters, and labels must be posted in all required areas.
- The employer must perform frequent and regular job site inspections of equipment.
- The employer must instruct all employees to recognize and avoid unsafe conditions and to know the regulations that pertain to the job so employees may control or eliminate any hazards.
- No one may use any tools, equipment, machines, or materials that do not comply with *29 CFR, Part 1926.*
- The employer must ensure that only qualified individuals operate tools, equipment, and machines.

- The employer must provide medical training and examinations when required by OSHA.
- Employers with more than than 10 employees must keep records of work-related injuries and illnesses. These records must be available to employees.
- Employers must not discriminate against employees who are exercising their rights under OSHA regulations.

Additional employer responsibilities are described in the Americans with Disabilities Act of 1990 (ADA). If a worker has a disability but is qualified to do the job, that worker has equal rights to employment. The US Equal Employment Opportunity Commission, along with state and local civil rights agencies, enforce ADA regulations.

Tips for a Positive Attitude

Here is a short checklist of things that you should keep in mind to help you develop and maintain a positive attitude:

- Remember that your attitude follows you wherever you go.
- Helpful suggestions and compliments are much more effective than negative ones.
- Look for the positive characteristics of your co-workers and supervisors. If it is necessary to stay home, phone the office early in the morning so that your supervisor can find another worker for the day.

GOING GREEN

Electricians Help Reduce Energy Consumption

Electricians are working to protect the environment and reduce energy consumption. Some companies specialize in energy efficiency in new installations and upgrading older facilities. One company helped the World Bank in Washington, DC to upgrade offices and reduce annual energy consumption by 4,500kW.

The World Bank offices occupy more than 3.5 million square feet in five separate buildings. Electricians upgraded more than 50,000 fluorescent light fixtures and 1,400 exit signs to improve energy efficiency and reduce lamp maintenance. This resulted in an annual energy savings exceeding $800,000 and eliminated the annual release of more than7 million pounds of oxides, which some persons believe causes global climate change and acid rain.

Additional Resources

29 *CFR* 1900–1910, *Standards for General Industry*, Occupational Safety and Health Administration US Department of Labor. **www.ecfr.gov**

29 *CFR* 1926, *Standards for the Construction Industry*, Occupational Safety and Health Administration US Department of Labor. **www.ecrf.gov**

3.0.0 Section Review

1. The minimum standards for the safe installation of electrical systems can be found in _____.

   a. *the National Electrical Code®(NEC®)*
   b. *29 CFR, Part 1910*
   c. *29 CFR, Part 1926*
   d. *NFPA 70E®*

2. OSHA standards for general industry are covered in _____.

   a. *29 CFR, Parts 1900–1910*
   b. *29 CFR, Part 1970*
   c. *29 CFR, Part 1926*
   d. *29 CFR, Parts 1965–1983*

SUMMARY

The electrical trade consists of three main areas of expertise: residential, commercial, and industrial. Each of these areas requires specific skills and training.

Regardless of the area of expertise, all electricians must understand the fundamentals of electricity, the hazards associated with electrical shock and arc flash, and the practices that must be applied in order to work safely.

Job opportunities in the electrical field can provide an above-average income and a rewarding yet challenging career with many potential levels of advancement.

1. Interior panels in residential wiring are typically installed during ______.

   a. rough-in
   b. trim-out
   c. service installation
   d. planning

2. A residential wiring system typically uses ______.

   a. medium-voltage cable
   b. tray cable
   c. flat conductor cable
   d. nonmetallic-sheathed cable

3. Commercial wiring is typically installed in ______.

   a. pairs
   b. air ducting
   c. metal raceways
   d. PVC raceways

4. Which of the following would most likely require special knowledge of hazardous locations?

   a. Residential wiring
   b. Industrial wiring
   c. Commercial wiring
   d. Service work

5. Tract homes are typically wired by ______.

   a. residential electricians
   b. commercial electricians
   c. homeowners
   d. utility workers

6. Minimum standards for apprentice training programs are established by ______.

   a. OSHA
   b. the DOL
   c. NCCER
   d. the employer

7. Which of the following is true with regard to your apprenticeship?

   a. Once you have finished your apprenticeship, your training is over.
   b. Licensing requirements vary from state to state.
   c. OSHA sets the minimum standards for training programs across the country.
   d. After completing your apprenticeship, you will automatically receive a license.

8. Which of the following is true with regard to tardiness and absenteeism?

   a. It is never acceptable to be absent, even when you are contagious.
   b. It is okay to be a little late for work as long as you make up the time.
   c. If you must be absent, call in early so that your employer can find a replacement.
   d. If you are not being paid, your absence does not cost the company anything.

9. If you see a safety violation at your job site, you should ______.

   a. ignore it unless it affects you directly
   b. make a mental note to avoid the area in future
   c. report it to your supervisor
   d. assume it is okay as long as no one has been injured

10. The primary mission of OSHA is to ______.

    a. inspect job sites for safety violations
    b. fine companies that violate safety regulations
    c. distribute safety equipment to workers
    d. ensure that employers maintain a safe workplace

Trade Terms Quiz

Fill in the blank with the correct term that you learned from your study of this module.

1. ___________ is the federal government agency established to ensure a safe and healthy environment in the workplace.

2. Job-related learning acquired while working is known as __________.

3. The main panelboard enclosure would probably be installed in the __________ stage.

4. The __________ connects the serving utility to the premises wiring system.

5. Devices and fixtures would be installed during __________.

6. An enclosure that houses conduit, boxes, and fittings in an electrical system is called a __________.

Trade Terms

Electrical service
Occupational Safety and Health Administration (OSHA)
On-the-job learning (OJL)

Raceway system
Rough-in
Trim-out

1. Phases of residential electrical wiring include

   ____

2. True or False? A major difference between residential and commercial wiring is that commercial wiring is usually installed in metal conduit.

3. Industrial electricians are usually split into two groups: __________ and __________.

4. Your apprenticeship program requires __________ hours of OJL per year.
   a. 500
   b. 1,000
   c. 2,000
   d. 4,000

5. True or False? Competency-based training means that you must demonstrate the skills necessary to perform hands-on tasks before advancing to the next stage of the curriculum.

6. A __________ allows you to complete as much as two years of your apprenticeship before you have finished high school.

7. True or False? It's okay to take scrap pieces home as long as you don't think they'll be needed on the job.

8. Many employees feel that sick days can be treated as floating holidays to be taken whenever they would like a day off. How is this unfair to an employer or co-workers?

   ____

9. True or False? OSHA requires that employers provide a safe and hazard-free job site.

10. In addition to the OSHA safety standards, name two other standards that relate to electrical systems and devices.

    ____

Tim Dean

Electrician/Electrical Trades Instructor
Central Ohio ABC/Madison Comprehensive High School

Provide a summary of how you got started in the construction industry.

Upon exiting The University of Toledo, I took a job with my brother-in-law, who worked as an electrician in Akron, Ohio. I knew little or nothing about electricity but needed a job to support myself and my wife.

Who inspired you to enter the industry? Why?

I suppose my brother-in-law, Tom Argenio, was my inspiration. He possessed a work ethic and craftsmanship that is very rare in our society today. He taught me not only how to be a good tradesman but also to understand the pride of quality workmanship.

What do you enjoy most about your job?

In the early days of my career, I simply appreciated having a job. But as time passed and my experience grew, I acquired a thirst for knowledge and understanding. Knowing the whys, whens, wheres, and hows brought new meaning to the skills I was attaining.

Do you think training and education are important in construction? If so, why?

The electrical trade is one of the most diverse and challenging of all the construction trades. Training and education are not only important but mandatory to stay safe, efficient, qualified, and prepared for the challenges of new technology.

How important are NCCER credentials to your career?

Being a part of NCCER has been one of the most rewarding experiences of my life. Acquiring credentials for performing as a Subject Matter Expert has benefited not only me personally by recognition and professional development but the organization I work for as well. I enjoy the challenges required to stay abreast of the continuous evolution of our industry, and NCCER provides the vehicle and resources for me to keep up with cutting-edge technological changes.

How has training/construction impacted your life and your career?

When I entered the trade, I was clueless. I had no idea what I was getting myself into, but as I continued I found a challenge in seeking an understanding of how and why things worked the way they did. I was engaged in the process and discovered that the more I knew and learned, the more I was worth to my employer.

Would you suggest construction as a career to others? If so, why?

I would recommend a career in the electrical trade to anyone who has a desire to learn and a thirst for understanding the new technologies that continue to evolve in our society. The electrical construction trade is both rewarding and challenging. Being a part of this great industry opens doors to new and exciting opportunities in many different areas. One only has to look around at the world we live in to know that being an electrician is much more than a job. Career opportunities abound, and there is truly no limit to what you can attain.

How do you define craftsmanship?

To me, craftsmanship is the result of acquiring knowledge, developing skills, infusing moral and ethical values, and blending personal pride to construct a product or process of recognizable and enduring quality.

Trade Terms Introduced in This Module

Electrical service: The electrical components that are used to connect the serving utility to the premises wiring system.

Occupational Safety and Health Administration (OSHA): The federal government agency established to ensure a safe and healthy environment in the workplace.

On-the-job learning (OJL): Job-related learning an apprentice acquires while working under the supervision of journey-level workers. Also called on-the-job training (OJT).

Raceway system: An enclosure that houses the conductors in an electrical system (such as fittings, boxes, and conduit).

Rough-in: The installation of the raceway system, wiring, or cable.

Trim-out: The installation and termination of devices and fixtures after rough-in.

SAMPLES OF NCCER TRAINING CREDENTIALS

13614 Progress Blvd • Alachua, Florida 32615 • p. 888.622.7320 f. 386.518.6255 • www.nccer.org

Official Training Transcript

Below are your credentials from NCCER's National Registry. These industry-recognized credentials give you flexibility in planning your career and ensure your achievements follow you wherever you go.

To access your training online via the Automated National Registry (ANR) web site, go to **https://anr.nccer.org**, click the **Individuals** button, and enter your NCCER card number and PIN *(Note: Your card number is below, and also on your NCCER wallet card. The default PIN is the last four digits of your SSN, you can change it after you log in.)* The first time you log in you'll need to answer a few security questions.

NCCER Card #: 01234567
Trainee Name: Sample Student
Sponsor: YouthBuild USA/YouthBuild
Address: 58 Day St
Somerville, MA 02144

Current Employer/School:

Module	Description	Instructor	Training Location	Completed
00101-04	Basic Safety	James Harris	YouthBuild USA/YouthBuild International Quad Area YouthBuild	2/26/2013
00102-04	Introduction to Construction Math	James Harris	YouthBuild USA/YouthBuild International Quad Area YouthBuild	2/26/2013
00103-04	Introduction to Hand Tools	James Harris	YouthBuild USA/YouthBuild International Quad Area YouthBuild	2/26/2013
00104-04	Introduction to Power Tools	James Harris	YouthBuild USA/YouthBuild International Quad Area YouthBuild	2/26/2013
00105-04	Introduction to Blueprints	James Harris	YouthBuild USA/YouthBuild International Quad Area YouthBuild	2/26/2013
00106-04	Basic Rigging	James Harris	YouthBuild USA/YouthBuild International Quad Area YouthBuild	2/26/2013
00107-04	Basic Communication Skills	James Harris	YouthBuild USA/YouthBuild International Quad Area YouthBuild	12/4/2012
00108-04	Basic Employability Skills	James Harris	YouthBuild USA/YouthBuild International Quad Area YouthBuild	2/26/2013

NO ENTRIES BELOW THIS LINE

nccer
Student 7 Samp
Certified Plus
4671784
nccer
Sample Student
2781481

Additional Resources

This module presents thorough resources for task training. The following reference material is recommended for further study.

29 *CFR* 1900–1910, *Standards for General Industry*, Occupational Safety and Health Administration US Department of Labor. **www.ecfr.gov**

29 *CFR* 1926, *Standards for the Construction Industry*, Occupational Safety and Health Administration US Department of Labor. **www.ecrf.gov**

National Electrical Code®*Handbook*, National Fire Protection Association. Latest Edition.

Figure Credits

Greenlee / A Textron Company, Module Opener
Tim Dean, Figure 1A
Hubbell Incorporated, Figures 1F, 2E and 2F
Al Hamilton, Figure 2B
Mike Powers, Figure 2C
Honeywell Fire Systems, Americas, Figure 2D

Section Review Answer Key

Answer	Section Reference	Objective
Section One		
1. b	1.1.0	1a
2. c	1.2.0	1b
3. a	1.3.0	1c
4. c	1.4.2	1d
Section Two		
1. d	2.1.0	2a
2. a	2.2.1	2b
Section Three		
1. a	3.1.9	3a
2. a	3.2.0	3b

NCCER CURRICULA — USER UPDATE

NCCER makes every effort to keep its textbooks up-to-date and free of technical errors. We appreciate your help in this process. If you find an error, a typographical mistake, or an inaccuracy in NCCER's curricula, please fill out this form (or a photocopy), or complete the online form at **www.nccer.org/olf**. Be sure to include the exact module ID number, page number, a detailed description, and your recommended correction. Your input will be brought to the attention of the Authoring Team. Thank you for your assistance.

Instructors – If you have an idea for improving this textbook, or have found that additional materials were necessary to teach this module effectively, please let us know so that we may present your suggestions to the Authoring Team.

NCCER Product Development and Revision

13614 Progress Blvd., Alachua, FL 32615

Email: curriculum@nccer.org
Online: www.nccer.org/olf

❏ Trainee Guide ❏ Lesson Plans ❏ Exam ❏ PowerPoints Other _______________________

Craft / Level: ___________________________________ Copyright Date: ______________

Module ID Number / Title: ____

Section Number(s): ____

Description: ____

____

____

____

Recommended Correction: ____

____

____

____

Your Name: ____

Address: ____

____

Email: _______________________________________ Phone: _________________________

Electrical Safety

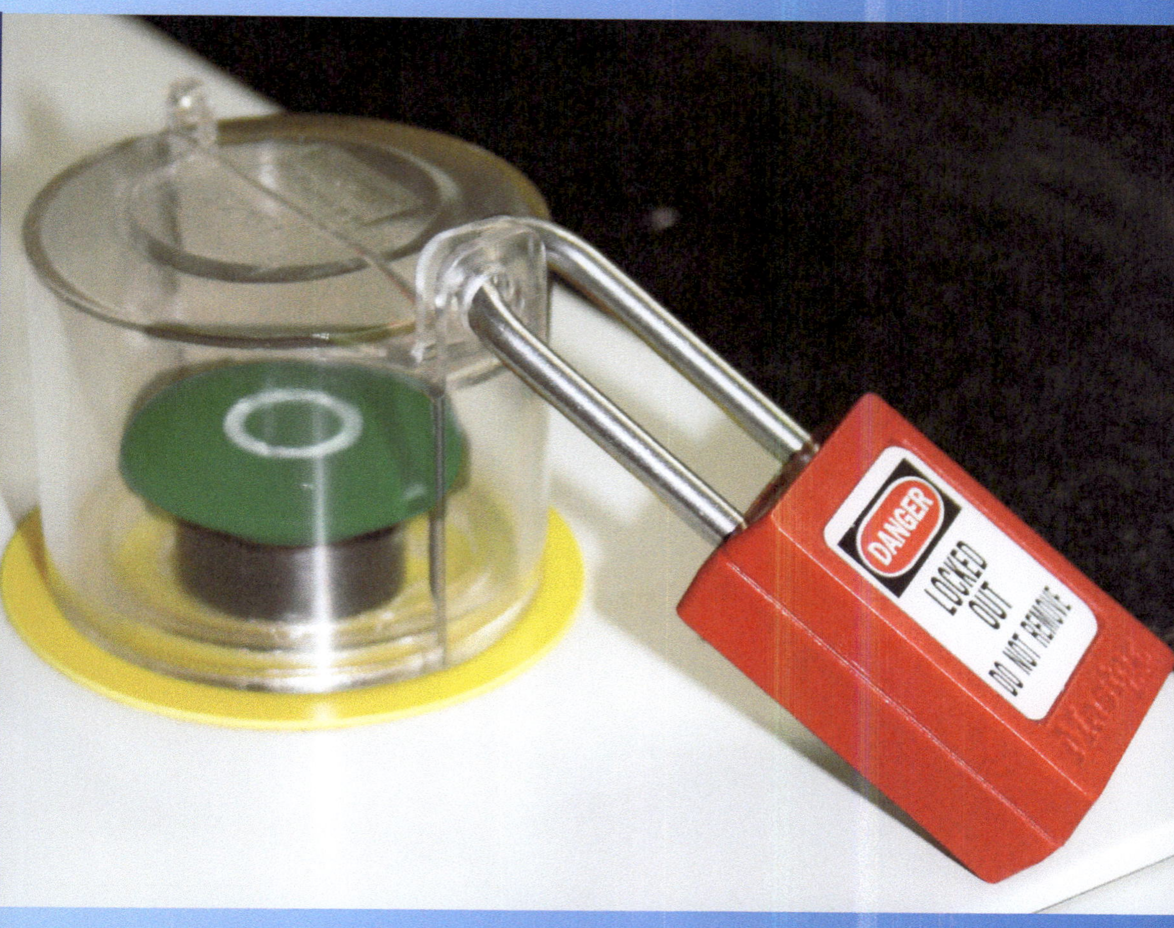

OVERVIEW

In order to be safe, electricians must be aware of potential hazards and stay constantly alert to them. This includes taking the proper precautions and practicing basic rules of safety. This module discusses electrical hazards and describes the various types of personal protective equipment (PPE) used to reduce injuries. It also covers the standards related to electrical safety and the OSHA lockout/tagout rule.

Module 26102-17

ELECTRICAL SAFETY

Objectives

When you have completed this module, you will be able to do the following:

1. Identify electrical hazards and their effects.
   a. Understand the effects of electrical shock on the human body.
   b. Verify that circuits are de-energized.
2. Use PPE to reduce the risk of injury.
   a. Identify OSHA requirements for protective equipment.
   b. Select and use protective equipment.
3. Identify the standards that relate to electrical safety.
   a. Apply OSHA requirements in the workplace.
   b. Understand the purpose of *NFPA 70E®*.
4. Recognize the safety requirements for various hazards.
   a. Identify the safety hazards associated with ladders, scaffolds, and lift equipment.
   b. Avoid back injuries by practicing proper lifting techniques.
   c. Demonstrate basic tool safety.
   d. Identify confined space entry procedures.
   e. Work safely with dangerous materials.
   f. Select and use appropriate fall protection.

Performance Tasks

Under the supervision of the instructor, you should be able to do the following:

1. Properly select and use PPE.
2. Describe the safety requirements for an instructor-supplied task, such as replacing the lights in your classroom.

   - Discuss the work to be performed and the hazards involved.
   - If a ladder is required, perform a visual inspection on the ladder and set it up properly.
   - Ensure that local emergency telephone numbers are either posted or known by you and your partner(s).
   - Plan an escape route from the location in the event of an accident.

Trade Terms

Double-insulated/ungrounded tools
Fibrillation
Grounded tool

Ground fault circuit interrupter (GFCI)
Polychlorinated biphenyls (PCBs)

Contents

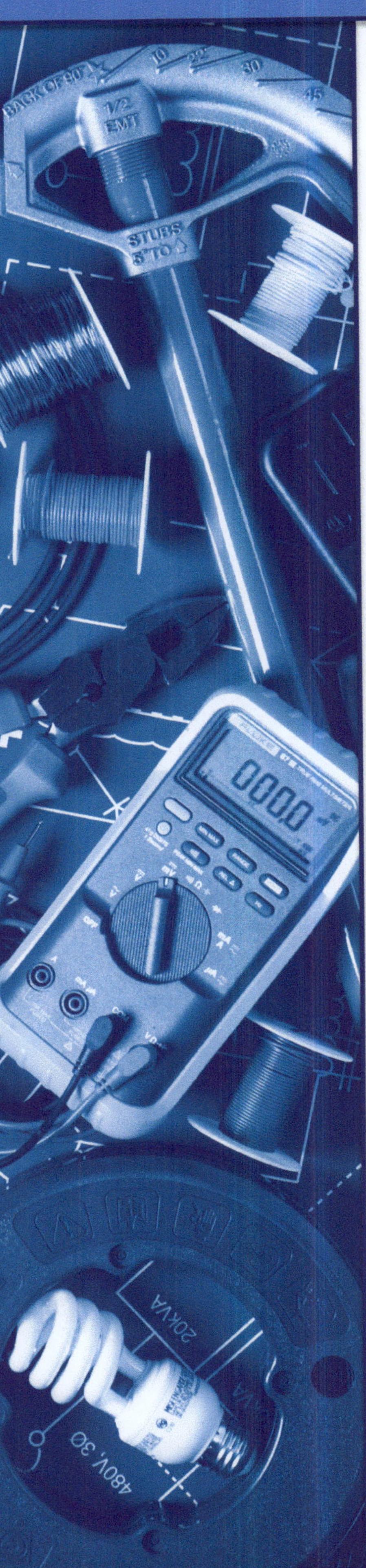

Figures and Tables

1.0.0 ELECTRICAL HAZARDS

Objective

Identify electrical hazards and their effects.
a. Understand the effects of electrical shock on the human body.
b. Verify that circuits are de-energized.

Trade Terms

Double-insulated/ungrounded tools: Electrical tools that are constructed so that the case is insulated from electrical energy. The case is made of a nonconductive material.

Fibrillation: Very rapid irregular contractions of the muscle fibers of the heart that result in the heartbeat and pulse going out of rhythm with each other.

Ground fault circuit interrupter (GFCI): A protective device that functions to de-energize a circuit or portion thereof within an established period of time when a current to ground exceeds some predetermined value. This value is less than that required to operate the overcurrent protective device of the supply circuit.

To be safe, you must understand potential hazards and stay alert to them. You must take the proper precautions and practice the basic rules of safety. You must be safety-conscious at all times and report any unsafe conditions to your supervisor and co-workers. Safety should become a habit. Keeping a safe attitude on the job will go a long way in reducing the number and severity of accidents. Remember that your safety is up to you.

As an apprentice electrician, you need to be especially careful. You should only work under the direction of experienced personnel who are familiar with the various job-site hazards and the means of avoiding them.

The most life-threatening hazards on a construction site are the following:

- Falls when you are working in high places
- The possibility of being crushed by falling materials or equipment
- Electric shock and arc-related burns caused by coming into contact with live electrical circuits

- The possibility of being struck by flying objects or moving equipment/vehicles such as trucks, forklifts, and other construction equipment

These exposures account for approximately 90 percent of fatalities on construction sites. Other hazards include cuts, burns, back sprains, and getting chemicals or objects in your eyes. Most injuries, from minor to deadly, are preventable if the proper precautions are taken.

1.1.0 Electrical Shock

Electricity can be described as a potential that results in the movement of electrons in a conductor. This movement of electrons is called electrical current. Some substances, such as silver, copper, steel, and aluminum, are excellent conductors. The human body is also a conductor. The conductivity of the human body greatly increases when the skin is wet or moistened with perspiration. Electrical current flows along any path in which the voltage can overcome the resistance. If the human body contacts an electrically energized point and is also in contact with the ground or another point in the circuit, the human body becomes a path for the current. *Table 1* shows the effects of current passing through the human body.

> **NOTE**
>
> The unit of measure represented by mA is known as a *milliampere* (or *milliamp*), and is equal to one one-thousandth of an ampere.

Table 1 Current Level Effects on the Human Body

Current Value	Typical Effects
1mA	Perception level. Slight tingling sensation.
5mA	Slight shock. Involuntary reactions can result in serious injuries such as falls from elevations.
6 to 30mA	Painful shock, loss of muscular control.
50 to 150mA	Extreme pain, respiratory arrest, severe muscular contractions. Death possible.
1000mA to 4300mA	Ventricular fibrillation, severe muscular contractions, nerve damage. Typically results in death.

Source: Occupational Safety and Health Administration

A primary cause of death from electrical shock is when the heart's rhythm is overcome by an electrical current. Normally, the heart's operation uses a very low-level electrical signal to cause the heart to contract and pump blood. When an abnormal electrical signal, such as current from an electrical shock, reaches the heart, the low-level heartbeat signals are overcome. The heart begins twitching in an irregular manner and goes out of rhythm with the pulse. This twitching is called **fibrillation**. The use of cardiopulmonary resuscitation (CPR) can keep oxygen flowing to the body, but unless the normal heartbeat is restored using an automated external defibrillator (AED), the individual will die. Other effects of electrical shock may include immediate heart stoppage and burns. In addition, the body's reaction to the shock can cause a fall or other accident. Delayed internal problems can also result. This is why it is critical for you to have a medical exam if you receive even a minor shock.

The amount of current measured in amperes that passes through a body determines the outcome of an electrical shock. The higher the voltage, the greater the chance for a fatal shock. In a one-year study in California, the following results were observed by the State Division of Industry Safety:

- Thirty percent of all electrical accidents were caused by contact with conductors. Of these accidents, 66 percent involved low-voltage conductors (those carrying 600 volts [V] or less).

- Portable, electrically operated hand tools made up the second largest number of injuries (15 percent). Almost 70 percent of these injuries happened when the frame or case of the tool became energized. These injuries could have been prevented by following proper safety practices, using properly maintained grounded or **double-insulated/ungrounded tools**, and using **ground fault circuit interrupter (GFCI)** protection.

In one ten-year study, investigators found 9,765 electrical injuries in the United States. A little more than 13 percent of the high-voltage injuries (over 600V) resulted in death. These high-voltage totals included limited-amperage contacts, which are often found on electronic equipment. When tools or equipment touch high-voltage overhead lines, the chance that a resulting injury will be fatal climbs to 28 percent. Of the low-voltage injuries, 1.4 percent were fatal.

1.1.1 Body Resistance

Electricity travels in closed circuits, and its normal route is through a conductor. Shock occurs when the body becomes part of the electric circuit (*Figure 1*). The current must enter the body at one point and leave at another. Shock normally occurs in one of three ways: the person must come in contact with both wires of the electric circuit; one wire of the electric circuit and the ground; or a metallic part that has become live by being in contact with an energized wire while the person is also in contact with the ground.

To understand the physical harm electrical shock does to the body, you need to understand something about the physiology of certain body parts: the skin, the heart, and muscles.

Think About It

Severity of Shock

In *Table 1*, how many milliamps separate a mild shock from a potentially fatal one? What is the fractional equivalent of this in amps? How many amps are drawn by a 60W light bulb?

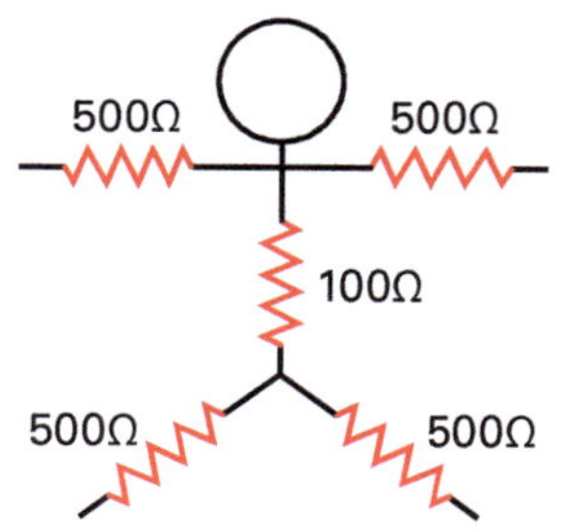

Figure 1 Body resistance.

Skin covers the body and is made up of three layers. The most important layer, as far as electric shock is concerned, is the outer layer of dead cells referred to as the horny layer. This layer is composed mostly of a protein called keratin, which provides the largest percentage of the body's electrical resistance. When it is dry, the outer layer of skin may have a resistance of several thousand

What's wrong with this picture?

Figure Credit: Mike Powers

Electrical Safety in the Workplace

Each year in the United States, approximately 20,000 electricity-related accidents occur at home and in the workplace. In a recent year, these accidents resulted in 700 deaths. Electrical accidents are the second leading cause of death in the workplace.

ohms. When it is moist, there is a radical drop in resistance, as is also the case if there is a cut or abrasion that pierces the horny layer. The amount of resistance provided by the skin will vary widely from person to person. A worker with a thick horny layer will have a much higher resistance than a child. The resistance will also vary widely at different parts of the body. For instance, the worker with high-resistance hands may have low-resistance skin on the back of his calf.

The heart is the pump that sends life-sustaining blood to all parts of the body. The blood flow is caused by the contractions of the heart muscle, which is controlled by electrical impulses. The electrical impulses are delivered by an intricate system of nerve tissue with built-in timing mechanisms, which make the chambers of the heart contract at exactly the right time. An outside electric current of as little as 75 mA can upset the rhythmic, coordinated beating of the heart by disturbing the nerve impulses. When this happens, the heart is said to be in fibrillation, and the pumping action stops. Death will occur quickly if the normal beat is not restored. Remarkable as it may seem, what is needed to defibrillate the heart is a shock of an even higher intensity.

The other muscles of the body are also controlled by electrical impulses delivered by nerves. Electric shock can cause loss of muscular control, resulting in the inability to let go of an electrical conductor. Electric shock can also cause injuries of an indirect nature in which involuntary muscle reaction from the electric shock can cause bruises, fractures, and even death resulting from collisions or falls.

The severity of shock received when a person becomes a part of an electric circuit is affected by three primary factors: the amount of current flowing through the body (measured in amperes), the path of the current through the body, and the length of time the body is in the circuit. Other factors that may affect the severity of the shock are the frequency of the current, the phase of the heart cycle when shock occurs, and the general health of the person prior to the shock. Effects can range from a barely perceptible tingle to immediate cardiac arrest. Although there are no absolute limits, or even known values that show the exact injury at any given amperage range, *Table 1* lists the general effects of electric current on the body for different current levels. As this table illustrates, a difference of only 100 mA exists between a current that is barely perceptible and one that is likely to kill you.

A severe shock can cause considerably more damage to the body than is visible. For example, a person may suffer internal hemorrhages and destruction of tissues, nerves, and muscle. In addition, shock is often only the beginning in a chain of events. The final injury may well be from a fall, cuts, burns, or broken bones.

1.1.2 Burns

The most common injury associated with electrical shock is a burn. Burns suffered in electrical accidents may be of three types: electrical burns, arc burns, and thermal contact burns.

Electrical burns are the result of electric current flowing through the tissues or bones. Tissue damage is caused by the heat generated by the current flow through the body. An electrical burn is one of the most serious injuries you can receive, and should be given immediate attention. Since the most severe burning is likely to be internal, a small surface wound could actually be an indication of severe internal burns.

Arc burns make up a substantial portion of the injuries from electrical malfunctions. The electric arc between metals can be up to 35,000°F, which is about four times hotter than the surface of the sun. Workers several feet from the source of the arc can receive severe or fatal burns. Since most electrical safety guidelines recommend safe working distances based on shock considerations, workers can be following these guidelines and still be at risk from arc. Electric arcs can occur due to poor electrical contact or failed insulation. Electrical arcing is caused by the passage of substantial amounts of current through the vaporized terminal material (usually metal or carbon).

> **WARNING!**
>
> Since the heat of the arc is dependent on the short circuit current available at the arcing point and the time it takes to clear (trip), arcs generated by low-voltage systems can be just as dangerous (or even more dangerous) than those at higher voltages.

The third type of burn is a thermal contact burn. It is caused by contact with heated objects thrown during the blast associated with an electric arc. This blast comes from the pressure developed by the near-instantaneous heating of the air surrounding the arc, and from the expansion of the metal as it is vaporized. (Copper expands by a factor in excess of 65,000 times in boiling.) The pressure wave can be great enough to hurl people, switchgear, and cabinets considerable distances. It can stop your heart, impale you with shrapnel, blow off limbs, cause deafness, and cause you to inhale vaporized metal. Another hazard associated with the blast is the hurling of molten metal droplets, which can also cause thermal contact burns and associated damage.

Many things can be done to reduce the chance of receiving an electrical shock. Always comply with your company's safety policy and all applicable rules and regulations, including jobsite rules. In addition, the Occupational Safety and Health Administration (OSHA) publishes the *Code of Federal Regulations (CFR)*. *CFR Part 1910* covers the OSHA standards for general industry and *CFR Part 1926* covers the OSHA standards for the construction industry.

Do not approach any electrical conductors closer than indicated in *Table 2* unless they are de-energized and your company has designated you as a qualified individual for that task. The values given in the table are minimum safe clearance distances; your company may have more restrictive requirements. *Table 2* is based on information provided in *NFPA 70E®, Standard for Electrical Safety in the Workplace, Table 130.4(D)(a)* for AC and *Table 130.4(D)(b)* for DC.

Table 2 Limited Approach Boundaries to Live Parts

Nominal System Voltage Range (Phase-to-Phase)	Limited Approach Boundary from Fixed Circuit Component
50 to 300	3 ft 6 in
301 to 750	3 ft 6 in
751 to 15kV	5 ft 0 in
15.1kV to 36kV	6 ft 0 in
36.1kV to 46kV	8 ft 0 in
46.1kV to 72.5kV	8 ft 0 in
72.6kV to 121kV	8 ft 0 in
138kV to 145kV	10 ft 0 in
161kV to 169kV	11 ft 8 in
230kV to 242kV	13 ft 0 in
345kV to 362kV	15 ft 4 in
500kV to 550kV	19 ft 0 in
765kV to 800kV	23 ft 9 in

1.1.3 Electrical Safety Precautions

There are several precautions you can take to help make your job safer. See the following, for example:

- Always remove all jewelry (e.g., rings, watches, exposed body piercings, bracelets, and necklaces) before working on electrical equipment. Most jewelry is made of conductive material and wearing it can result in a shock and other injuries if the jewelry is caught in moving components.
- When working on energized equipment, it is safer to work in pairs. In doing so, if one of the workers experiences a harmful electrical shock, the other worker can release the victim and call for help.
- Plan each job before you begin it. Make sure you understand exactly what it is you are going to do. If you are not sure, ask your supervisor.

> **WARNING!**
>
> The first time you perform a specific task, such as installing a bolt-in breaker, it cannot be done energized. You must demonstrate how to safely perform a task on a de-energized system first before attempting energized work.

- You will need to look over the appropriate prints and drawings to locate isolation devices and potential hazards. Never defeat safety interlocks. Remember to plan your escape route before starting work. Know where the nearest phone is and the emergency number to dial for assistance.
- If you realize that the work will go beyond the scope of what was planned, stop and get instructions from your supervisor before continuing. Do not attempt to plan as you go.

- It is critical that you stay alert. Workplaces are dynamic, and situations relative to safety are always changing. If you leave the work area to pick up material, take a break, or have lunch, reevaluate your surroundings when you return. Always remember to plan ahead. Complete a checklist or pre-task safety plan and/or permit before beginning work.

1.2.0 Verifying That Circuits Are De-Energized

You should always assume that all the circuits are energized until you have verified that the circuit is de-energized. This is called a live-dead-live test. Follow these steps to verify that a circuit is de-energized:

Step 1 Ensure that the circuit is properly tagged and locked out (*CFR 1910.333/1926.417*).

Step 2 Verify the test instrument operation on a known source using the appropriately rated tester.

Step 3 Using the test instrument, check the circuit to be de-energized. The voltage should be zero.

Step 4 Verify the test instrument operation, once again on a known power source.

Think About It

Bodily Harm

What factors affect the amount of damage to the body during an electric shock?

Additional Resources

National Electrical Code® Handbook, Latest Edition. Quincy, MA: National Fire Protection Association.

1.0.0 Section Review

1. Which of the following is true regarding electrical shock?

    a. Most electrical accidents involve contact with high-voltage conductors (over 600V).
    b. The majority of electrical accidents involving electrically operated tools happen when the frame or case of the tool becomes energized.
    c. Construction workers are no more likely to be electrocuted than workers in other industries.
    d. CPR can be used to correct heart fibrillation following a shock.

2. A live-dead-live test is used to verify ______.

    a. that a circuit is operating properly
    b. the operation of a test meter
    c. that a circuit is de-energized
    d. the voltage of a known power source

2.0.0 USING PPE TO REDUCE THE RISK OF INJURY

Objective

Use PPE to reduce the risk of injury.
a. Identify OSHA requirements for protective equipment.
b. Select and use protective equipment.

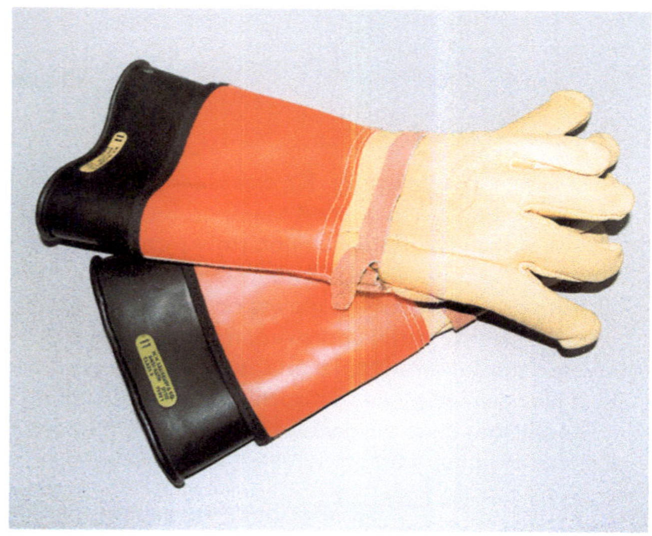

You must also become familiar with common personal protective equipment. In particular, know the voltage rating of each piece of equipment. Rubber gloves are used to prevent the skin from coming into contact with energized circuits. A separate leather cover protects the rubber glove from punctures and other damage (*Figure 2*). The leather protectors also provide the wearer with a certain level of protection against arc flash. Glove dust can be used to absorb moisture and prevent the gloves from sticking.

2.1.0 OSHA Requirements for Protective Equipment

OSHA addresses the safeguards for personal protection in *CFR 1910.335*. The design requirements for specific types of electrical protective equipment are covered in *CFR 1910.137*.

CFR 1910.137(a)137 presents the design requirements for electrical protective equipment, while *CFR 1910.335* presents safeguards for the use of this equipment. Some of these requirements include the following:

- Employees working in areas where there are potential electrical hazards shall be provided with, and shall use, electrical protective equipment that is appropriate for the specific parts of the body to be protected and for the work to be performed.
- Protective equipment shall be maintained in a safe, reliable condition and shall be periodically inspected or tested, as required by *CFR 1910.137*.
- If the insulating capability of protective equipment may be subject to damage during use, the insulating material shall be protected.

Figure 2 Rubber gloves, leather protectors, and glove dust.

- Employees shall wear nonconductive head protection wherever there is a danger of head injury from electric shock or burns due to contact with exposed energized parts.
- Employees shall wear protective equipment for the eyes and face wherever there is danger of injury to the eyes or face from electric arcs or flashes or from flying objects resulting from an electrical explosion.
- When working near exposed energized conductors or circuit parts, each employee shall use insulated tools or handling equipment if the tools or handling equipment might make contact with such conductors or parts. If the insulating capability of insulated tools or handling equipment is subject to damage, the insulating material shall be protected.
- Fuses must only be removed using appropriate fuse-handling equipment that is insulated for the circuit voltage.

Arc Blast

An electrician in Louisville, KY was cleaning a high-voltage switch cabinet. He removed the padlock securing the switch enclosure and opened the door. He used a voltage meter to verify absence of voltage on the three load phases at the rear of the cabinet. However, he did not test all potentially energized parts within the cabinet, and some components were still energized. He was wearing standard work boots and safety glasses, but was not wearing protective rubber gloves or arc-rated apparel. As he used a paintbrush to clean the switch, an arc blast occurred, which lasted approximately one-sixth of a second. The electrician was knocked down by the blast. He suffered third-degree burns and required skin grafts on his arms and hands. The investigation determined that the blast was caused by debris, such as a cobweb, falling across the open switch.

The Bottom Line: Always wear the appropriate personal protective equipment and test all components for the absence of voltage before working in or around electrical devices.

WARNING!

Fuses should never be installed or removed when energized. This could cause an electrical arc, which can result in death or a serious injury.

- Ropes and handlines used near exposed energized parts shall be nonconductive.
- Protective shields, protective barriers, or insulating materials shall be used to protect each employee from shock, burns, or other electrically related injuries while that employee is working near exposed energized parts that might be accidentally contacted or where dangerous electric heating or arcing might occur. When normally enclosed live parts are exposed for maintenance or repair, they shall be guarded to protect unqualified persons from contact with the live parts.

The types of electrical safety equipment, protective apparel, and protective tools available for use are quite varied. This module will discuss the most common types of safety equipment. These include the following:

- Currently tested rubber protective equipment, including gloves and blankets
- Arc-rated apparel
- Natural fiber clothing
- Electrical hot sticks
- Fuse pullers
- Footwear rated to protect against electrical hazards
- Safety glasses
- Face shields

2.2.0 Selecting and Using Protective Equipment

All electrical workers may be exposed to energized circuits or equipment. Two of the most important articles of protection for electrical workers are insulated rubber gloves and rubber blankets, which must be matched to the voltage rating for the circuit or equipment. Rubber protective equipment is designed for the protection of the user. If it fails during use, a serious injury could occur.

Rubber protective equipment is available in two types. Type 1 designates rubber protective equipment that is manufactured of natural or synthetic rubber that is properly vulcanized, and Type 2 designates equipment that is ozone resistant, made from any elastomer or combination of elastomeric compounds. Ozone is a form of oxygen that is produced from electricity and is present in the air surrounding a conductor under high voltages. Normally, ozone is found at voltages of 10kV and higher, such as those found in electric utility transmission and distribution systems. Type 1 protective equipment can be damaged by corona cutting, which is the cutting action of ozone on natural rubber when it is under mechanical stress. Type 1 rubber protective equipment can also be damaged by ultraviolet rays. However, it is very important that the rubber protective equipment in use today be natural rubber (Type 1). Type 2 rubber protective equipment is very stiff and is not as easily worn as Type 1 equipment.

2.2.1 Classes of Protective Equipment

The American National Standards Institute (ANSI) and the American Society for Testing and Materials International (ASTM) have designated a specific classification system for rubber protective equipment. The maximum AC use voltage and equipment tag colors are shown in *Table 3*.

Before rubber protective equipment can be worn by personnel in the field, all equipment must have a current test date stenciled on the equipment. Insulating gloves must be inspected each day by the user before they can be used. They must also be electrically tested every six months and any time the insulating value is in question. Because rubber protective equipment is used for personal protection and serious injury could result from its misuse or failure, it is important that an adequate safety factor be provided between the voltage on which it is to be used and the voltage at which it was tested.

All rubber protective equipment must be marked with the appropriate voltage rating and last inspection date. The markings that are required to be on rubber protective equipment must be applied in a manner that will not interfere with the protection that is provided by the equipment.

2.2.2 Rubber Gloves

Both high- and low-voltage rubber gloves are of the gauntlet type and are available in various sizes. For the best possible protection and service life, observe the following general guidelines/considerations when using rubber gloves in electrical work:

- Always wear leather protectors over your gloves. Leather protectors shield the gloves against contact with sharp or pointed objects that may cut, snag, or puncture the rubber. Leather protectors also provide burn protection not provided by the gloves themselves.
- Always wear rubber gloves right side out (serial number and size to the outside).
- Always keep the gauntlets up. Rolling them down sacrifices a valuable area of protection. Tuck the sleeves of your shirt or protective suit under the glove cuffs to prevent an arc blast from coming inside your clothing.

Table 3 Rubber Protective Equipment Classification

Classification (Tag Color)	Max Voltage
Class 00 (beige tag)	500V
Class 0 (red tag)	1,000V
Class 1 (white tag)	7,500V
Class 2 (yellow tag)	17,000V
Class 3 (green tag)	26,500V
Class 4 (orange tag)	36,000V

- Always inspect and field check gloves before using them. Always check the inside for any debris.
- Use light amounts of manufacturer-approved glove dust or cotton liners with the rubber gloves. This helps to absorb some of the perspiration that can damage the gloves over years of use.
- Wash the rubber gloves in lukewarm, clean, fresh water after each use. Dry the gloves inside and out prior to returning to storage. Never use any type of cleaning solution on the gloves.
- Once the gloves have been properly cleaned, inspected, and tested, they must be properly stored. Store them in a cool, dry, dark place that is free from ozone, chemicals, oils, solvents, or other materials that could damage the gloves. Do not store gloves near hot pipes or in direct sunlight. Store both gloves and sleeves in their natural shape in a bag or box inside their leather protectors. They should be undistorted, right side out, and unfolded. Storing gloves inside out will reduce the life of the gloves.
- Gloves can be damaged by many different chemicals, especially petroleum-based products such as oils, gasoline, hydraulic fluid inhibitors, hand creams, pastes, and salves. If contact is made with these or other petroleum-based products, the contaminant should be wiped off immediately. If any signs of physical damage or chemical deterioration are found (e.g., swelling, softness, hardening, stickiness, ozone deterioration, or sun checking), the protective equipment must not be used.
- Never wear watches or rings while wearing rubber gloves; this can cause damage from the inside out and defeats the purpose of using rubber gloves. Never wear anything conductive.
- Rubber gloves must be electrically tested every six months by a certified testing laboratory. Always check the inspection date before using gloves.

- Use rubber gloves only for their intended purpose, not for handling chemicals or other work. This also applies to the leather protectors.

Before rubber gloves are used, a visual inspection and an air test should be made. This should be done prior to use and as many times during the day as you feel necessary. To perform a visual inspection, stretch a small area of the glove, checking to see that no defects exist, such as the following:

- Embedded foreign material
- Deep scratches
- Pinholes or punctures
- Snags or cuts

Gloves and sleeves can be inspected by rolling the outside and inside of the protective equipment between the hands. This can be done by squeezing together the inside of the gloves or sleeves to bend the outside area and create enough stress to the inside surface to expose any cracks, cuts, or other defects. When the entire surface has been checked in this manner, the equipment is then turned inside out, and the procedure is repeated. It is very important not to leave the rubber protective equipment inside out as that places stress on the preformed rubber.

Remember that even the smallest amount of damage to a rubber glove compromises its insulating ability. Look for signs of deterioration from age, such as hardening and slight cracking. Also, if the glove has been exposed to petroleum products, it should be considered suspect because deterioration can be caused by such exposure. If the gloves are suspect, turn them in for evaluation. If the gloves are defective, turn them in for disposal. Never leave a damaged glove lying around; someone may think it is a good glove and not perform an inspection prior to using it.

After visually inspecting the glove, other defects may be observed by applying the air test (*Figure 3*). The following general steps can be used to perform an air test:

Step 1 Stretch the glove and look for any defects.

Step 2 Twirl the glove around quickly or roll it down from the glove gauntlet to trap air inside.

Step 3 Trap the air by squeezing the gauntlet with one hand. Use the other hand to squeeze the palm, fingers, and thumb to check for weaknesses and defects.

Step 4 Hold the glove up to your ear to try to detect any escaping air.

Step 5 If the glove does not pass this inspection, it must be turned in for disposal.

Gloves may also be tested using a glove tester (*Figure 4*). To use a glove tester, stretch the glove cuff over the housing and secure it in place, then inflate and check for leaks.

Never use compressed gas for the air test or inflate gloves beyond the manufacturer's recommendations, as this can damage the gloves.

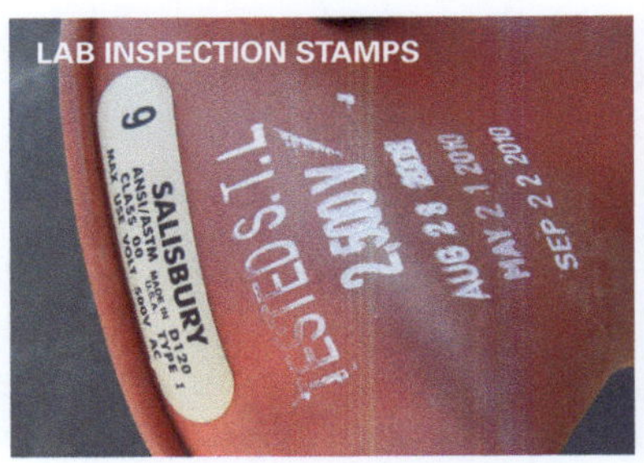

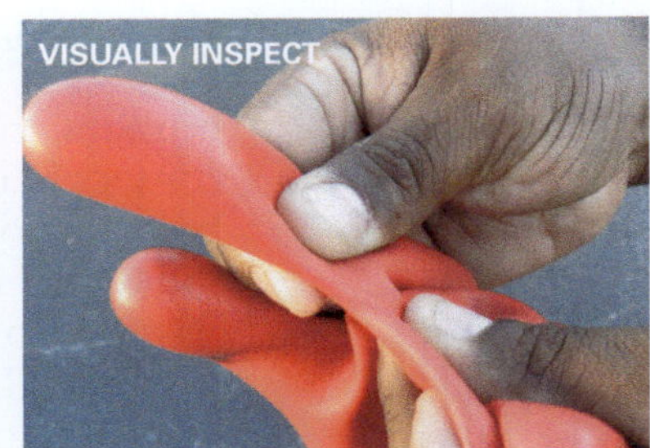

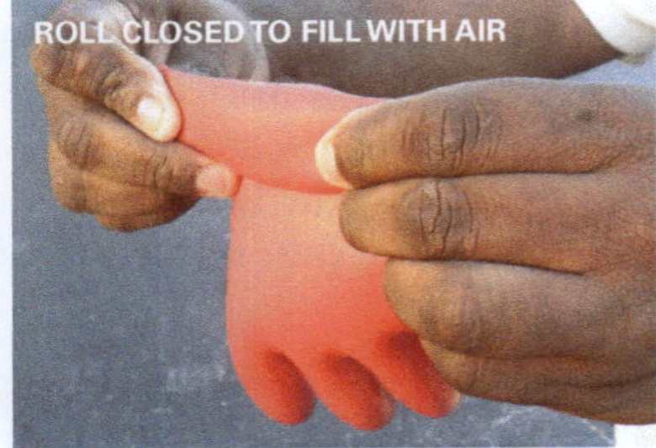

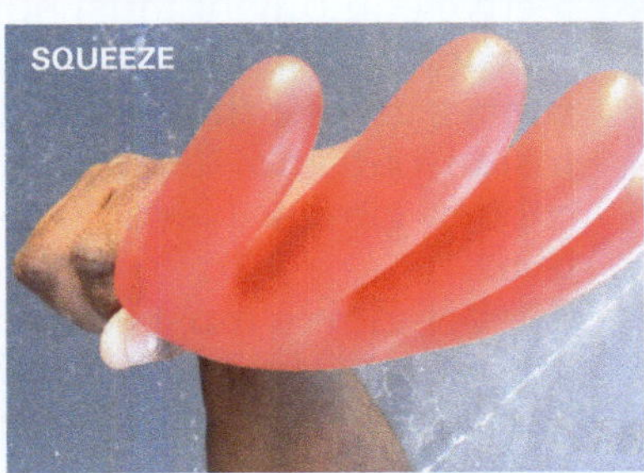

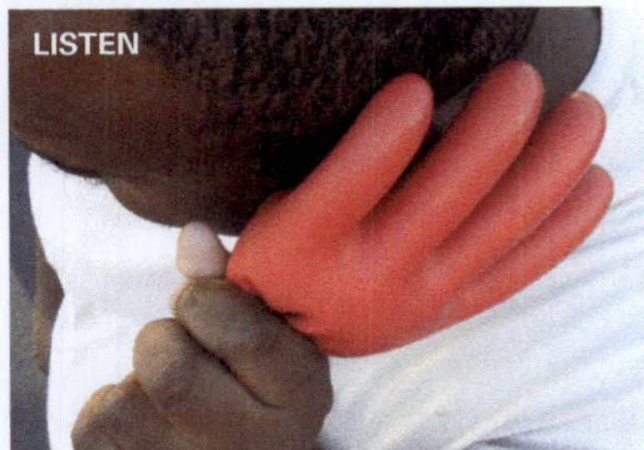

Figure 3 Glove inspection.

NCCER – *Maritime Electrical*

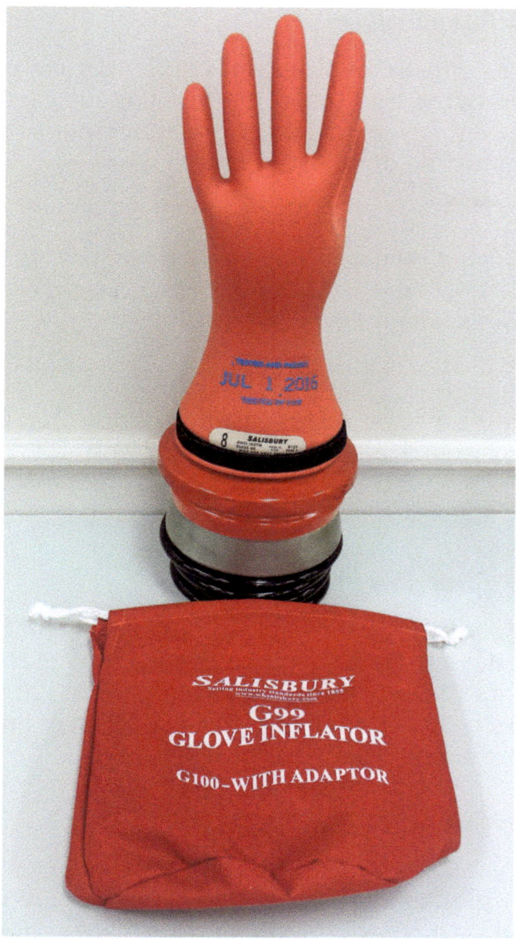

Figure 4 Glove tester.

2.2.3 Insulating Blankets

An insulating blanket is a versatile cover-up device best suited for the protection of maintenance technicians against accidental contact with energized electrical equipment.

These blankets are designed and manufactured to provide insulating quality and flexibility for use in covering. Insulating blankets are designed only for covering equipment and should not be used on the floor. Special rubber floor mats, called switchboard matting, are available for floor use. Use caution when installing these over sharp edges or when covering pointed objects.

Blankets must be tested yearly and inspected before each use. To check rubber blankets, place the blanket on a flat surface and roll the blanket from one corner to the opposite corner. If there are any irregularities in the rubber, this method will expose them. After the blanket has been rolled from each corner, it should then be turned over and the procedure repeated.

Insulating blankets are cleaned in the same manner as rubber gloves. Once the protective equipment has been properly cleaned, inspected, and tested, it must be properly stored. It should be stored in a cool, dry, dark place that is free from ozone, chemicals, oils, solvents, or other materials that could damage the equipment. Avoid storage near hot pipes or in direct sunlight. Blankets may be stored rolled in containers that are designed for this use; the inside diameter of the roll should be at least 2" (50 mm).

2.2.4 Other Protective Apparel

Besides rubber gloves, there are other types of special-application protective apparel, such as arc-rated suits, face shields, and rubber sleeves. Manufacturing plants should have other types of special-application protective equipment available for use, such as high-voltage sleeves, high-voltage boots, nonconductive protective helmets, nonconductive eyewear and face protection, and switchboard blankets.

All equipment must be inspected before use and during use, as necessary. The equipment used and the extent of the precautions taken depend on each individual situation; however, it is better to be overprotected than underprotected when you are trying to prevent electric shock, arc blast, and burns.

When working with energized equipment, arc-rated suits with rated face and head protection may be required in some applications.

Arc-Flash Hazard Warnings

In other than dwelling units, *NEC Section 110.16* requires that all switchboards, switchgear, panelboards, industrial control panels, meter socket enclosures, and motor control centers be clearly marked to warn qualified persons of potential electric arc flash hazards. *NEC Section 110.16(B)* requires service equipment at these locations to be marked with the nominal system voltage, available fault current at the service overcurrent protective devices, clearing time of the service overcurrent protective devices based on the available fault current at the service equipment, and the date the label was applied.

The clothes worn beneath arc-rated protective apparel must also be carefully selected. Never wear synthetic-fiber clothing during energized work; these types of materials will melt when exposed to high temperatures and because they burn hotter, will actually increase the severity of a burn. Wear cotton clothing, leather boots or shoes, safety glasses, and hard hats. Use hearing protection where needed.

2.2.5 Hot Sticks

Hot sticks are insulated tools designed for the manual operation of disconnecting switches, fuse removal and insertion, and the application and removal of temporary grounds.

A hot stick is made up of two parts, the head or hood and the insulating rod. The head can be made of metal or hardened plastic, while the insulating section may be wood, plastic, laminated wood, or other effective insulating materials. Telescoping sticks are also available.

Most plants have hot sticks available for different purposes. Select a stick of the correct type and size for the application and inspect it before use. Look for signs of obvious damage, deep scratches, dust, or surface contaminants. Never use a damaged hot stick.

Storage of hot sticks is important. They should be hung up vertically on a wall to prevent any damage. They should also be stored away from direct sunlight and prevented from being exposed to petroleum products.

2.2.6 Fuse Pullers

Use the plastic or fiberglass style of fuse puller for removing and installing low-voltage cartridge fuses. All fuse pulling and replacement operations must be done using fuse pullers.

> **WARNING!**
>
> Fuses should only be pulled when they are safely de-energized. Failure to do so can result in a serious injury or death.

The best type of fuse puller is one that has a spread guard installed. This prevents the puller from opening if resistance is met when installing fuses.

2.2.7 Shorting Probes

Before working on de-energized circuits that have capacitors installed, you must discharge the capacitors using a safety shorting probe. This procedure requires special training and may only be performed by qualified individuals.

2.2.8 Eye and Face Protection

NFPA 70E® and OSHA require that you wear safety glasses at all times whenever you are working on or near energized circuits. Face protection is worn over your safety glasses, and is required whenever there is danger of electrical arcs or flashes, or from flying or falling objects resulting from an electrical explosion. *NFPA 70E*® is discussed in detail later in this module.

Think About It

Dressing for Safety

How could minor flaws in clothing cause harm? What about metal components, such as the rivets in jeans, or synthetic materials, such as polyester? How are these dangerous? What about rings, watches, earrings, or other body piercings? How does protective apparel prevent accidents?

NCCER – *Maritime Electrical*

Using a Phasing Tester

Using a phasing tester or other voltage detection equipment can expose you to a considerable arc flash hazard. Always take the time to use the appropriate personal protective equipment. Remember, you have only one chance to protect yourself.

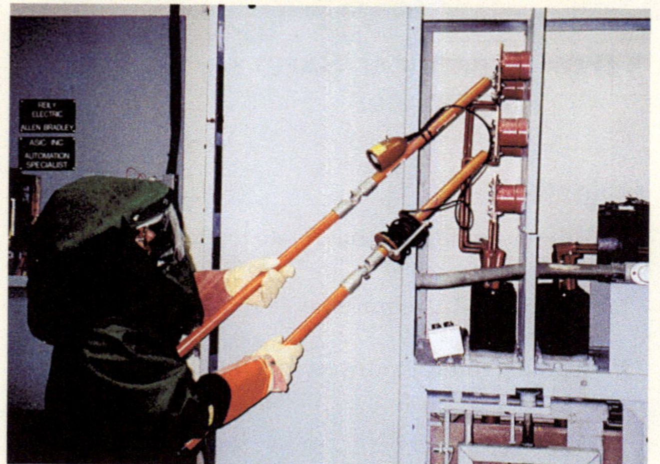

Figure Credit: Tim Ely

Additional Resources

National Electrical Code® Handbook, Latest Edition. Quincy, MA: National Fire Protection Association.

2.0.0 Section Review

1. Periodic inspection and testing of protective equipment is required by _____.

    a. ANSI
    b. OSHA
    c. NFPA
    d. NEMA

2. Rubber protective equipment with a beige tag is rated for _____.

    a. 500V
    b. 1,000V
    c. 7,500V
    d. 17,000V

3.0.0 ELECTRICAL SAFETY STANDARDS

Objective

Identify the standards that relate to electrical safety.

a. Apply OSHA requirements in the workplace.
b. Understand the purpose of *NFPA 70E®*.

Trade Term

Grounded tool: An electrical tool with a three-prong plug at the end of its power cord or some other means to ensure that stray current travels to ground without passing through the body of the user. The ground plug is bonded to the conductive frame of the tool.

All electricians must become familiar with both national safety standards and local requirements. The three main national standards applicable to electrical work are the *National Electrical Code (NEC®)*, *NFPA 70E®*, and *OSHA 29 CFR, Parts 1910 and 1926*. This module provides a brief overview of *NFPA 70E®* and OSHA requirements. The *NEC®* is discussed in detail in a later module.

3.1.0 Applying OSHA Requirements in the Workplace

The purpose of OSHA is "to ensure safe and healthful working conditions for working men and women." OSHA is authorized to enforce standards and assist and encourage the states in their efforts to ensure safe and healthful working conditions. OSHA assists states by providing for research, information, education, and training in the field of occupational safety and health.

The law that established OSHA specifies the duties of both the employer and employee with respect to safety. Some of the key requirements are outlined below; note that this list does not include everything, nor does it override the procedures called for by your employer:

- Employers shall provide a place of employment free from recognized hazards likely to cause death or serious injury.
- Employers shall comply with OSHA standards.

- Employers shall be subject to fines and other penalties for violation of those standards.

> **CAUTION**
>
> OSHA states that employees have a duty to follow the safety rules laid down by the employer. Additionally, some states can reduce the amount of benefits paid to an injured employee if that employee was not following known, established safety rules. Your company may also terminate you if you violate an established safety rule.

The OSHA standards are split into several sections. As discussed earlier, the two that affect you the most are *CFR 1926*, construction specific, and *CFR 1910*, which is the standard for general industry. Either or both may apply depending on where you are working and what you are doing. Your company may also have its own policies and procedures. In addition, you will be required to follow the safety procedures of any plant or facility where you are working.

3.1.1 OSHA Requirements for Electrical Safety

The most important piece of safety equipment required when performing work in an electrical environment is common sense. All areas of electrical safety precautions and practices draw upon common sense and attention to detail. One of the most dangerous conditions in an electrical work area is a poor attitude toward safety.

As stated in *CFR 1910.333(a)/1926.416*, safety-related work practices shall be employed to prevent electric shock or other injuries resulting from either direct or indirect electrical contact when work is performed near or on equipment or circuits that are or may be energized. The following are considered some of the basic and necessary attitudes and electrical safety precautions that lay the groundwork for a proper safety program:

> **WARNING!**
>
> Only qualified individuals may work on or near electrical equipment. Your employer will determine who is qualified. Remember, your employer's safety rules must always be followed.

- OSHA requires the posting of hard hat areas. Be alert to those areas and always wear your hard hat properly, with the bill in front. Hard hats should be worn whenever overhead hazards exist, or there is the risk of exposure to electric shock or burns.

Safety on the Job Site

Uncovered openings present several hazards, including the following:

- Workers may trip over them.
- If they are large enough, workers may fall through them.
- If there is a work area below, tools or other objects may fall through them, causing serious injury to workers below.

Any hole deeper than 6' (1.8 m) and more than 2" (50 mm) in any direction must be protected with a cover or with guardrails, as shown here. A cover must be able to handle twice the anticipated load, be secured in place, and have the word "HOLE" or "COVER" on it.

Figure Credit: Mike Powers

- Wear safety shoes on all job sites. Keep them in good condition.
- Do not wear clothing with exposed metal zippers, buttons, or other metal fasteners. Avoid wearing loose-fitting or torn clothing.
- Always wear safety glasses with full side shields. In addition, the job may also require protective equipment such as face shields or goggles.
- All electrical work shall be in compliance with the latest *NEC*® and OSHA standards.
- The noncurrent-carrying metal parts of fixed, portable, and plug-connected equipment shall be grounded. Choose either a grounded tool or a double-insulated tool. An example of a double-insulated tool is shown in *Figure 5*.
- Extension cords shall be the three-wire type, shall be protected from damage, and if hung overhead, shall not be fastened with staples or bare wire or hung in a manner that could cause damage to the outer jacket or insulation. Never run an extension cord through a doorway or window that can pinch the cord. Also, never allow vehicles or equipment to drive over cords.
- Exposed lamps in temporary lights shall be guarded to prevent accidental contact, except where lamps are deeply recessed in the reflector. Temporary lights must be supported in accordance with their listed labeling.
- All receptacles for attachment plugs shall be of an approved type and properly installed. Installation of the receptacle will be in accordance with the listing and labeling for each receptacle.
- *NEC Section 590.6* states that all 125V, single-phase, 15A, 20A, and 30A receptacle outlets for temporary or permanent power shall be ground fault protected.
- Each disconnecting means for motors and appliances and each service feeder or branch circuit at the point where it originates shall be legibly marked with its purpose and voltage.

Working Around Openings

Did you know that OSHA could cite your company for working around the open pipe in the photo shown here, even if you weren't responsible for creating it? If you are working near uncovered openings, you have only two options:

- Cover the opening properly (either you may do it or you may have the responsible contractor cover it).
- Stop work and leave the job site until the opening has been covered.

Figure Credit: Mike Powers

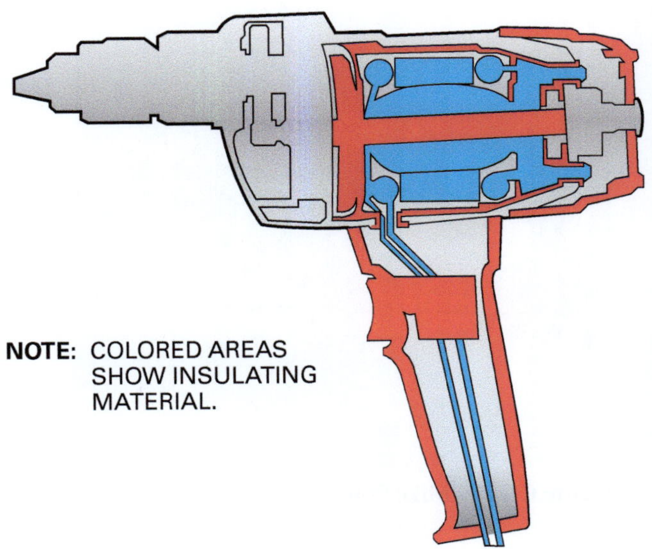

Figure 5 Double-insulated electric drill.

- Flexible cords shall be used in continuous lengths (no splices) and shall be of a type listed in *NEC Table 400.4*.
- All receptacles shall be ground fault protected if possible or special-purpose ground fault circuit interrupter for personnel (SPGFCI) shall be used per *NEC Section 590.6(B)(2)*. If that is not possible, a written assured equipment grounding conductor program must be continuously enforced at the site for all cord sets per *NEC Section 590.6(B)(3)*. A typical ground fault circuit interrupter (GFCI) is shown in *Figure 6*.

All work on electrical equipment should be done with circuits de-energized and cleared or grounded. Work on energized electrical equipment should be avoided if at all possible. *CFR 1910.333(a)(1)* states that live parts to which an employee may be exposed shall be de-energized before the employee works on or near them, unless the employer can demonstrate that de-energizing introduces additional or increased hazards or is not possible because of equipment design or operational limitations. Live parts that operate at less

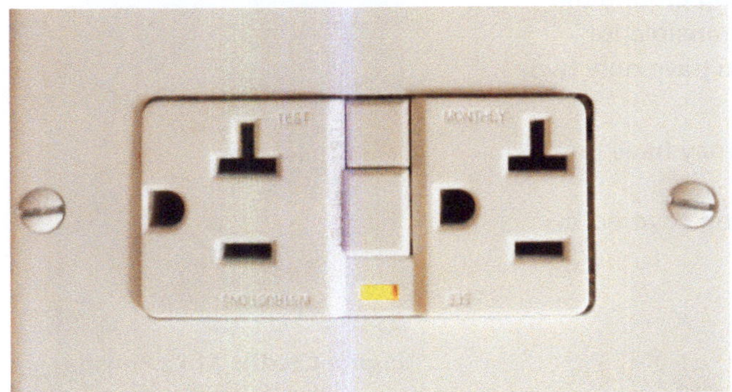

Figure 6 Typical GFCI receptacle.

than 50 volts to ground need not be de-energized if there will be no increased exposure to electrical burns or to explosion due to electric arcs.

All conductors, buses, and connections should be considered energized until proven otherwise. As stated in *1910.333(b)(1)*, conductors and parts of electrical equipment that have not been locked out or tagged out in accordance with this section should be considered energized. Routine operation of the circuit breakers and disconnect switches contained in a power distribution system can be hazardous if not approached in the right manner. Several basic precautions that can be observed in switchgear operations are:

- Wear proper clothing made of natural fiber or fire-resistant fabric in accordance with *NFPA 70E®*.
- Wear eye, face, and head protection.
- Whenever operating circuit breakers in low-voltage or medium-voltage systems, always stand off to the side of the unit.
- Always try to operate disconnect switches and circuit breakers under a no-load condition.
- Never intentionally force an interlock on a system or circuit breaker.

Often, a circuit breaker or disconnect switch is used for providing lockout on an electrical system. Perform the following procedures when using the device as a lockout point:

- Breakers must always be locked out and tagged whenever you are working on a circuit that is tied to an energized breaker. Always follow the standard rack-out and removal procedures that were supplied with the switchgear. Once removed, a sign must be hung on the breaker identifying its use as a lockout point, and approved safety locks must be installed when the breaker is used for isolation. Breakers equipped with closing springs should be discharged to release all stored energy in the breaker mechanism.
- Some of the circuit breakers used are equipped with keyed interlocks (Kirk locks) for protection during operation. These locks are used to ensure proper sequence of operation only. They are not used to lock out a circuit or system. When opening or closing a disconnect manually, it should be done quickly with a positive force. Lockout/tagout should be used when the disconnects are open.
- Whenever performing switching or fuse replacements, always use the protective equipment necessary to ensure personnel safety. Always prepare for the worst-case accident when performing switching.

Potential Hazards

A self-employed builder was using a metal cutting tool on a metal carport roof and was not using GFCI protection. The male and female plugs of his extension cord partially separated, and the active pin touched the metal roofing. When the builder grounded himself on the gutter of an adjacent roof, he received a fatal shock.

The Bottom Line: Always use GFCI protection and be on the lookout for potential hazards.

- Whenever re-energizing circuits following maintenance or removal of a faulted component, use extreme care. Always verify that the equipment is in a condition to be re-energized safely. All connections should be insulated and all covers and screws installed. Have all personnel stand clear of the area for the initial re-energization. Never assume everything is in perfect condition. Verify the conditions.

The following procedure is provided as a guideline for ensuring that equipment and systems will not be damaged by reclosing low-voltage circuit breakers into faults. If a low-voltage circuit breaker has opened for no apparent reason, perform the following:

Step 1 Verify that the equipment being supplied is not physically damaged and shows no obvious signs of overheating or fire.

Step 2 Make all appropriate tests to locate any faults.

Step 3 Reclose the feeder breaker. Stand off to the side when closing the breaker. Whenever possible, avoid reaching across the front of the device.

Step 4 If the circuit breaker trips again, do not attempt to reclose the breaker. In a plant environment, Electrical Engineering should be notified, and the cause of the trip must be isolated and repaired.

Think About It

Working on Energized Systems

Some electricians commonly work on energized systems because they think it's too much trouble to turn off the power. What practices have you seen around your home or workplace that could be deadly?

The same general procedure should be followed for fuse replacement, with the exception of transformer fuses. If a transformer fuse blows, the transformer and feeder cabling should be inspected and tested before re-energizing. A blown fuse to a transformer is very significant because it normally indicates an internal fault. Transformer failures are catastrophic in nature and can be extremely dangerous. If applicable, contact the in-plant Electrical Engineering Department prior to commencing any effort to re-energize a transformer.

Power must always be removed from a circuit when removing and installing fuses. The air break disconnects (or quick disconnects) provided on the upstream side of a large transformer must be opened prior to removing the transformer's fuses. Otherwise, severe arcing will occur as the fuse is removed or installed. This arcing can result in personnel injury and equipment damage.

> **WARNING!**
>
> When replacing fuses, always wear appropriate PPE, which may include arc-rated suits and rated face and head protection.

To replace fuses servicing circuits *below* 600 volts, observe the following steps:

Step 1 Turn off the power to the disconnect.

Step 2 Verify that the fuses are de-energized.

Step 3 Remove the blown fuse.

Step 4 Install the new fuse. Push it in firmly and verify that it is seated properly.

Step 5 Turn the power back on.

To replace fuses servicing systems *above* 600 volts, observe the following steps:

Step 1 Open and lock out the disconnect switches.

Step 2 Unlock the fuse compartment.

Step 3   Verify that the fuses are de-energized.

Step 4   Attach the fuse removal hot stick to the fuse and remove it.

3.1.2 OSHA Lockout/Tagout Rule

The OSHA lockout/tagout rule covers the specific procedure to be followed for the "servicing and maintenance of machines and equipment in which the unexpected energization or startup of the machines or equipment, or releases of stored energy, could cause injury to employees." This standard establishes minimum performance requirements for the control of such hazardous energy.

The first step to be completed before working on a circuit is to ensure that equipment is isolated from all potentially hazardous energy (for example, electrical, mechanical, hydraulic, chemical, or thermal), and tagged and locked out before employees perform any servicing or maintenance activities in which the unexpected energization, startup, or release of stored energy could cause injury. All employees shall be instructed in the lockout/tagout procedure.

> **CAUTION**
>
> Although 99 percent of your work may be electrical, be aware that you may also need to lock out mechanical and other types of energy equipment. The following is an example of a lockout/tagout procedure. Make sure to use the procedure that is specific to your employer or job site.

> **WARNING!**
>
> This procedure is provided for your information only. The OSHA procedure provides only the minimum requirements for lockouts/tagouts. Consult the lockout/tagout procedure for your company and the plant or job site at which you are working. Remember that your life could depend on the lockout/tagout procedure. It is critical that you use the correct procedure for your site. The *NEC*® requires that remote-mounted motor disconnects be permanently equipped with a lockout feature.

Think About It

GFCIs

Explain how GFCIs protect people. Where should a GFCI be installed in the circuit to be most effective?

I. Introduction

   A. This lockout/tagout procedure has been established for the protection of personnel from potential exposure to hazardous energy sources during construction, installation, service, and maintenance of electrical energy systems.

   B. This procedure applies to and must be followed by all personnel who may be potentially exposed to the unexpected startup or release of hazardous energy (e.g., electrical, mechanical, pneumatic, hydraulic, chemical, thermal, kinetic, or other).

   Exceptions:

   1. This procedure does not apply to process and/or utility equipment or systems with cord and plug power supplies when the cord and plug are the only source of hazardous energy, are removed from the source, and remain under the exclusive control of the authorized employee.

   2. This procedure does not apply to troubleshooting (diagnostic) procedures and installation of electrical equipment and systems when the energy source cannot be de-energized because continuity of service is essential or shutdown of the system is impractical. Additional personal protective equipment for such work is required and the safe work practices identified for this work must be followed.

II. Definitions

   - *Affected employee* – Any person working on or near equipment or machinery when maintenance or installation tasks are being performed by others during lockout/ tagout conditions.

   - *Appointed authorized employee* – Any person appointed by the job-site supervisor to coordinate and maintain the security of a group lockout/tagout condition.

   - *Authorized employee* – Any person authorized by the job-site supervisor to use lockout/tagout procedures while working on electrical equipment.

   - *Authorized supervisor* – The assigned job-site supervisor who is in charge of coordination of procedures and maintenance of security of all lockout/ tagout operations at the job site.

NCCER – *Maritime Electrical*

- *Energy isolation device* – An approved electrical disconnect switch capable of accepting approved lockout/tagout hardware for the purpose of isolating and securing a hazardous electrical source in an open or safe position.
- *Lockout/tagout hardware* – A combination of padlocks, danger tags, and other devices designed to attach to and secure electrical isolation devices.

III. Training

A. Each authorized supervisor, authorized employee, and appointed authorized employee shall receive initial and as-needed user-level training in lockout/tagout procedures.

B. Training is to include recognition of hazardous energy sources, the type and magnitude of energy sources in the workplace, and the procedures for energy isolation and control.

C. Retraining will be conducted on an as-needed basis whenever lockout/tagout procedures are changed or there is evidence that procedures are not being followed properly.

IV. Protective Equipment and Hardware

A. Lockout/tagout devices shall be used exclusively for controlling hazardous energy sources.

B. All padlocks must be numbered and assigned to one employee only.

C. No duplicate or master keys will be made available to anyone except the site supervisor.

D. A current list with the lock number and authorized employee's name must be maintained by the site supervisor.

E. Danger tags must be of the standard white, red, and black DANGER—DO NOT OPERATE design and shall include the authorized employee's name, the date, and the appropriate network company (use permanent markers).

F. Danger tags must be used in conjunction with padlocks, as shown in *Figure 7*.

V. Procedures

A. Preparation for lockout/tagout:

1. Check the procedures to ensure that no changes have been made since you last used a lockout/tagout.

(A) ELECTRICAL LOCKOUT

(B) PNEUMATIC LOCKOUT

Figure 7 Lockout/tagout devices.

2. Identify all authorized and affected employees involved with the pending lockout/tagout.

B. Sequence for lockout/tagout:

1. Notify all authorized and affected personnel that a lockout/tagout is to be used and explain the reason why.

2. Shut down the equipment or system using the normal Off or Stop procedures.

3. Lock out energy sources and test disconnects to be sure they cannot be moved to the On position and open the control cutout switch. If there is no cutout switch, block the magnet in the Switch Open position before working on electrically operated equipment/ apparatus such as motors, relays, etc. Remove the control wire.

4. Lock and tag the required switches in the Open position. Each authorized employee must affix a separate lock and tag. An example is shown in *Figure 8*.

5. Dissipate any stored energy by attaching the equipment or system to ground.

6. Verify that the test equipment is functional via a known power source.

7. Confirm that all switches are in the Open position and use test equipment to verify that all parts are de-energized.

8. If it is necessary to temporarily leave the area, upon returning, retest to ensure that the equipment or system is still de-energized.

Figure 8 Multiple lockout/tagout device.

C. Restoration of energy:

1. Confirm that all personnel and tools, including shorting probes, are accounted for and removed from the equipment or system.

2. Completely reassemble and secure the equipment or system.

3. Replace and/or reactivate all safety controls.

4. Remove locks and tags from isolation switches. Authorized employees must remove their own locks and tags.

5. Notify all affected personnel that the lockout/tagout has ended and the equipment or system is energized.

6. Operate or close isolation switches to restore energy.

VI. Emergency Removal Authorization

A. In the event a lockout/tagout device is left secured, and the authorized employee is absent, or the key is lost, the authorized supervisor can remove the lockout/tagout device.

B. The authorized employee must be informed that the lockout/tagout device has been removed.

C. Written verification of the action taken, including informing the authorized employee of the removal, must be recorded in the job journal.

3.2.0 NFPA 70E®

In addition to the *NEC®*, the National Fire Protection Association also publishes the *Standard for Electrical Safety in the Workplace (NFPA 70E®)*. The *NEC®* specifies the minimum installation provisions necessary to safeguard persons and property from electrical hazards, while *NFPA 70E®* addresses the hazards that arise during the installation, operation, and maintenance of electrical equipment. In other words, the *NEC®* applies to installations, while *NFPA 70E®* applies to workplaces. *NFPA 70E®* is a national consensus standard, which means that it is a standard developed by the same people it affects and is then adopted by a nationally recognized institution. Other national consensus standards include those developed by ASTM International and the American National Standards Institute (ANSI).

The electrical safety requirements in *CFR 1910.331* to *1910.335* are to use appropriate safe work practices, including de-energization of equipment, and the use of PPE appropriate to the

NCCER – *Maritime Electrical*

What's wrong with this picture?

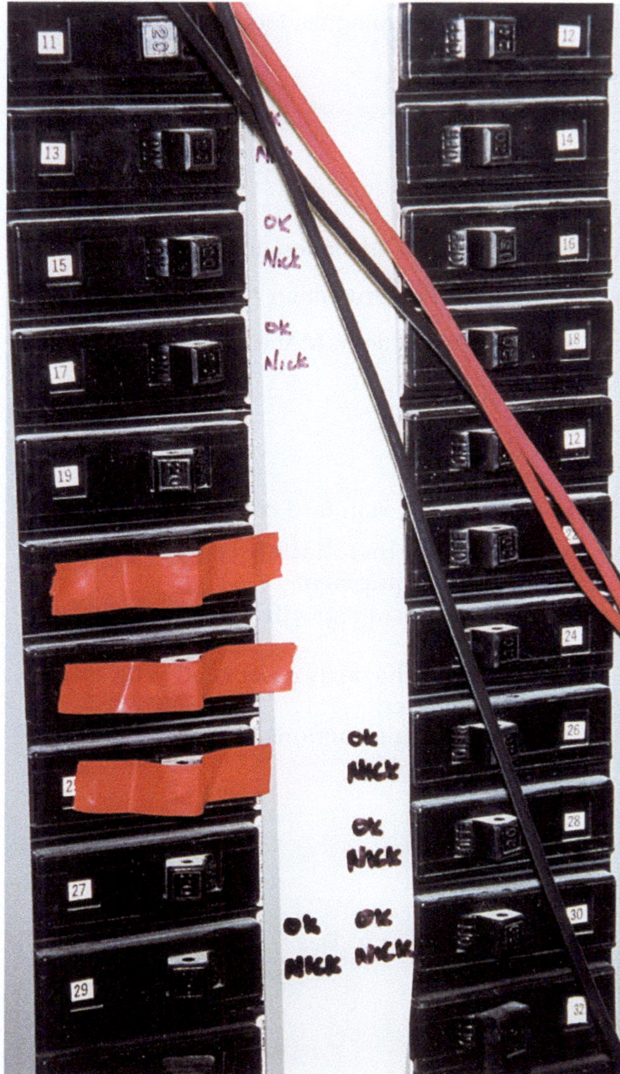

Figure Credit: Mike Powers

hazard. *NFPA 70E®* is really the first consensus standard to provide practical guidance in hazard assessment, training of qualified persons, and management of electrical hazards. For many, *NFPA 70E®* provided the first functional understanding of the electrical arc flash hazard, direction for establishing flash protection boundaries, and levels of protection when qualified persons must work within the flash protection boundary. *NFPA 70E®* is arranged as follows:

- *Chapter 1, Safety-Related Work Practices* – This chapter covers the basic safety requirements for working on or near electrical equipment. It covers safety training programs, explains the difference between qualified and unqualified individuals, and includes requirements for analysis of both the electrical shock hazard and the electrical arc flash/blast hazard for a given task. The analysis includes assessment of the presence and extent of hazardous voltage and the available fault current for arc flash. The result of each hazard analysis is also selection of safe work practices and appropriate PPE to be used in performing the task.
- *Chapter 2, Safety-Related Maintenance Requirements* – This chapter covers the maintenance requirements for various types of electrical equipment. This includes premises wiring, various types of electric equipment and devices, and personal safety and protective equipment.
- *Chapter 3, Safety Requirements for Special Equipment* – This chapter covers electrolytic cells, batteries, lasers, and various types of power electronic equipment, such as radio and television transmitters, UPS systems, arc welding equipment, and lighting controllers.
- *Annexes A through P* – The annexes provide various technical resources. These include safe approach distances, determination of flash protection boundaries and incident energy exposure from arc flash, a sample electrical lockout/tagout procedure, and other useful information.

Lockout/Tagout Dilemma

In Georgia, electricians found energized switches after the lockout of a circuit panel in an older system that had been upgraded several times. The existing wiring did not match the current site drawings. A subsequent investigation found many such situations in older facilities.

The Bottom Line: Never rely solely on drawings. It is mandatory that the circuit be tested after lockout to verify that it is de-energized.

Additional Resources

Managing Electrical Hazards, Latest Edition. Upper Saddle River, NJ: Pearson Education, Inc.

National Electrical Code® Handbook, Latest Edition. Quincy, MA: National Fire Protection Association.

Standard for Electrical Safety in the Workplace (NFPA 70E®), Latest Edition. Quincy, MA: National Fire Protection Association.

3.0.0 Section Review

1. The OSHA lockout/tagout rule _____.
   a. applies only to those employees performing electrical work
   b. applies to any stored energy that could cause injury (electrical, mechanical, hydraulic, chemical, thermal, kinetic, or other)
   c. represents the most stringent requirements and supersedes site procedures
   d. only applies to government work sites

2. The *NFPA 70E®* standard covers the _____.
   a. minimum installation requirements for electrical equipment
   b. manufacturing specifications for electrical equipment
   c. workplace hazards associated with electrical equipment
   d. operating instructions for electrical equipment

4.0.0 HAZARDS AND SAFETY REQUIREMENTS

Objective

Recognize the safety requirements for various hazards.

a. Identify the safety hazards associated with ladders, scaffolds, and lift equipment.
b. Avoid back injuries by practicing proper lifting techniques.
c. Demonstrate basic tool safety.
d. Identify confined space entry procedures.
e. Work safely with dangerous materials.
f. Select and use appropriate fall protection.

Performance Tasks

1. Properly select and use PPE.
2. Describe the safety requirements for an instructor-supplied task, such as replacing the lights in your classroom.
   - Discuss the work to be performed and the hazards involved.
   - If a ladder is required, perform a visual inspection on the ladder and set it up properly.
   - Ensure that local emergency telephone numbers are either posted or known by you and your partner(s).
   - Plan an escape route from the location in the event of an accident.

Trade Term

Polychlorinated biphenyls (PCBs): Toxic chemicals that may be contained in liquids used to cool certain types of large transformers and capacitors.

There are many hazards on a job site, including fall hazards, equipment hazards, chemical hazards, and tool hazards. The best prevention is safety awareness. If an accident occurs, know where first aid is available and follow company procedures for reporting it. Emergency first aid telephone numbers must be readily available to everyone on the job site. Refer to *CFR 1910.151/1926.23* and *1926.50* for specific requirements.

4.1.0 Ladders, Scaffolds, and Lift Equipment

About half of the injuries that result from contact with energized components involve ladders and scaffolds. In addition to electrical injuries, the involuntary recoil that can occur when a person is shocked can result in a fall from a ladder or elevated surface.

4.1.1 Ladders

Many job-site accidents involve the misuse of ladders. Be sure to use the following general rules every time you use any ladder—they can prevent serious injury or even death:

- Before using any ladder, inspect it. Look for loose or missing rungs, cleats, bolts, or screws. Also check for cracked, bent, broken, or badly worn rungs, cleats, or side rails. Various types of ladder damage is shown in *Figure 9.*
- Before you climb a ladder, make sure you clear any debris from the base of the ladder so you do not trip over it when you descend.
- If you find a ladder in poor condition, do not use it. Report it and tag it for repair or disposal.
- Never modify a ladder by cutting it or weakening its parts.
- Do not set up ladders where they may be run into by others, such as in doorways or walkways. If it is absolutely necessary to set up a ladder in such a location, protect the ladder with barriers.
- Do not increase a ladder's reach by putting it in a mechanical lift or standing it on boxes, barrels, or anything other than a flat, solid surface.
- Check your shoes for grease, oil, or mud before climbing a ladder. These materials could make you slip.
- Always face the ladder and maintain three-point contact with the ladder (either have both feet and one hand on the ladder or both hands and one foot as you climb).
- Never lean out from the ladder. Keep your belt buckle centered between the rails. If something is out of reach, get down and move the ladder.

WARNING!

When performing electrical work, always use ladders made of nonconductive material.

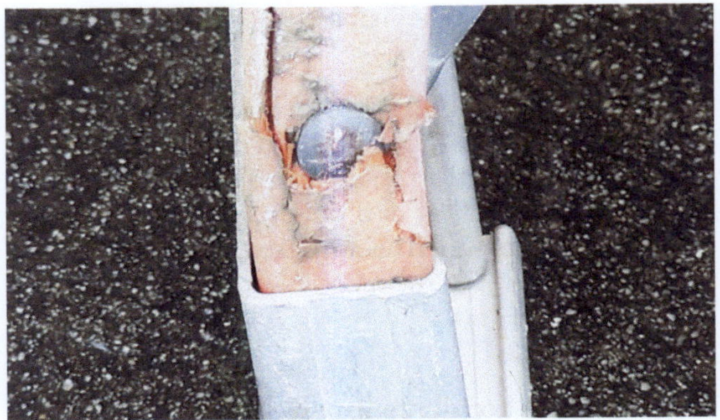

(A) CRUMBLING RAIL

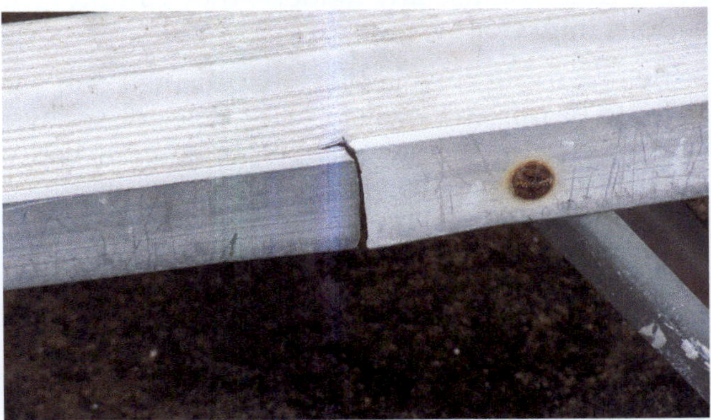

(B) CRACKED STEP

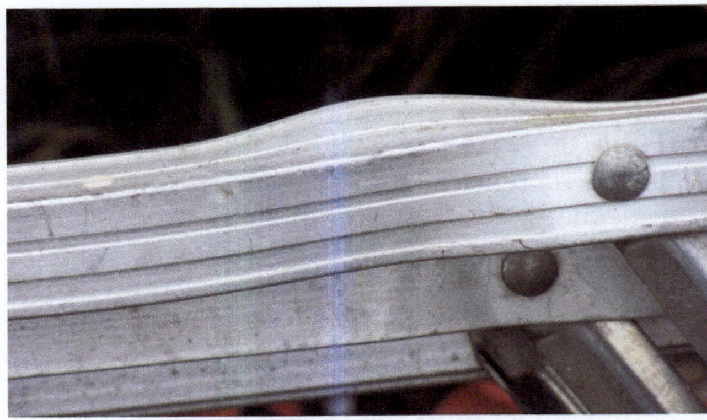

(C) BENT BACK BRACE

Figure 9 Types of ladder damage.

Straight and extension ladders – There are also some specific hazards related to straight ladders and extension ladders. For this reason, the following rules must be observed when working with them:

- Always place a straight ladder at the proper angle. The horizontal distance from the ladder feet to the base of the wall or support should be about one-fourth the working height of the ladder (*Figure 10*).
- Secure straight ladders to prevent slipping. Use ladder shoes or hooks at the top and bottom.

Another method is to secure a board to the floor against the ladder feet. For brief jobs, someone can hold the straight ladder.

- Side rails should extend above the top support point by at least 36" (0.9 m). This provides stability as well as a handhold when stepping around the ladder.
- It takes two people to safely extend and raise an extension ladder. Extend the ladder only after it has been raised to an upright position.
- Never carry an extended ladder.
- Never use two ladders spliced together.
- Ladders should not be painted because paint can hide defects.

Stepladders – The following specific rules must be followed when use working with stepladders:

- Always open the stepladder all the way and lock the spreaders to avoid collapsing the ladder accidentally.
- Use a stepladder that is high enough for the job so that you do not have to reach. It is a good idea to have someone hold the ladder if it is more than 10' (3 m) high.
- Never use a stepladder as a straight ladder.
- Never stand on or straddle the top two rungs of a stepladder.

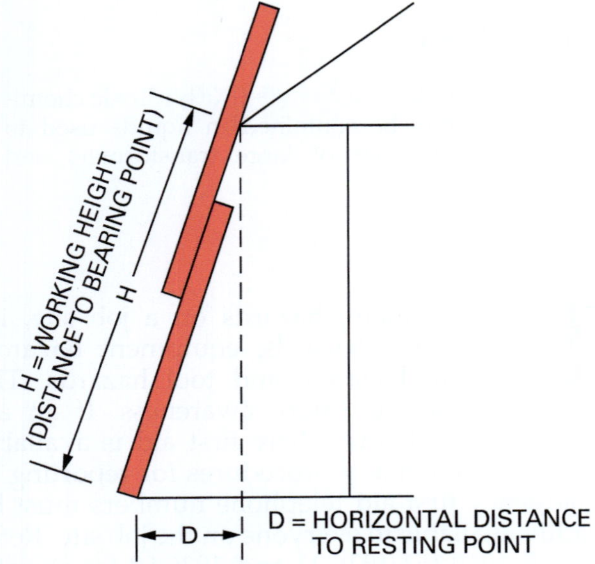

Figure 10 Straight ladder positioning.

NCCER – *Maritime Electrical*

Fireman's Rule

To set up a straight ladder, place your feet against the side rails, stand straight up, and put your hands at right angles to your body directly in front of you. If you can grab the side rails, the ladder is at the correct angle. This is called the Fireman's Rule.

Figure Credit: Mike Powers

- Do not use ladders as shelves.

Do not leave tools or materials on a stepladder.

Sometimes you will need to move or remove protective equipment, guards, or guardrails to complete a task using a ladder. Always wear appropriate fall protection and replace what you moved or removed before leaving the area.

4.1.2 Scaffolds

Working on scaffolds (*Figure 11*) also involves being safe and alert to hazards. In general, keep scaffold platforms clear of unnecessary material or scrap. These can become deadly tripping hazards or falling objects. Carefully inspect each part of the scaffold as it is erected. Your life may depend on it! Makeshift scaffolds have caused many injuries and deaths on job sites. Use only scaffolding and planking materials designed and marked for their specific use. When working on a scaffold, follow the established specific requirements set by OSHA for the use of fall protection. When appropriate, wear an approved harness with a lanyard properly anchored to the structure.

What's wrong with this picture?

Figure Credit: Mike Powers

Figure 11 Scaffolding.

The following are some of the basic OSHA rules for working safely on scaffolds:

- Scaffolds must be erected on sound, rigid footing (referred to as a mud sill) that can carry the maximum intended load. If the scaffolding is not erected on concrete or another firm surface, you can use 2 × 10 or 2 × 12 scaffold-grade lumber to support the base plates of the scaffolding.
- Scaffolds must be erected straight and plumb with no bent or deformed pieces. Correctly erected scaffolding will be symmetrical with the same parts on both sides (unless it has an outrigger attachment).
- Guardrails and toe boards must be installed on the open sides and ends of platforms higher than 10' (3 m) above the ground or floor. Note that many companies lower this to 6' (1.8 m) or even 4' (1.2 m).
- There must be a screen of $\frac{1}{2}$" (13 mm) maximum openings between the toe board and the mid rail where persons are required to work or pass under the scaffold.
- Scaffold planks must extend over their end supports not less than 6" (150 mm) nor more than 12" (300 mm).
- If the scaffold does not have built-in ladders that meet the standard, then it must have an attached ladder access.
- All employees must be trained to erect, dismantle, and use scaffold(s).
- Unless it is impossible, fall protection must be worn while building or dismantling all scaffolding.
- Work platforms must be completely decked for use by employees.
- Use trash containers or other similar means to keep debris from falling and never throw or sweep material from above.

4.1.3 Lift Equipment

On the job, you may be working in the operating area of lifts, hoists, or cranes. The following safety rules are for those who are working in the area with overhead equipment but are not directly involved in its operation:

- Stay alert and pay attention to the warning signals from operators.
- Never stand or walk under a load, regardless of whether it is moving or stationary.
- Always warn others of moving or approaching overhead loads.
- Never attempt to distract signal persons or operators of overhead equipment.
- Obey warning signs.
- Do not use equipment that you are not qualified to operate.

Case History

Scaffolds and Electrical Hazards

Remember that scaffolds are excellent conductors of electricity. Recently, a maintenance crew needed to move a scaffold and although time was allocated in the work order to dismantle and rebuild the scaffold, the crew decided to push it instead. They did not follow OSHA recommendations for scaffold clearance and did not perform a job site survey. During the move, the five-tier scaffold contacted a 12,000V overhead power line. All four members of the crew were killed and the crew chief received serious injuries.

The Bottom Line: Never take shortcuts when it comes to your safety and the safety of others. Trained safety personnel should survey each job site prior to the start of work to assess potential hazards. Safe working distances should be maintained between scaffolding and power lines.

NCCER – *Maritime Electrical*

What's wrong with this picture?

Figure Credit: Mike Powers

- Cranes that are operated in areas with places in which a person can become trapped or pinched must have barricades placed around them to warn away workers.
- Hoists rigged in shafts or the outside of buildings must be secured to prevent them from being pulled down.
- Never overload a hoist, lift, or crane. Always follow lift ratings—there is no such thing as a safety factor that can be used to cheat a load.
- Cranes, lifts, and hoists must be inspected daily.
- Never lift a person using a crane, hoist, or material lift.
- Only people who have been trained in rigging should ever do rigging for a lift. It is extremely easy for a load to fall out of inappropriately rigged lifts.
- Personnel hoists require gates on every landing that can only be opened from the hoist side. This ensures that they are not opened onto a fall hazard.

Vertical Towers

This electrician is installing a light fixture using a vertical tower or Genie® lift. This lift is designed to fold up small enough so that it can be maneuvered through house doorways. Some models can be driven.

Figure Credit: Mike Powers

4.2.0 Proper Lifting Techniques

Back injuries cause many lost working hours every year. That is in addition to the misery felt by the person with the hurt back! Learn how to lift properly and size up the load. To lift, first stand close to the load. Then, squat down and keep your back straight and your chin up. Get a firm grip on the load and keep the load close to your body. Lift by straightening your legs. Make sure that you lift with your legs and not your back. Do not be afraid to ask for help if you feel the load is too heavy. An example of proper lifting is shown in *Figure 12*.

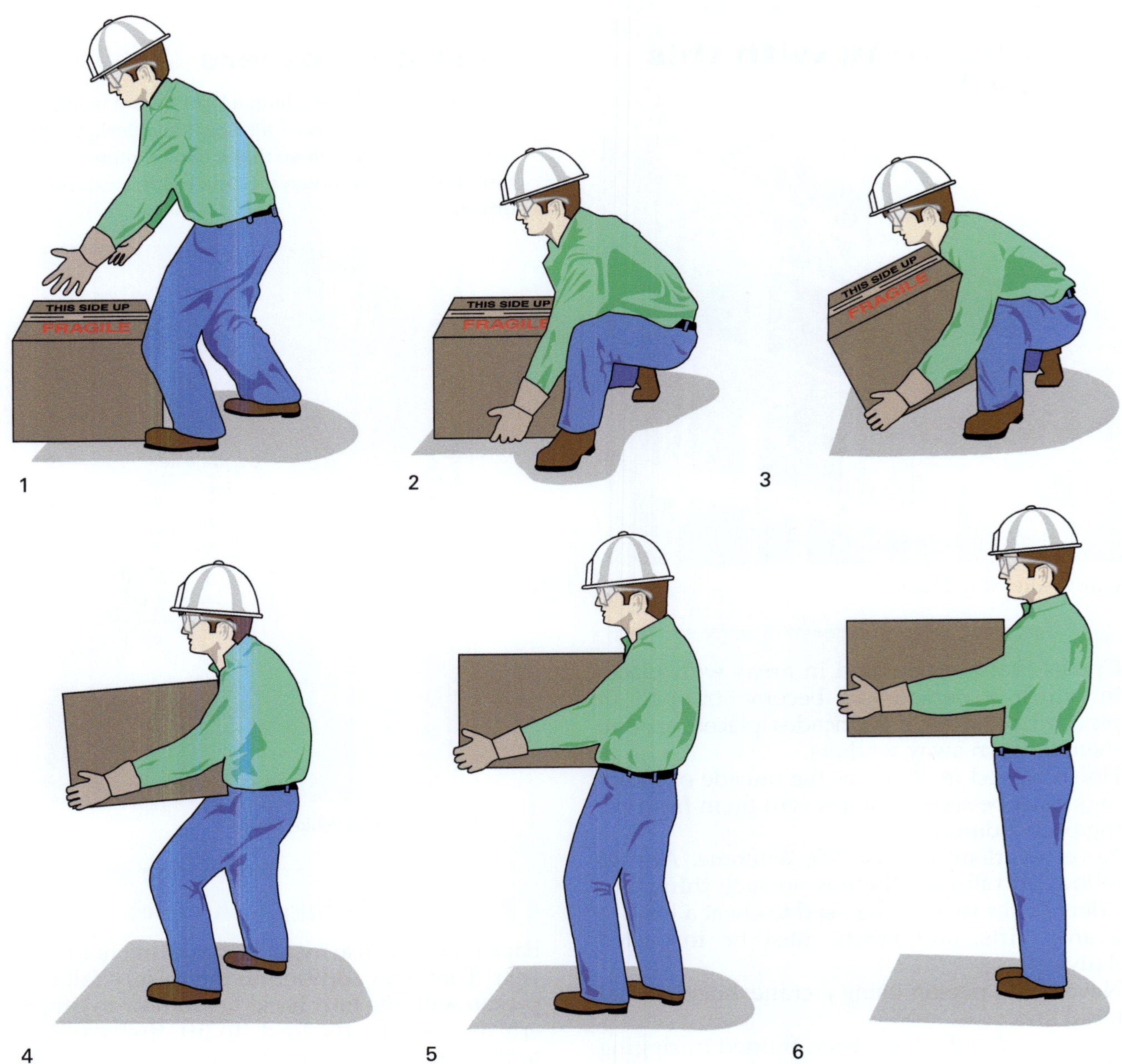

Figure 12 Proper lifting.

Lift Safely to Preserve Your B.A.C.K.

B – Balance: keep your stance wide, and get a good grip on the object.

A – Alignment: keep your back relaxed and upright.

C – Contract and close: contract your stomach muscles and hold loads close.

K – Knees: make sure you bend them, not your waist.

NCCER – *Maritime Electrical*

What's wrong with this picture?

Figure Credit: Mike Powers

Keep the following precautions in mind when lifting:

- Make the lift smoothly and under control.
- Move your feet to pivot; do not twist or you may injure yourself.
- Constantly scan the path ahead for obstructions. If you cannot see your path over or around the object being carried, then you must have help to transport the object.
- Avoid lifting objects over your head.
- Never lift over the side or tailgate of a pickup truck.
- Don't twist your body when lifting an object or setting it down.
- Never reach over an obstacle to lift a load.
- Don't step over objects in your way.

Lifting puts an extraordinary amount of pressure on your back. For example, if you bend from the waist to pick up an object weighing X, you are applying 10 times the amount of pressure (10X) to your lower back. Lower back injuries are one of the most common workplace injuries because it's so easy to be careless about lifting, especially when you are in a hurry. Remember, it is much easier to ask for help than it is to nurse an injured back.

4.3.0 Basic Tool Safety

When using any tools for the first time, read the operator's manual to learn the recommended safety precautions. If you are not certain about the operation of any tool, ask the advice of a more experienced worker. Before using a tool, you should know its function and how it works.

4.3.1 Hand Tool Safety

Hand tools are non-powered tools and may include anything from screwdrivers to cable strippers (*Figure 13*). Hand tools are dangerous if they are misused or improperly maintained.

Keep the following precautions in mind when using hand tools:

- Use tools only for their designated purpose.
- Always maintain your tools properly. If the wooden handle on a tool such as an axe or hammer is loose, splintered, or cracked, the head of the tool may fly off and strike the user or another person. Tag the tool out of service and do not use it.
- Repair or replace damaged or worn tools. A wrench with its jaws sprung might easily slip, causing hand injuries. If the wrench flies, it may strike the user or another person.
- Impact tools such as chisels, wedges, and drift pins are unsafe if they have mushroomed heads. The heads might shatter when struck, sending sharp fragments flying.
- Use bladed tools with the blades and points aimed away from yourself and other people.
- Store bladed tools properly; use the sheath or protective covering if there is one.
- Keep blades sharp and inspect them regularly. Dull blades are difficult to use and control and can be far more dangerous than well-maintained blades.
- Never leave tools on top of ladders or scaffolding.

In general, your risks are greatly reduced by inspecting and maintaining tools regularly and always wearing appropriate personal protective equipment such as safety goggles, hard hats, and filtering masks. Also, keep floors dry and clean to prevent accidental slips that can result in injuries caused by the tools you may be using.

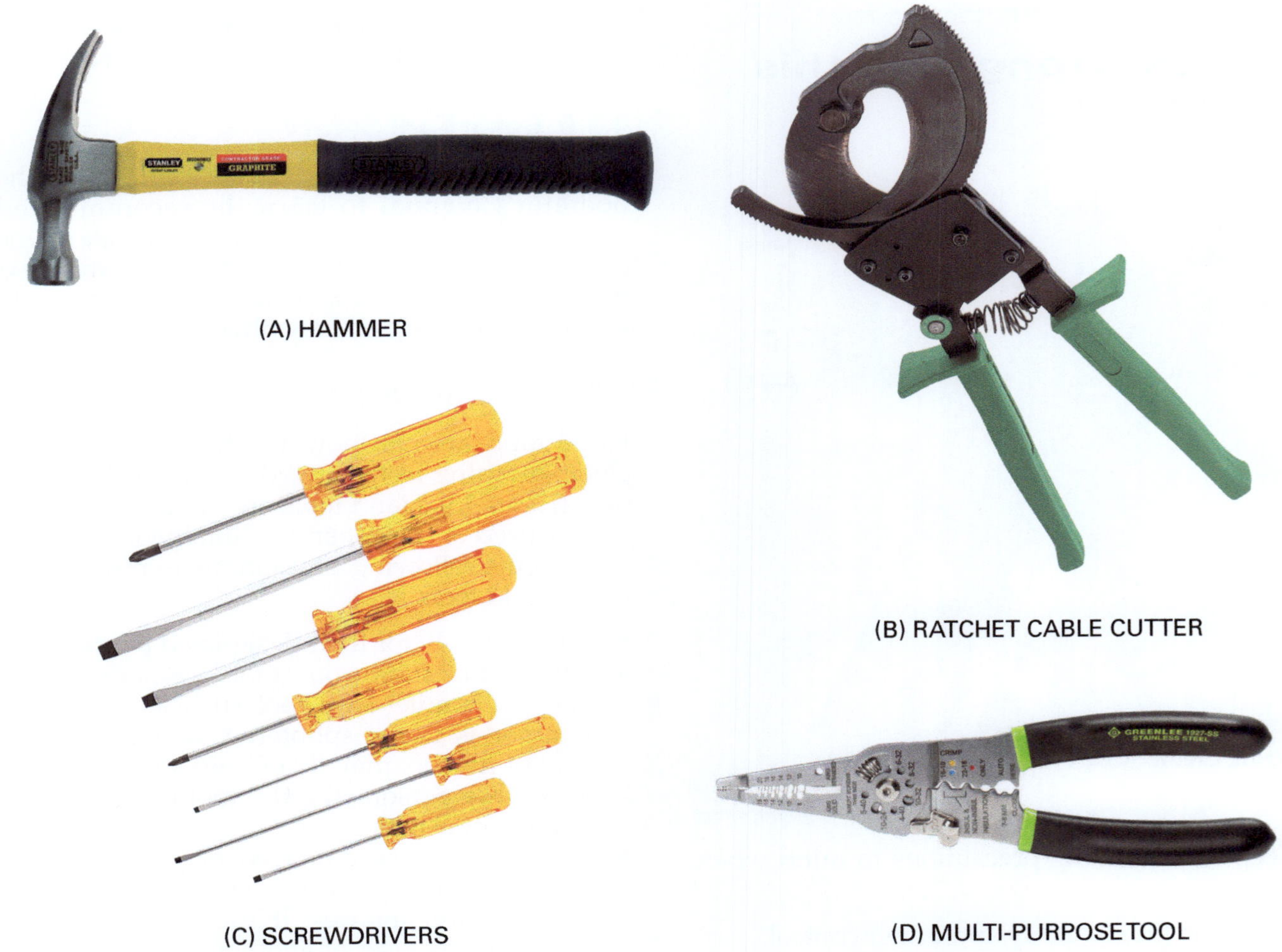

(A) HAMMER

(B) RATCHET CABLE CUTTER

(C) SCREWDRIVERS

(D) MULTI-PURPOSE TOOL

Figure 13 Hand tools.

4.3.2 *Power Tool Safety*

Power tools can be hazardous when they are improperly used or not well maintained. Most of the risks associated with hand tools are also risks when using power tools. When you add a power source to a tool, however, the risk factors increase. Power tools are powered by different sources. Some examples of power sources for power tools include the following:

- Electricity
- Pneumatics (air pressure)
- Liquid fuel (gasoline or propane)
- Hydraulics (fluid pressure)

You must know the safety rules and proper operating procedures for each tool you use. Specific operating procedures and safety rules for using a tool are provided in the operator's/user's manual supplied by the manufacturer. Before operating any power tool for the first time, always read the manual to familiarize yourself with the tool. If the manual is missing, contact the manufacturer for a replacement.

WARNING!

Never use a tool if you are unsure of how to use it correctly.

Follow these general guidelines to prevent accidents and injury:

- Inspect all tools for damage before use. Remove damaged tools from use and tag DO NOT USE.
- Never carry or lower a tool by the cord or hose.
- Keep cords and hoses away from heat, oil, and sharp edges.
- Do not attempt to operate any power tool before being checked out by your instructor on that particular tool.
- Always wear eye protection, a hard hat, and any other required personal protective equipment when operating power tools.
- Wear face and hearing protection when required.
- Wear proper respiratory equipment when necessary.

- Wear the appropriate clothing for the job being done. Wear close-fitting clothing that cannot become caught in moving tools. Roll up or button long sleeves, tuck in shirttails, and tie back long hair. Do not wear any jewelry, including watches or rings.
- Do not distract others or let anyone distract you while operating a power tool.
- Do not engage in horseplay.
- Do not run or throw objects.
- Consider the safety of others, as well as yourself. Observers should be kept at a safe distance away from the work area.
- Never leave a power tool running while it is unattended.
- Assume a safe and comfortable position before using a power tool. Be sure to maintain good footing and balance in order to respond to kickbacks, jumps, or sudden shifts.
- Secure work with clamps or a vise, freeing both hands to safely operate the tool.
- To avoid accidental starting, never carry a tool with your finger on the switch.
- Be sure that a power tool is properly grounded and connected to a ground fault circuit interrupter (GFCI) before using it.
- Ensure that power tools are disconnected before performing maintenance or changing accessories.
- Use a power tool only for its intended use.
- Keep your feet, fingers, and hair away from the blade and/or other moving parts of a power tool.
- Never use a power tool with guards or safety devices removed or disabled.
- Never operate a power tool if your hands or feet are wet.
- Keep the work area clean at all times.
- Become familiar with the correct operation and adjustments of a power tool before attempting to use it.
- Keep tools sharp and clean for the best performance.
- Follow the instructions in the user's manual for lubricating and changing accessories.
- Keep a firm grip on the power tool at all times.
- Use electric extension cords of sufficient size to service the particular power tool you are using.
- Many jobs require extension cords to be suspended. Use only nonconductive supports or listed products.
- Do not run extension cords across walkways where they will pose a tripping hazard.
- Do not run extension cords in areas of vehicle traffic or where wheeled carts are in use.
- Report unsafe conditions to your instructor or supervisor.
- Tools that shoot nails (*Figure 14*), rivets, or staples, and operate at pressures greater than 100 pounds per square inch (psi) or 689.5 kilopascals (kPa), must be equipped with a safety device that won't allow fasteners to be shot unless the muzzle is pressed against a work surface.
- Compressed-air guns should never be pointed toward anyone and the muzzle should never be pressed against a person.
- The use of stud-type guns (including both powder-actuated tools and tools powered by gasoline, rocket fuel, and battery/spring power) requires special training and certification.
- Never use explosive or flammable materials around stud-type guns.
- Never point stud-type guns at anybody.
- Never pick up an unattended stud-type gun. Instead, tell your supervisor that the tool has been left unattended.
- Never play with stud-type guns. These tools are as dangerous as a loaded firearm.

Figure 14 Pneumatic nail gun.

4.4.0 Confined Space Entry Procedures

Occasionally, you may be required to do your work in a confined space. If this is the case, you need to be aware of some special safety considerations. For details on the subject of working in manholes and vaults, refer to *CFR 1926.1200* through *1926.1213*. The general precautions are listed in the following paragraphs.

4.4.1 General Guidelines

A confined space includes (but is not limited to) any of the following: a manhole (*Figure 15*), boiler, tank, tunnel, hopper, bin, sewer, vat, pipeline, vault, pit, air duct, or vessel. A confined space is identified as follows:

- It has restricted entry and exit.
- It is not intended for continued human occupancy.
- It has the potential for other hazards, such as electrical, explosive, and animal.
- It has the potential for entrapment/engulfment.
- It has the potential for accumulating a dangerous atmosphere.

Figure 15 Manhole.

Entry into a confined space occurs when any part of the body crosses the plane of entry. No employee shall enter a confined space unless the employee has been trained in confined space entry procedures. Other requirements for confined spaces include the following:

- All hazards must be eliminated or controlled before a confined space entry is made.
- The air quality in the confined space must be continually monitored.
- All appropriate personal protective equipment shall be worn at all times during confined space entry and work.
- Ladders used for entry must be secured.
- A rescue retrieval system must be in use when entering confined spaces and while working in permit-required confined spaces (discussed later). Each employee must be capable of being rescued by the retrieval system, or a trained, practiced rescue squad must be available and able to respond immediately.
- Only no-entry rescues will be performed by company personnel. Entry rescues will be performed by trained rescue personnel identified on the entry permit.
- The area outside the confined space must be properly barricaded, and appropriate warning signs must be posted. *Figure 16* shows a confined space cover with a warning sign.
- Entry permits can only be issued and signed by a qualified person such as the job-site supervisor. Permits must be kept at the confined space while work is being conducted. At the end of the shift, the entry permits must be made part of the job journal and retained for one year.

4.4.2 Confined Space Hazard Review

Before determining the proper procedure for confined space entry, a hazard review shall be performed. The hazard review shall include, but not be limited to, the following conditions:

Figure 16 Ventilated confined space cover with warning sign.

- The past and current uses of the confined space
- The physical characteristics of the space including size, shape, air circulation, etc.
- Proximity of the space to other hazards
- Existing or potential hazards in the confined space, such as atmospheric conditions (oxygen levels, flammable/explosive levels, and/or toxic levels), presence/potential for liquids, presence/potential for particulates, or potential for engulfment
- Potential for mechanical/electrical hazards in the confined space (including work to be done)

Once the hazard review is completed, the supervisor, in consultation with the project managers and/or safety manager, shall classify the confined space as one of the following:

- A non-permit confined space
- A permit-required confined space

Permit-required confined spaces must be posted. Once the confined space has been properly classified, the appropriate entry and work procedures must be followed.

The controlling contractor must list all confined spaces on site and designate them as permit-required or non-permit confined spaces.

Only qualified and trained individuals may enter a confined space.

4.5.0 Dangerous Materials

Be prepared in case an accident occurs on the job site. First aid training that includes certification classes in CPR and artificial respiration could be the best insurance you and your fellow workers ever receive. Make sure that you know where first aid is available at your job site. Also, make sure you know the accident reporting procedure. Each job site should also have a first aid manual or booklet giving easy-to-find emergency treatment procedures for various types of injuries. Emergency first aid telephone numbers should be readily available to everyone on the job site. Refer to *CFR 1910.151/1926.23* and *1926.50* for specific requirements.

4.5.1 Solvents

The solvents that are used by electricians may give off vapors that are toxic enough to make people temporarily ill or even cause permanent injury. Many solvents are skin and eye irritants. Solvents can also be systemic poisons when they are swallowed or absorbed through the skin.

Solvents in spray or aerosol form are dangerous in another way. Small aerosol particles or solvent vapors mix with air to form a combustible mixture with oxygen. The slightest spark could cause an explosion in a confined area because the mix is perfect for fast ignition. There are procedures and methods for using, storing, and disposing of most solvents and chemicals. These procedures are normally found in the safety data sheets (SDSs) available at your facility.

An SDS is required for all materials that could be hazardous to personnel or equipment. These sheets contain information on the material, such as the manufacturer and chemical makeup. As much information as possible is kept on the hazardous material to prevent a dangerous situation; or, in the event of a dangerous situation, the information is used to rectify the problem in as safe a manner as possible. See *Figure 17* for an example of SDS information you may find on the job.

_______________ **Section 1 – Product & Company Identification** _______________

Product Name..........................:   RIDGID Dark Thread Cutting Oil
Product Catalog No...............:   41590, 70830, 41610, 41600

Recommended Use..............:   Thread Cutting

Company Name.....................:   Ridge Tool Company
Address.................................:   400 Clark Street
    Elyria, Ohio 44035-6001
Telephone.............................:   1-800-519-3456 (USA) (8:00 am – 5:00 pm EST, M-F)
Emergency Telephone..........:   call 9-1-1 or local emergency number
Website.................................:   www.RIDGID.com

Issue Date.............................:   May 29, 2015

_______________ **Section 2 – Hazards Identification** _______________

This product is classified as not hazardous per US OSHA 29CFR 1910.1200 (HazCom 2012) and Canada's Hazardous Products Regulations (WHMIS 2015).

GHS Label Elements: Not applicable

_______________ **Section 3 – Composition / Information On Ingredients** _______________

Component:	CAS #	% By Weight
Mineral Oil	Confidential	40-100%

This product does not contain silicone or chlorinated additives.

Specific chemical identities and/or exact percentages have been withheld as trade secrets

_______________ **Section 4 – First Aid Measures** _______________

INGESTION:
   Rinse mouth thoroughly. Call a Poison Center or doctor if you feel unwell. Do NOT induce vomiting.

INHALATION:
   Move to fresh air. Call a Poison Center or doctor if you feel unwell.

Figure 17 Portion of an SDS.

It is always best to use a nonflammable, nontoxic solvent whenever possible. However, any time solvents are used, it is essential that your work area be adequately ventilated and that you wear the appropriate personal protective equipment:

- Wear a chemical face shield with chemical goggles to protect the eyes and skin from sprays and splashes.
- Wear a chemical apron to protect your body from sprays and splashes. Remember that some solvents are acid-based. If they come into contact with your clothes, solvents can eat through your clothes to your skin.
- A paper filter mask does not stop vapors; it is used only for nuisance dust. In situations where a paper mask does not supply adequate protection, chemical cartridge respirators might be needed. These respirators can stop many vapors if the correct cartridge is selected. In areas where ventilation is a serious problem, a self-contained breathing apparatus (SCBA) must be used.
- Make sure that you have been given a full medical evaluation and that you are properly trained in using respirators at your site.

SDS Centers

SDSs are required to be in an accessible location. Many job sites provide SDS centers, such as that shown here.

Figure Credit: ZingGreen Safety Products

Chemical Safety

The first line of defense with chemicals is to read and follow the directions found on the container. If you follow these instructions, you should be safe from chemical exposure. Be aware that everyone reacts differently to chemicals and you may be hypersensitive to a particular chemical that does not bother your co-workers. Leave the area at the first sign of an allergic reaction and seek medical attention.

The best respiratory protection when using solvents (or other materials that present inhalation hazards) is to avoid the hazard entirely. Off-shift work, ventilation, and/or rescheduled work schedules should always be used to eliminate the need for working in areas with poor air quality. For example, in an area where hazardous solvents are used, the electrical work can be done off schedule when the solvents are not being used. When this cannot be done, protection against high concentrations of dust, mist, fumes, vapors, and gases is provided by appropriate respirators.

Appropriate respiratory protective devices should be used for the hazardous material involved and the extent and nature of the work performed.

An air-purifying respirator is, as its name implies, a respirator that removes contaminants from air inhaled by the wearer. The respirators may be divided into the following types: particulate-removing (mechanical filter), gas- and vapor-removing (chemical filter), and a combination of particulate-removing and gas- and vapor-removing.

Particulate-removing respirators are designed to protect the wearer against the inhalation of particulate matter in the ambient atmosphere. They may be designed to protect against a single type of particulate, such as pneumoconiosis-producing and nuisance dust, toxic dust, metal fumes or mist, or against various combinations of these types.

Gas- and vapor-removing respirators are designed to protect the wearer against the inhalation of gases or vapors in the ambient atmosphere. They are designated as gas masks, chemical-cartridge respirators (nonemergency gas respirators), and self-rescue respirators. They may be designed to protect against a single gas such as chlorine; a single type of gas, such as acid gases; or a combination of types of gases, such as acid gases and organic vapors.

If you are required to use a respiratory protective device, you must be evaluated by a physician to ensure that you are physically fit to use a respirator. You must then be fitted and thoroughly instructed in the respirator's use.

Any employee whose job entails having to wear a respirator must keep his face free of facial hair in the seal area.

> **WARNING!**
>
> Do not use any respirator unless you have been fitted for it and thoroughly understand its use. As with all safety rules, follow your employer's respiratory program and policies.
>
> Respiratory protective equipment must be inspected regularly and maintained in good condition. Respiratory equipment must be properly cleaned on a regular basis and stored in a sanitary, dustproof container.

4.5.2 Asbestos

Asbestos is a mineral-based material that is resistant to heat and corrosive chemicals. Depending on the chemical composition, asbestos fibers may range in texture from coarse to silky. The properties that make asbestos fibers so valuable to industry are its high tensile strength, flexibility, heat and chemical resistance, and good frictional properties. Asbestos fibers enter the body by inhalation of airborne particles or by ingestion and can become embedded in the tissues of the respiratory or digestive systems. Exposure to asbestos can cause numerous disabling or fatal diseases. Among these diseases are asbestosis, an emphysema-like condition; lung cancer; mesothelioma, a cancerous tumor that spreads rapidly in the cells of membranes covering the lungs and body organs; and gastrointestinal cancer. The use of asbestos was banned in 1978.

Case History

Altered Respiratory Equipment

A self-employed man applied a solvent-based coating to the inside of a tank. Instead of wearing the proper respirator, he used nonstandard air supply hoses and altered the face mask. All joints and the exhalation slots were sealed with tape. He collapsed and was not discovered for several hours.

The Bottom Line: Never alter or improvise safety equipment.

Because asbestos was still in the manufacturing pipeline for a while after it was banned, you need to assume that any facility constructed before 1980 has asbestos in it. The owner must have a survey with any work rules needed to work safely around the asbestos. Common products that contain asbestos include thermal pipe insulation, mastic for ducts and insulation, spray-on fireproofing, floor tiles, ceiling tiles, roof insulation, exterior building sheathing, old wire insulation, and even pipe. As an electrician, you must not drill through or otherwise work with asbestos—you can only be trained to work around it when it can be done safely. Asbestos work is a trade that requires special training and protective equipment.

The signs shown in *Figure 18* must be placed in areas containing asbestos.

4.5.3 Batteries

Working around wet cell batteries can be dangerous if the proper precautions are not taken. Batteries often give off hydrogen gas as a byproduct. When hydrogen mixes with air, the mixture can be explosive in the proper concentration. For this reason, smoking is strictly prohibited in battery rooms, and only insulated tools should be used. Proper ventilation also reduces the chance of explosion in battery areas. Follow your company's procedures for working near batteries. Also, ensure that your company's procedures are followed for lifting heavy batteries.

> **WARNING!**
>
> Battery-powered scissor lifts may have unsealed batteries that require water levels to be checked and topped off. Never use a flame to look inside a battery—it can cause an explosion. Charging batteries with low water levels can damage them.

4.5.4 Acids

Batteries also contain acid, which will eat away human skin and many other materials. Personal protective equipment for battery work typically includes chemical aprons, sleeves, gloves, face shields, and goggles to prevent acid from contacting skin and eyes. Follow your site procedures for dealing with spills of these materials. Also, know the location of first aid when working with these chemicals.

Because of the chance that battery acid may contact someone's eyes or skin, wash stations are located near battery rooms. Do not connect or disconnect batteries without proper supervision.

> **DANGER**
> ASBESTOS
> CANCER AND LUNG DISEASE HAZARD
> AUTHORIZED PERSONNEL ONLY
> RESPIRATORS AND PROTECTIVE
> CLOTHING ARE REQUIRED IN THIS AREA

> **DANGER**
> CONTAINS ASBESTOS FIBERS
> AVOID CREATING DUST
> CANCER AND LUNG DISEASE HAZARD

Figure 18 Danger signs for areas containing asbestos

Everyone who works in the area should know where the nearest wash station is and how to use it. Battery acid should be flushed from the skin and eyes with large amounts of water or with a neutralizing solution.

> **CAUTION**
>
> If you come in contact with battery acid, flush the affected area with water and report it immediately to your supervisor.

4.5.5 PCBs and Vapor Lamps

Polychlorinated biphenyls (PCBs) are chemicals that were marketed under various trade names as a liquid insulator/cooler in older transformers. In addition to being used in older transformers, PCBs are also found in some large capacitors and in the small ballast transformers used in street lighting and ordinary fluorescent light fixtures. Disposal of these materials is regulated by the Environmental Protection Agency (EPA) and must be done through a regulated disposal company; use extreme caution and follow your facility procedures.

> **WARNING!**
>
> Do not come into contact with PCBs. They present a variety of serious health risks, including lung damage and cancer.

In addition, any vapor lamps, such as fluorescent, halide, or mercury vapor lamps, must be recycled. The tubes must be packaged and handled carefully to avoid breakage.

4.5.6 Lead Safety

In 2010, the Environmental Protection Agency (EPA) enacted the Renovation, Repair, and Painting (RRP) rule. This standard is designed primarily to protect young children living in or spending time in buildings containing lead-based paint. It requires lead-safe certification and work practices to be followed during renovation work in all homes or other child-occupied facilities built prior to 1978. The rule is triggered by disturbing more than 6 sq ft (0.6 sq m) per room of an interior painted surface or more than 20 sq ft (1.8 sq m) of an exterior painted surface.

Housing and Urban Development (HUD) projects have a lower trigger of only 2 sq ft (0.2 sq m) of interior lead-based paint, more than 20 sq ft (1.8 sq m) of exterior lead-based paint, or 10 percent of the total surface area on an interior or exterior painted component that contains lead-based paint. This can include relatively small areas such as window sills, baseboard, and trim.

The RRP rule contains the following basic requirements:

- At least one RRP certified renovator is required at each job site. Certification involves lead-safe work training by an EPA-accredited training provider.
- In addition to individual certification, the contracting firm or agency must also be certified.
- Contractors must give the client a copy of the "Renovate Right" pamphlet (available for download at **www.epa.gov**).

4.5.7 *Silica Safety*

Exposure to silica can cause lung cancer, silicosis, chronic obstructive pulmonary disease, and kidney disease. According to OSHA, approximately two million construction workers are exposed to respirable crystalline silica on the job each year. This includes those who drill, cut, crush, or grind materials such as concrete and stone. To combat the adverse health effects of occupational exposure to silica, OSHA has introduced a silica rule. This rule reduces the permissible exposure limit (PEL) for respirable crystalline silica to 50 micrograms per cubic meter of air, averaged over an 8-hour shift. It also requires employers to use the following controls to limit worker exposure:

- Implement engineering controls (for example, tools equipped with shrouds or dust collection systems and tools equipped with integrated water delivery systems) to limit worker exposure to the PEL.
- Provide respirators when engineering controls cannot adequately limit exposure.
- Limit worker access to high exposure areas.
- Develop a written exposure control plan.
- Offer medical exams to monitor highly exposed workers and supply information about their lung health.
- Train workers on silica risks and how to limit exposures.

OSHA estimates that this rule will save over 600 lives and prevent more than 900 new cases of silicosis each year. *Table 4A* and *Table 4B* provide OSHA exposure controls for various types of equipment/tasks. Following the guidelines in this table will ensure compliance with the standard.

4.6.0 Fall Protection

All employees must receive documented training before working in any area where there is the possibility of exposure to a fall of 6' (1.8 m) or more. This rule does not apply to ladders, scaffolds, stairways, and mechanical lifts, which have their own standards.

Fall protection must be used when employees are on a walking or working surface that is 6' (1.8 m) or more above a lower level and has an unprotected edge or side. The areas covered include, but are not limited to the following:

> **WARNING!**
>
> An edge or side where there is no guardrail system at least 39" (1 m) high is considered unprotected.

- Equipment
- Finished and unfinished floors or mezzanines
- Temporary or permanent walkways/ramps
- Finished or unfinished roof areas
- Elevator shafts and hoistways
- Floor, roof, or walkway holes
- Work areas that are 6' (1.8 m) or more above dangerous equipment (*Exception*: If the dangerous equipment is unguarded, fall protection must be used at all heights regardless of the fall distance.)

Figure 19 shows a simple temporary guardrail installed on the stairway of a residence under construction. *Figure 20* shows a complex guardrail system used during the construction of a department store. The large central opening will eventually house the building escalators.

Figure 19 Temporary guardrail on stairs.

Figure 20 Complex guardrail system.

According to OSHA, an employee can never be exposed to a fall of more than 6' (1.8 m). This is called 100 percent fall protection. The first order of protection is always to eliminate the hazard if possible. If it is not possible to eliminate the hazard, the employee must be protected by one of the following in this order of preference:

1. Guardrail system
2. Fall restraint (limiting the length of a lanyard so that it does not reach the edge of a building or roof)
3. Personal fall arrest system (PFAS)
4. Controlled access zone or other administrative system

4.6.1 Guardrail System

Guardrail systems must be constructed as follows:

- The top rails must be 42" (+/–3") or 1.1 m (+/–75 mm) and must be capable of withstanding 200 lbs (90.7 kg).
- The mid-rails must be at 21" (+/–3") or 530 mm (+/–75 mm) and must be capable of withstanding 150 lbs (68.0 kg).
- The toe board must be 4" tall (100 mm) and no more than $\frac{1}{4}$" (6 mm) above the floor for drainage.
- Guardrails can be made of 2 × 4s, pipes, chains, or cables.
- Chains and cables require flags every 6' (1.8 m). If cables are used, they must be secured to avoid deflection greater than 3" (75 mm) in, out, or down from the 42" (1.1 m) requirement.
- Banding material is not allowed for guardrail construction. Cable guardrails require the use of cable clamps on the cable. Clamps must be forged and not malleable. They must be torqued and installed properly. Ensure that the clamp does not damage the loadbearing line.

Warning lines, signs, or barricades must be installed back from the edge at least 6' (1.8 m). That way, even if someone falls over the barrier, they will not fall to the level below.

Table 4A OSHA Exposure Controls for Silica

Equipment/Task	Engineering and Work Practice Control Methods	Required Respiratory Protection and Minimum Assigned Protection Factor (APF)	
		≤ 4 hours/shift	**> 4 hours/shift**
Stationary masonry saws	Use saw equipped with integrated water delivery system that continuously feeds water to the blade. Operate and maintain tool in accordance with manufacturer's instructions to minimize dust emissions.	None	None
Handheld power saws (any blade diameter)	Use saw equipped with integrated water delivery system that continuously feeds water to the blade. Operate and maintain tool in accordance with manufacturer's instructions to minimize dust emissions.	None when used outdoors APF 10 when used indoors or in an enclosed area	None when used outdoors APF 10 when used indoors or in an enclosed area
Handheld power saws for cutting fiber-cement board (with blade diameter of 8 inches or less)	For tasks performed outdoors only: Use saw equipped with commercially available dust collection system. Operate and maintain tool in accordance with manufacturer's instructions to minimize dust emissions. Dust collector must provide the air flow recommended by the tool manufacturer, or greater, and have a filter with 99% or greater efficiency.	None	None
Walk-behind saw	Use saw equipped with integrated water delivery system that continuously feeds water to the blade. Operate and maintain tool in accordance with manufacturer's instructions to minimize dust emissions.	None when used outdoors APF 10 when used indoors or in an enclosed area	None when used outdoors APF 10 when used indoors or in an enclosed area
Rig-mounted core saws or drills	Use tool equipped with integrated water delivery system that supplies water to cutting surface. Operate and maintain tool in accordance with manufacturer's instructions to minimize dust emissions.	None	None
Handheld and stand-mounted drills (including impact and rotary hammer drills)	Use drill equipped with commercially available shroud or cowling with dust collection system. Operate and maintain tool in accordance with manufacturer's instructions to minimize dust emissions. Dust collector must provide the air flow recommended by the tool manufacturer, or greater, and have a filter with 99% or greater efficiency and a filter-cleaning mechanism. Use a HEPA-filtered vacuum when cleaning holes.	None	None

Table 4B OSHA Exposure Controls for Silica

Equipment/Task	Engineering and Work Practice Control Methods	Required Respiratory Protection and Minimum Assigned Protection Factor (APF)	
		≤ 4 hours/shift	> 4 hours/shift
Jackhammers and handheld powered chipping tools	Use tool with water delivery system that supplies a continuous stream or spray of water at the point of impact. OR Use tool equipped with commercially available shroud and dust collection system. Operate and maintain tool in accordance with manufacturer's instructions to minimize dust emissions. Dust collector must provide the air flow recommended by the tool manufacturer, or greater, and have a filter with 99% or greater efficiency and a filter-cleaning mechanism.	None when used outdoors APF 10 when used indoors or in an enclosed area	APF 10 when used outdoors or indoors/in an enclosed area
Handheld grinders for uses other than mortar removal	For tasks performed outdoors only: Use grinder equipped with integrated water delivery system that continuously feeds water to the grinding surface. Operate and maintain tool in accordance with manufacturer's instructions to minimize dust emissions. OR Use grinder equipped with commercially available shroud and dust collection system. Operate and maintain tool in accordance with manufacturer's instructions to minimize dust emissions. Dust collector must provide 25 cubic feet per minute (cfm) or greater of airflow per inch of wheel diameter and have a filter with 99% or greater efficiency and a cyclonic pre-separator or filter-cleaning mechanism.	None	None when used outdoors APF 10 when used indoors or in an enclosed area
Heavy equipment and utility vehicles for tasks such as grading and excavating but not including: demolishing, abrading, or fracturing silica-containing materials	Apply water and/or dust suppressants as necessary to minimize dust emissions. OR When the equipment operator is the only employee engaged in the task, operate equipment from within an enclosed cab.	None	None

What's wrong with this picture?

Figure Credit: Mike Powers

4.6.2 *Personal Fall Arrest System*

A personal fall arrest system (PFAS) provides fall arrest after a worker falls. This equipment must be selected, inspected, donned, anchored, and maintained to be effective. The complete system usually consists of a full-body harness, lanyard, and anchorage device.

Full-body harnesses – Full-body harnesses are the only acceptable equipment to wear for PFAS. Select the appropriate harness based on size and gender. Inspect the equipment before use. Harnesses must be worn snug (but not tight) with all required straps attached. When properly applied, you should be able to slide two fingers under the straps with little difficulty. The D-ring in the back of the harness must be centered between the shoulder blades. After donning the harness, have a co-worker pull sharply up on the D-ring. You should feel the grab around the thighs, chest, and buttocks. Jobs that require positioning must be accomplished using a full-body harness with side D-rings. Safety belts are not allowed.

Lanyards – Lanyards are used to connect the harness to the attachment point. As no employee can be exposed to a fall of more than 6' (1.8 m), standard lanyards must be no longer than 6' (1.8 m). You can be exposed to 1,800 lbs of force (816.5 kg) in a properly worn harness. The use of shock absorbing lanyards or retractable lanyards can reduce that force to as low as 400 to 600 lbs (181.4 to 272.2 kg). Shock absorbers work by slowing the employee to a stop by ripping stitches while elongating up to 42" (1.1 m). All lanyards must have locking snap hooks. Never attach two locking snap hooks to the same D-ring as they can foul each other, causing the relatively weak gates to break. *Figure 21* shows an electrician making an adjustment to an outdoor wallwasher lighting fixture while hanging out of a 15th-story window. For extra protection, he is attached to two lanyards on two separate D-rings.

A twin-tailed lanyard is required when climbing. While climbing, you cannot unhook your lanyard to move it to another anchorage and still have 100 percent fall protection. Thus, with two lanyards, you can "walk" to where you are working.

Retractable lanyards come in a variety of sizes from 6' to over 150' (1.8 m to 45.7 m). *Figure 22* shows a worker tied off to a retractable lanyard

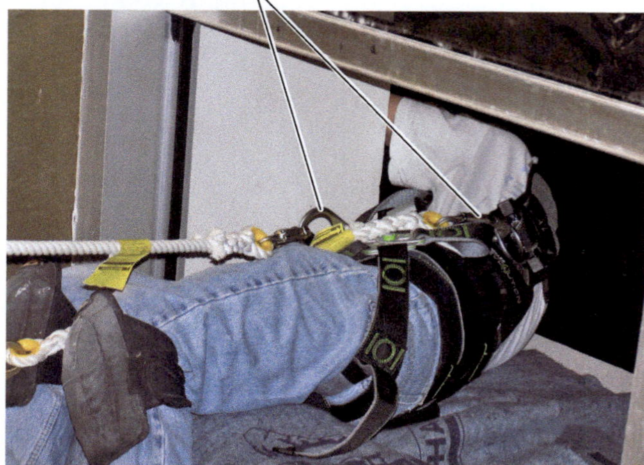

Figure 21 Electrician tied off to two lanyards.

when working on a boom lift. Retractable lanyards are used where movement or close proximity with the ground will render standard lanyards ineffective or inefficient. Retractable lanyards can be longer than 6' (1.8 m) because when a fall occurs they grab hold within 2' (0.6 m). This quick reaction also eliminates the need for a shock absorber.

> **WARNING!**
> Do not use a shock absorber in line with a retractable lanyard unless it is manufactured into the system because it may interfere with the response of the lanyard.

OSHA standards require PFAS to control falls for any working surface 6' (1.8 m) or more above the ground. If you are forced to attach below the D-ring on your harness, you will fall that distance plus the 6' (1.8 m) length of a standard lanyard. This could result in a fall of up to 12' (3.7 m). A PFAS is not designed to meet those forces and will transfer much higher impacts when it arrests the fall. To combat this, the American National Standards Institute (ANSI) has changed the equipment manufacturing standards for fall protection. The new standard is commonly referred to as the 11' (3.3 m) lanyard. The lanyard is still only 6' (1.8 m) long, but the allowable deployment (extendable length) of the shock absorber has been increased. This lessens the force transferred to the body through the harness, reducing the likelihood of injury.

Before using a PFAS, you must examine the space below any potential fall point. Make sure it is clear of any hazards or obstructions. When assessing the required free length of a PFAS, you

Figure 22 Proper fall protection on a boom lift.

What's wrong with this picture?

Figure Credit: Mike Powers

must account for the deployment of the shock absorber. In addition, PFAS manufacturers require up to a 3' (0.9 m) safety factor. Taking this safety factor into account, *Table 5* indicates the minimum distance required between any hazard/obstruction and the worker or attachment point.

As you can see, a typical single-story building may not provide enough height to use standard lanyards. Retractable lanyards that engage and stop falls in less than 2' (0.6 m) are usually the best choice for fall protection when used below 20' (6.1 m) or on any type of lift equipment.

ANSI has also addressed the locking mechanism (gate) on snap hooks. OSHA requires the gate to be rated for 350 lbs (158.8 kg) when the rest of the system is rated at 5,000 lbs (2,268 kg). This creates a weak link in the PFAS. ANSI issued a new standard increasing the strength of these gates to 3,600 lbs (1,632.9 kg).

Anchorage devices – Anchorage devices and points are the interface between the PFAS and the structure to which they are attached. This point must hold 5,000 lbs (2,268 kg). This is the equivalent of a full-size extended cab pickup truck.

A lanyard cannot be wrapped around an anchorage and then attached to itself unless it is specially designed with cross arm straps. Cross arm straps are made of webbing, 2" (50 mm) wide, of any necessary length, with two different size D-rings. The cross-arm straps are passed over whatever object you are going to attach them to and wrapped around to reduce the length of the lanyard. Beam clamps, wire hangers, trolleys, and other manufactured devices are also used for specific applications.

Equipment inspection process – All PFAS must be inspected when received and before each use. Check the manufacturer's tag for the manufacturer's inspection date. If the fall equipment has no date, it should be disposed of immediately. Carefully look over the webbing. If you observe any burns, ripped stitches, color marker threads, distorted grommets, bent or cracked buckle tongues, distorted D-rings, or bent, cracked, or pulled fabrics, remove the PFAS from service. Retractable lanyards must undergo the same inspection process as other PFAS. In addition, you must also pull out the entire lanyard for inspection, let it back in, and then pull out 2' to 4' (0.6 to 1.2 m) of lanyard and give it a swift tug to see if it properly engages. If any of these inspections fail, the equipment must be destroyed immediately or tagged DO NOT USE and returned to the shop.

> **WARNING!**
> All fall protection equipment that is involved in a fall must be taken out of service and destroyed. Any employees involved in a fall must receive medical attention, even if they do not feel they have been injured. Falls can cause internal injuries that are not readily apparent to the victim.

Rescue – Never use fall protection equipment to pull anyone up; always use ladders or equipment to rescue from below. If the standard equipment is not available to provide rescue, a plan must be created before work can proceed. Rescues must be accomplished from below using ladders, lifts, and/or scaffolds.

> **WARNING!**
> Unless a fallen person is in immediate danger, never attempt to lift the worker up by the lanyard. This could cause an additional drop for the fallen worker and/or injure the rescuers.

Table 5 Required Free Lengths for Various Lanyards

Lanyard Type	Lanyard Length	Shock Absorber Length	Average Worker Height	Safety Factor	Required Free Length
Standard 6' lanyard with 3.5' shock absorber	6'	3.5'	6'	3'	18.5'
New ANSI 6' lanyard with 4' shock absorber	6'	4'	6'	3'	19'
New ANSI 11' lanyard with 5' shock absorber	6' (11' freefall)	5'	N/A (accounted for by lanyard)	3'	20'

What's wrong with this picture?

Figure Credit: Mike Powers

Immediately summon the fire department to assist in the rescue effort unless you can rescue the person without assistance. Rescue must take place as quickly as possible, as hanging from a harness presents additional hazards. If you fall, continue to move your limbs while awaiting rescue. This will help maintain circulation in your lower extremities.

> **WARNING!**
>
> Per OSHA, any employee whose weight, including tools, exceeds 310 lbs (140.6 kg) cannot wear fall protection.

PFAS selection – The type of system selected depends on the fall hazards associated with the work to be performed. First, a hazard analysis must be conducted by the job-site supervisor prior to the start of work. Based on the hazard analysis, the job-site supervisor and project manager, in consultation with the safety manager, will select the appropriate fall protection system. All employees must be instructed in the use of the fall protection system before starting work.

4.6.3 Controlled Access Zones

There are times when a guardrail cannot be attached to the building. In these cases, a controlled access zone may be the solution. A controlled access zone may consist of guards, barricades, badge systems, or other administrative measures. Roofers are allowed to be within 6' (1.8 m) of the edge of a roof. All other workers must remain 15' (4.6 m) from the edge. *Figure 23* shows a controlled access zone on a roof. With a controlled access zone, even a worker who trips over the barricade won't fall off the roof.

Figure 23 Controlled access zone.

Additional Resources

29 CFR Parts 1900–1910, Standards for General Industry. Occupational Safety and Health Administration, U.S. Department of Labor.

29 CFR Part 1926, Standards for the Construction Industry. Occupational Safety and Health Administration, U.S. Department of Labor.

4.0.0 Section Review

1. Maintaining three-point contact on a ladder means that ______.

   a. you must either have both feet and one hand on the ladder or both hands and one foot as you climb
   b. at least three of the four rail points (two side rails and two feet) must be resting on a solid surface
   c. you may reach as far to the left and right as required as long as you have one hand and both feet on the ladder
   d. all splices must be attached at three points

2. Which of the following is true with regard to lifting?

   a. Step carefully over objects rather than going around them.
   b. To avoid foot injuries, stand as far away as possible when lifting a load.
   c. Lift with your back muscles.
   d. Lift by straightening your legs.

3. Which of the following is true with regard to hand tools?

   a. Point the blade of a sharp tool away from yourself to avoid injury if it slips during use.
   b. A dull tool is safer than a sharp one.
   c. The side of a wedge can be used as a hammer if necessary.
   d. Mushroomed heads on impact tools are preferred because they offer a larger surface area for striking.

4. Which of the following is considered a confined space?

   a. Narrow alleyway between buildings
   b. Shallow trench
   c. Grain hopper
   d. Crawl space under open decking

5. Fluorescent light fixtures may contain ______.

   a. acid
   b. lead
   c. CFCs
   d. PCBs

6. Fall protection is required when working at elevations of ______.

   a. 6' (1.8 m) or more
   b. 7' (2.1 m) or more
   c. 8' (2.4 m) or more
   d. 10' (3 m) or more

Summary

Safety must be your concern at all times so that you do not become either the victim of an accident or the cause of one. Safety requirements and safe work practices are provided by OSHA and your employer. It is essential that you adhere to all safety requirements and follow your employer's safe work practices and procedures. Also, you must be able to identify the potential safety hazards of your job site. The consequences of unsafe job-site conduct can often be expensive, painful, or even deadly. Report any unsafe act or condition immediately to your supervisor. You should also report all work-related accidents, injuries, and illnesses to your supervisor immediately. Remember, proper construction techniques, common sense, and a good safety attitude will help to prevent accidents, injuries, and fatalities.

1. The most life-threatening hazards on a construction site typically include all of the following *except* _____.

    a. falls
    b. electric shock
    c. being crushed or struck by falling or flying objects
    d. chemical burns

2. If a person's heart begins to fibrillate due to an electrical shock, the solution is to _____.

    a. leave the person alone until the fibrillation stops
    b. immerse the person in ice water
    c. use the Heimlich maneuver
    d. have a qualified person use an AED

3. Low-voltage conductors _____.

    a. are not powerful enough to cause death from electrocution
    b. must exceed 480V to cause death
    c. are responsible for most electrocution deaths due to the frequency of contact
    d. are unlikely to cause injury if rubber-soled shoes are worn

4. Class 0 rubber gloves are used when working with voltages less than _____.

    a. 500 volts
    b. 1,000 volts
    c. 5,000 volts
    d. 7,500 volts

5. An important use of a hot stick is to _____.

    a. replace busbars
    b. test for voltage
    c. replace fuses
    d. test for continuity

6. Which of these statements correctly describes a double-insulated power tool?

    a. There is twice as much insulation on the power cord.
    b. It can safely be used in place of a grounded tool.
    c. It is made entirely of plastic or other non-conducting material.
    d. The entire tool is covered in rubber.

7. Which of the following applies in a lockout/tagout procedure?

    a. Only the supervisor can install lockout/tagout devices.
    b. If several employees are involved, the lockout/tagout equipment is applied only by the first employee to arrive at the disconnect.
    c. Lockout/tagout devices applied by one employee can be removed by another employee as long as it can be verified that the first employee has left for the day.
    d. Lockout/tagout devices are installed by every authorized employee involved in the work.

8. The *NEC*® covers the _____.

    a. minimum installation requirements for electrical equipment
    b. manufacturing specifications for electrical equipment
    c. workplace hazards associated with electrical equipment
    d. operating instructions for electrical equipment

9. What is the proper distance from the feet of a straight ladder to the wall?

    a. one-fourth the working height of the ladder
    b. one-half the height of the ladder
    c. a distance equal to the height of the first three rungs
    d. one-fourth of the square root of the height of the ladder

10. What are the minimum and maximum distances (in inches) that a scaffold plank can extend beyond its end support?

    a. 4; 8
    b. 6; 10
    c. 6; 12
    d. 8; 12

11. What happens to the permits used to enter a confined space?

   a. They are reviewed at the end of each shift and discarded at the completion of the job.
   b. They are recorded in the job journal at the end of each shift and retained for a year.
   c. They are recorded in the job journal at the end of each shift and submitted to OSHA.
   d. They are reviewed at the end of each shift and retained for three months.

12. The best way to protect yourself from solvent hazards is to ______.

   a. always wear vinyl gloves and a paper filter mask
   b. ask a co-worker for instructions
   c. ask the supplier
   d. read and follow all instructions on the product's SDS

13. It is safe to assume that asbestos may be found in any facility constructed before ______.

   a. 1980
   b. 1985
   c. 1990
   d. 1995

14. Which of the following applies to vapor lamps?

   a. You may throw vapor lamps out with regular trash as long as you wrap them carefully to avoid breakage.
   b. Disposal of these materials is regulated by the Environmental Protection Agency (EPA).
   c. You may dispose of up to six vapor lamps per year in regular trash.
   d. These lamps do not contain harmful materials.

15. A PFAS anchorage point for a 6' (1.8 m) lanyard must be able to hold ______.

   a. 250 lbs (113.4 kg)
   b. 500 lbs (226.8 kg)
   c. 1,000 lbs (453.6 kg)
   d. 5,000 lbs (2,268 kg)

Trade Terms Quiz

Fill in the blank with the correct term that you learned from your study of this module.

1. A life-threatening condition of the heart in which the muscle fibers contract irregularly is called __________.

2. Tools that have a case made of nonconductive material and have been constructed so that the case is insulated from electrical energy are called __________.

3. Chemicals known as __________ are often used to cool certain types of large transformers and capacitors.

4. A __________ will de-energize a circuit or a portion of it if the current to ground exceeds some predetermined value.

5. A __________ has a three-prong plug at the end of its power cord or some other means to ensure that stray current travels to ground without passing through the body of the operator.

Trade Terms

Double-insulated/ungrounded tools
Fibrillation
Grounded tool

Ground fault circuit interrupter (GFCI)
Polychlorinated biphenyls (PCBs)

1. *CFR Part 1910* covers the

    ____

2. *CFR Part 1926* covers the

    ____

3. True or False? You should use a compressor for an air test on high-voltage gloves because it has more pressure and is much faster.

    ____

4. There are six classes of rubber protective equipment. List the six classes and their voltage ratings.

    ____

5. All conductors, buses, and connections should be considered ___________ until proven otherwise.

6. The distance from the ladder feet to the base of the wall or support should be about ___________ the vertical distance from the bottom to the top of the ladder.

7. As it is erected, each part of a scaffolding shall be carefully ___________

8. List the general guidelines used in identifying a confined space.

    ____

9. Before determining the proper procedure for confined space entry, a(n) ___________ must be performed.

10. Fall protection must be used when employees are on a walking or working surface that is ___________ or more above a lower level and has an unprotected edge or side.

Michael J. Powers
Tri-City Electrical Contractors, Inc.

How did you choose a career in the electrical field?

My father was an electrician and after I "burned out" with a career in fast-food management, I decided to choose a completely different field.

Tell us about your apprenticeship experience.

It was excellent! I worked under several very knowledgeable electricians and had a pretty good selection of teachers. Over my four-year apprenticeship, I was able to work on a variety of jobs, from photomats to kennels to colleges.

What positions have you held and how did they help you to get where you are now?

I have been an electrical apprentice, a licensed electrician, a job-site superintendent, a master electrician, and am currently a corporate safety and training director. The knowledge I acquired in electrical theory in apprenticeship school and preparing for my licensing exams, as well as the practical on-the-job experience over thirty years in the trade, were wonderful training for my current position.

I also serve on the authoring team for NCCER's Electrical curricula, which has provided me with not only the opportunity to share what I have learned, but is also a great way to keep current in other areas by meeting with electricians from a variety of disciplines (we have commercial, residential, and industrial electricians on the team, as well as instructors).

What would you say was the single greatest factor that contributed to your success?

Choosing a company that recognized and rewarded competent, hard workers and provided them with the support and guidance to allow them to develop and succeed in the industry.

What does your current job entail?

I am responsible for safe work practices and procedures through the job-site management team at Tri-City Electrical Contractors, Inc. I also assist in developing, delivering, and administering the training program, from apprenticeship to in-house to outsourced training.

What advice do you have for trainees?

Training in all its aspects is the key to your success and advancement in the industry. Any time you are given a training opportunity, take it, even if it might not appear relevant at the time. Eventually, all knowledge can be applied to some situation.

Most importantly, have fun! The construction industry is composed of good people. I firmly believe that construction workers, as a group, are much more honest and direct than any other comparable group. Wait—did I say comparable group? That's a misstatement—there is no comparable group. Construction workers build America!

Trade Terms Introduced in This Module

Double-insulated/ungrounded tools: Electrical tools that are constructed so that the case is insulated from electrical energy. The case is made of a nonconductive material.

Fibrillation: Very rapid irregular contractions of the muscle fibers of the heart that result in the heartbeat and pulse going out of rhythm with each other.

Grounded tool: An electrical tool with a three-prong plug at the end of its power cord or some other means to ensure that stray current travels to ground without passing through the body of the user. The ground plug is bonded to the conductive frame of the tool.

Ground fault circuit interrupter (GFCI): A protective device that functions to de-energize a circuit or portion thereof within an established period of time when a current to ground exceeds some predetermined value. This value is less than that required to operate the overcurrent protective device of the supply circuit.

Polychlorinated biphenyls (PCBs): Toxic chemicals that may be contained in liquids used to cool certain types of large transformers and capacitors.

Additional Resources

This module presents thorough resources for task training. The following resource material is suggested for further study.

29 CFR Parts 1900–1910, Standards for General Industry. Occupational Safety and Health Administration, U.S. Department of Labor.

29 CFR Part 1926, Standards for the Construction Industry. Occupational Safety and Health Administration, U.S. Department of Labor.

Managing Electrical Hazards, Latest Edition. Upper Saddle River, NJ: Pearson Education, Inc.

National Electrical Code®Handbook, Latest Edition. Quincy, MA: National Fire Protection Association.

Standard for Electrical Safety in the Workplace (NFPA 70E®), Latest Edition. Quincy, MA: National Fire Protection Association.

Figure Credits

The Master Lock Company, Module Opener, Figures 8, 16

U.S. Department of Labor, Tables 1, 2, and 4

Mike Powers, Figures 3, 9, 10, 11, 15, 18–22

Brady Corporation, Figure 7B

©Stanley Black & Decker, Figure 13A

Greenlee / A Textron Company, Figure 13B, 13D

Klein Tools, Inc., Figure 13C

Courtesy of RIDGID®

RIDGID® is the registered trademark of RIDGID, Inc., Figures 14, 17

Section Review Answer Key

Answer	Section Reference	Objective
Section One		
1. b	1.1.0	1a
2. c	1.2.0	1b
Section Two		
1. b	2.1.0	2a
2. a	2.2.1	2b
Section Three		
1. b	3.1.2	3a
2. c	3.2.0	3b
Section Four		
1. a	4.1.1	4a
2. d	4.2.0	4b
3. a	4.3.1	4c
4. c	4.4.1	4d
5. d	4.5.5	4e
6. a	4.6.0	4f

NCCER CURRICULA — USER UPDATE

NCCER makes every effort to keep its textbooks up-to-date and free of technical errors. We appreciate your help in this process. If you find an error, a typographical mistake, or an inaccuracy in NCCER's curricula, please fill out this form (or a photocopy), or complete the online form at **www.nccer.org/olf**. Be sure to include the exact module ID number, page number, a detailed description, and your recommended correction. Your input will be brought to the attention of the Authoring Team. Thank you for your assistance.

Instructors – If you have an idea for improving this textbook, or have found that additional materials were necessary to teach this module effectively, please let us know so that we may present your suggestions to the Authoring Team.

NCCER Product Development and Revision

13614 Progress Blvd., Alachua, FL 32615

Email: curriculum@nccer.org
Online: www.nccer.org/olf

❏ Trainee Guide   ❏ Lesson Plans   ❏ Exam   ❏ PowerPoints   Other ________________

Craft / Level: Copyright Date:

Module ID Number / Title:

Section Number(s):

Description:

Recommended Correction:

Your Name:

Address:

Email: Phone:

Electrical Theory

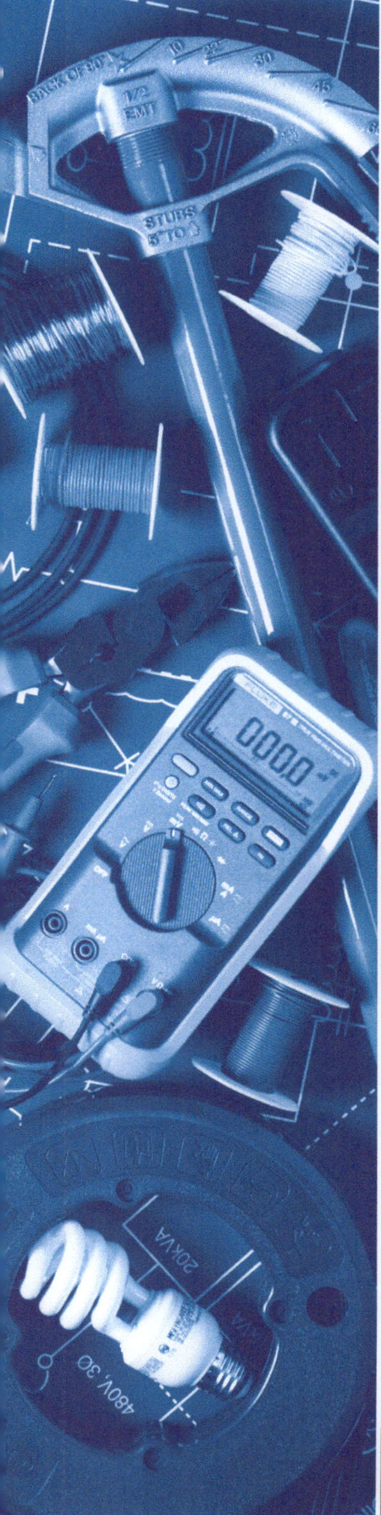

OVERVIEW

Knowledge of electrical circuits is essential in the electrical field. Sound understanding of basic circuits, as well as the methods for calculating the electrical energy within them, forms the foundation for utilizing these principles in practical applications. This module explains how to apply Ohm's law to series, parallel, and series-parallel circuits. It also covers Kirchhoff's voltage and current laws.

Module 26104-17

ELECTRICAL THEORY

Objectives

When you have completed this module, you will be able to do the following:

1. Calculate values in resistive circuits.
   a. Identify resistances in series.
   b. Identify resistances in parallel.
   c. Simplify series-parallel circuits.
   d. Apply Ohm's law to various types of circuits.
2. Apply Kirchhoff's laws to various types of circuits.
   a. Use Kirchhoff's current law.
   b. Use Kirchhoff's voltage law.

Performance Tasks

1. This is a knowledge-based module. There are no performance tasks.

Trade Terms

Kirchhoff's current law
Kirchhoff's voltage law
Parallel circuits
Series circuit
Series-parallel circuits

Industry Recognized Credentials

If you are training through an NCCER-accredited sponsor, you may be eligible for credentials from NCCER's Registry. The ID number for this module is 26104-17. Note that this module may have been used in other NCCER curricula and may apply to other level completions. Contact NCCER's Registry at 888.622.3720 or go to **www.nccer.org** for more information.

Note

NFPA 70®, *National Electrical Code*® and *NEC*® are registered trademarks of the National Fire Protection Association, Quincy, MA.

Contents

Figures

1.0.0 RESISTIVE CIRCUITS

Objective

Calculate values in resistive circuits.
  a. Identify resistances in series.
  b. Identify resistances in parallel.
  c. Simplify series-parallel circuits.
  d. Apply Ohm's law to various types of circuits.

Trade Terms

Parallel circuits: Circuits containing two or more parallel paths through which current can flow.

Series circuit: A circuit that has only one path for current flow.

Series-parallel circuits: Circuits that contain both series and parallel current paths.

Resistance, which is measured in ohms (Ω), is calculated in different ways depending on the type of circuit. Different equations are used for series and parallel circuits. Resistance calculations are often used to determine other circuit characteristics, such as voltage and current.

> **NOTE**
>
> There are two theories of electron flow. Most electricians use *electron flow theory*, which describes current flow from negative to positive. However, electronics engineers frequently use *conventional electron flow*, which describes current flow in the opposite direction (from positive to negative). To many, this makes it easier to analyze complex electronics. This curriculum uses electron flow theory (from negative to positive). It is important to be aware of this to avoid communication issues and miscalculations.

1.1.0 Resistances in Series

A **series circuit** is a circuit that has only one path for current flow. In the series circuit shown in *Figure 1*, the current (I) is the same in all parts of the circuit. This means that the current flowing through R_1 is the same as the current flowing through R_2 and R_3, and it is also the same as the current supplied by the battery. Unknown values can be found using any variation of Ohm's law

or the power equation. The three forms of Ohm's law (used to find voltage, resistance, or current) are as follows:

To find voltage:

$$E = I \times R$$

To find resistance:

$$R = E \div I$$

To find current:

$$I = E \div R$$

Where:

  E = voltage in volts (V)
  I = current in amperes, or amps (A)
  R = resistance in ohms (Ω)

The three forms of the power equation (used to find power, current, or voltage) are as follows:

To find power:

$$P = I \times E$$

To find voltage:

$$E = P \div I$$

To find current:

$$I = P \div E$$

Where:

  P = power in watts (W)
  I = current in amperes
  E = voltage in volts

When resistances are connected in series, like the example shown in *Figure 1*, the total resistance in the circuit is equal to the sum of all the resistances in the circuit:

$$R_T = R_1 + R_2 + R_3$$

Where:

  R_T = total resistance
  R_1, R_2, and R_3 = each of the resistances in series (There can be any number of these.)

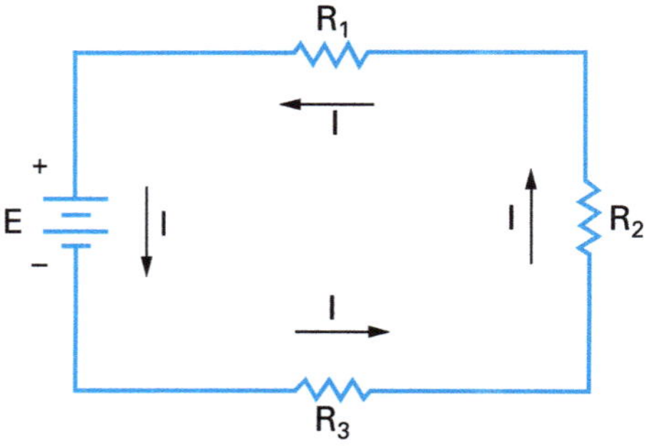

Figure 1 Series circuit.

Example 1:

The circuit shown in *Figure 2 (A)* has 50Ω, 75Ω, and 100Ω resistors in series. Find the total resistance of the circuit.

Add the values of the three resistors in series:

$$R_T = R_1 + R_2 + R_3$$
$$R_T = 50Ω + 75Ω + 100Ω$$
$$R_T = 225Ω$$

The total resistance is 225Ω.

Example 2:

The circuit shown in *Figure 2 (B)* has three lamps connected in series with the resistances shown. Find the total resistance of the circuit.

Add the values of the three lamp resistances in series:

$$R_T = R_1 + R_2 + R_3$$
$$R_T = 20Ω + 40Ω + 60Ω$$
$$R_T = 120Ω$$

The total resistance is 120Ω.

1.2.0 Resistances in Parallel

Circuits that contain two or more parallel paths through which current can flow are called **parallel circuits**. The total resistance in a parallel circuit is given by the formula:

$$R_T = \cfrac{1}{\cfrac{1}{R_1} + \cfrac{1}{R_2} + \cfrac{1}{R_3}}$$

Where:

R_T = total resistance in parallel
R_1, R_2, and R_3 = each of the branch resistances
(There can be any number of these.)

Example 1:

Find the total resistance of the 2Ω, 4Ω, and 8Ω resistors in parallel shown in *Figure 3*.

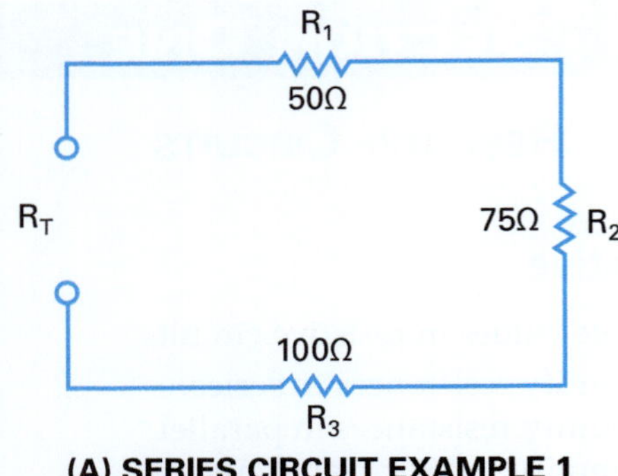

(A) SERIES CIRCUIT EXAMPLE 1

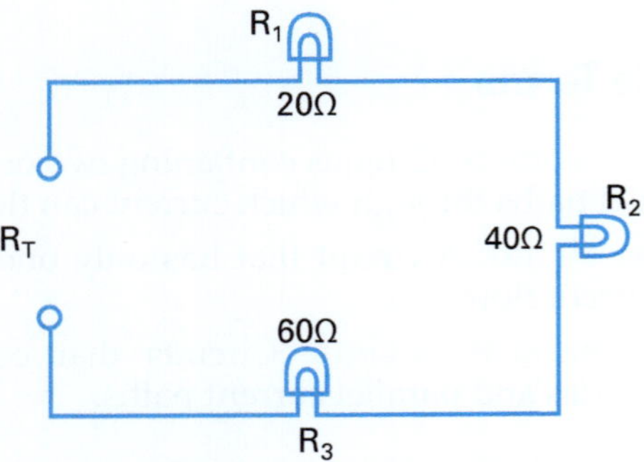

(B) SERIES CIRCUIT EXAMPLE 2

Figure 2 Example series circuits.

To find the total resistance, use the following steps:

Step 1 Write the formula for the three resistances in parallel:

$$R_T = \cfrac{1}{\cfrac{1}{R_1} + \cfrac{1}{R_2} + \cfrac{1}{R_3}}$$

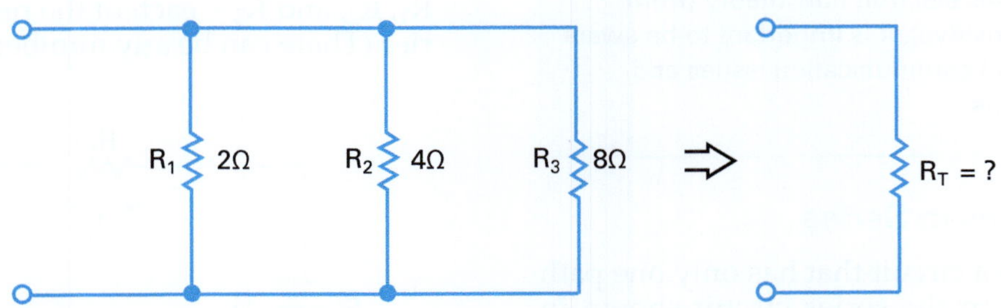

Figure 3 Parallel branch.

NCCER – *Maritime Electrical*

Series Circuits

Simple series circuits are often used as voltage dividers, but are seldom encountered in practical wiring. The only simple series circuit you may recognize is an older strand of Christmas lights, in which the entire string went dead when one lamp burned out. Think about what the actual wiring of a series circuit would look like in household receptacles. How would the circuit physically be wired? What kind of illumination would you get if you wired your household receptacles in series and plugged half a dozen lamps into those receptacles?

Step 2 Substitute the resistance values:

$$R_T = \cfrac{1}{\frac{1}{2} + \frac{1}{4} + \frac{1}{8}}$$

$$R_T = \cfrac{1}{0.5 + 0.25 + 0.125}$$

$$R_T = \frac{1}{0.875} = 1.14\Omega$$

Note that when resistances are connected in parallel, the total resistance is always less than the resistance of any single branch. In this example:

$$R_T = 1.14\Omega \;<\; R_1 = 2\Omega$$
$$R_T = 1.14\Omega \;<\; R_2 = 4\Omega$$
$$R_T = 1.14\Omega \;<\; R_3 = 8\Omega$$

Example 2:

Add a fourth parallel resistor of 2Ω to the circuit in *Figure 3*. What is the new total resistance, and what is the net effect of adding another resistance in parallel?

To find the total resistance, use the following steps:

Step 1 Write the formula for four resistances in parallel:

$$R_T = \cfrac{1}{\frac{1}{R_1} + \frac{1}{R_2} + \frac{1}{R_3} + \frac{1}{R_4}}$$

Step 2 Substitute the resistance values:

$$R_T = \cfrac{1}{\frac{1}{2} + \frac{1}{4} + \frac{1}{8} + \frac{1}{2}}$$

$$R_T = \cfrac{1}{0.5 + 0.25 + 0.125 + 0.5}$$

$$R_T = \frac{1}{1.375} = 0.73\Omega$$

The net effect of adding another resistance in parallel is a reduction of the total resistance from 1.14Ω to 0.73Ω.

1.2.1 Simplified Formulas

When there are only two resistors in a parallel circuit, it is often easier to calculate the total resistance by multiplying the two resistances and dividing the product by the sum of the resistances. This is sometimes called the *product-over-sum method*, and it is shown in the following equation (this formula only works if there are two resistors in the circuit):

$$R_T = \frac{R_1 \times R_2}{R_1 + R_2}$$

Where:

R_T = total resistance of unequal resistors in parallel
R_1, R_2 = two unequal resistors in parallel

When resistors in parallel are all equal to one another, their total resistance is equal to the resistance of one resistor divided by the number of resistors in the circuit, as shown in the following equation:

$$R_T = \frac{R}{N}$$

Where:

R_T = total resistance of equal resistors in parallel
R = resistance of one of the equal resistors
N = number of equal resistors

If two resistors with the same resistance are connected in parallel, their equivalent resistance is equal to one resistor of half of that value, as shown in *Figure 4*.

The two 200Ω resistors in parallel are the equivalent of one 100Ω resistor; the two 100Ω resistors are the equivalent of one 50Ω resistor; and the two 50Ω resistors are the equivalent of one 25Ω resistor.

Example 1:

What is the total resistance of a 6Ω (R_1) resistor and an 18Ω (R_2) resistor in parallel?

Because there are only 2 resistors, the product-over-sum method can be used:

$$R_T = \frac{R_1 \times R_2}{R_1 + R_2}$$

$$R_T = \frac{6 \times 18}{6 + 18}$$

$$R_T = \frac{108}{24} = 4.5\Omega$$

Example 2:

Find the total resistance of a 100Ω (R_1) resistor and a 150Ω (R_2) resistor in parallel.

Because there are only 2 resistors, the product-over-sum method can be used:

$$R_T = \frac{R_1 \times R_2}{R_1 + R_2}$$

$$R_T = \frac{100 \times 150}{100 + 150}$$

$$R_T = \frac{1500}{250} = 60\Omega$$

1.3.0 Series-Parallel Circuits

Finding current, voltage, and resistance in series circuits and parallel circuits is fairly easy. When working with either type of arrangement (series or parallel), use only the rules that apply to that type. In **series-parallel circuits**, some parts of the circuit are connected in series, and other parts are connected in parallel. Thus, in some parts the rules for series circuits apply, and in other parts, the rules for parallel circuits apply.

To analyze or solve a problem involving a series-parallel circuit, it is necessary to recognize which parts of the circuit are series connected and which parts are parallel connected. This can be obvious if the circuit is simple. However, more complex circuits must be redrawn, putting it into a form that is easier to recognize.

In a series circuit, the current is the same at all points. A parallel circuit has one or more points where the current divides and flows in separate

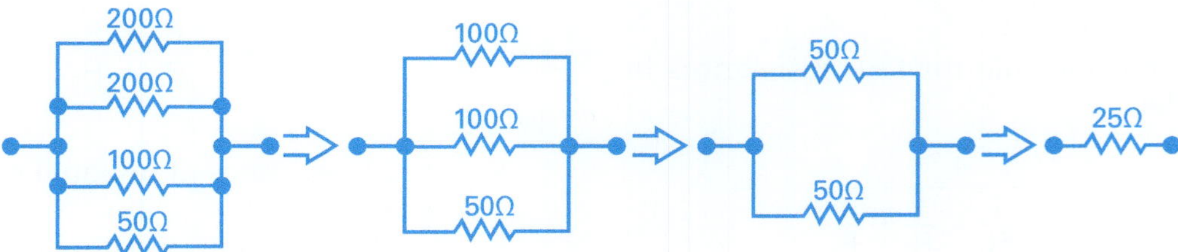

Figure 4 Equal resistances in a parallel circuit.

Think About It

Parallel Circuits

An interesting fact about circuits is the drop in resistance in a parallel circuit as more resistors are added. But this fact does not mean that you can add an endless number of devices, such as lamps, in a parallel circuit. Why not?

branches. A series-parallel circuit has both separate branches and series loads. The easiest way to find out whether a circuit is a series, parallel, or series-parallel circuit is to start at the negative terminal of the power source and trace the path of current through the circuit back to the positive terminal of the power source. If the current does not divide anywhere, it is a series circuit. If the current divides into separate branches, but there are no series loads, it is a parallel circuit. If the current divides into separate branches and there are also series loads, it is a series-parallel circuit. *Figure 5* shows electric lamps connected in series, parallel, and series-parallel circuits.

After determining that a circuit is series-parallel, redraw the circuit so that the branches and the series loads are more easily recognized. This is especially helpful when computing the total resistance of the circuit. *Figure 6* (*A*) shows resistors connected in a series-parallel circuit, and *Figure 6* (*B*) shows the equivalent circuit redrawn in a simpler form.

1.3.1 Reducing Series-Parallel Circuits

Most of the time, all that is known about a series-parallel circuit is the applied voltage and the values of the individual resistances. To find the voltage drop across any of the loads or the current in any of the branches, the total circuit current must also be known. To find the total current, the total resistance of the circuit must be known.

To find the total resistance, reduce the circuit to its simplest form. Usually, this means simplifying the values of all the resistors into a single,

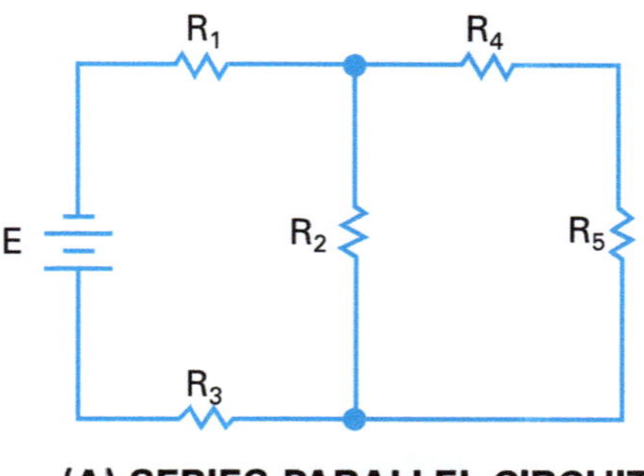

(A) SERIES-PARALLEL CIRCUIT

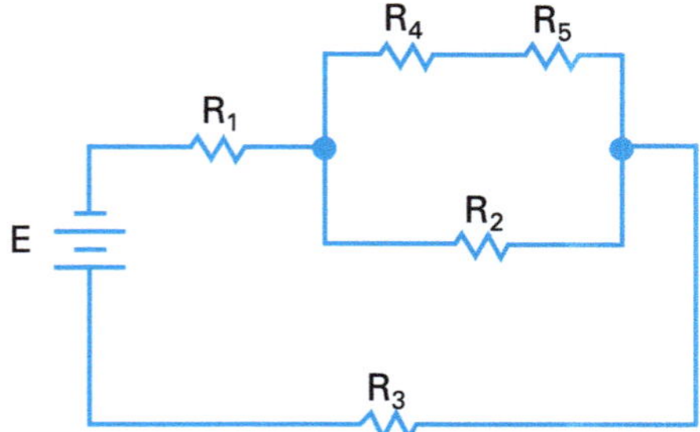

(B) REDRAWN VERSION OF THE CIRCUIT

Figure 6 Redrawing a series-parallel circuit.

Think About It

Series-Parallel Circuits

Explain *Figure 6*. Which resistors are in series and which are in parallel?

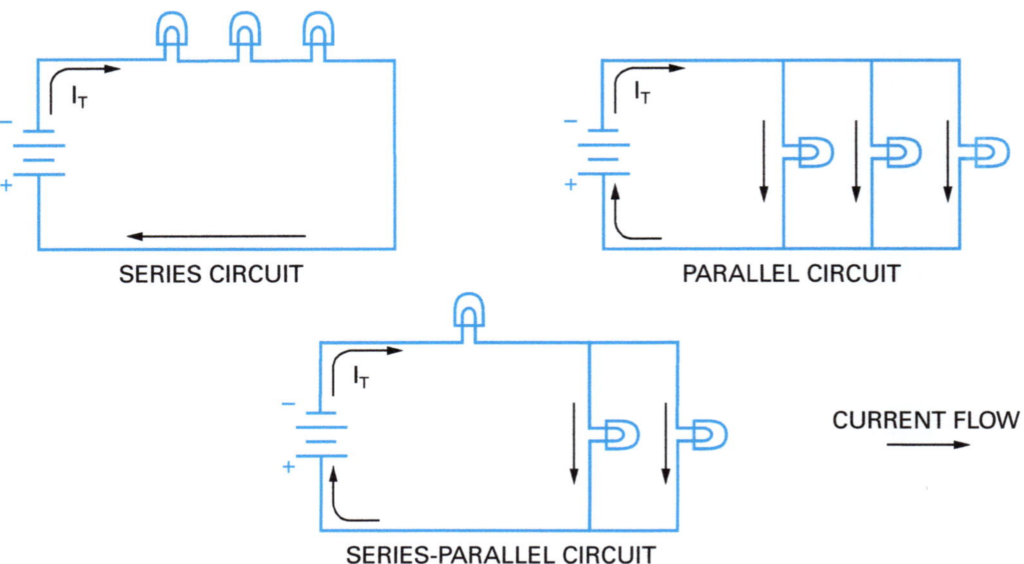

Figure 5 Series, parallel, and series-parallel circuits.

equivalent (effective) value. This simple series circuit has the equivalent resistance of the series-parallel circuit it was derived from, and also has the same total current. There are four basic steps in reducing a series-parallel circuit:

Step 1   If necessary, redraw the circuit so that all parallel combinations of resistances and series resistances are easily recognized.

Step 2   For each parallel combination of resistances, calculate their effective resistance (their equivalent series resistance).

Step 3   Replace each of the parallel combinations with one resistance whose value is equal to the effective resistance of that combination. This provides a circuit with all series loads.

Step 4   Find the total resistance of this circuit by adding the resistances of all the series loads.

The following are four general steps for finding voltage drops across individual resistors; this step-by-step process will be further explained and illustrated in the following sections:

Step 1   Reduce the circuit into a simpler form by converting any parallel resistances into an equivalent series resistance value.

Step 2   Find the total resistance by adding all of the series resistances in the simplified circuit.

Step 3   Now that you have the total resistance (R_T), you can use this along with the applied voltage (E_T) to find the total current (I_T). Do this by using Ohm's law to solve for I_T ($I_T = E_T \times R_T$).

Step 4   Use Ohm's law with the individual resistances and total current in order to find the voltage drops across each resistor.

Examine the series-parallel circuit shown in *Figure 7* and reduce it to an equivalent series circuit.

Parallel Circuits

Most practical circuits are wired in parallel, like the pole lights shown here.

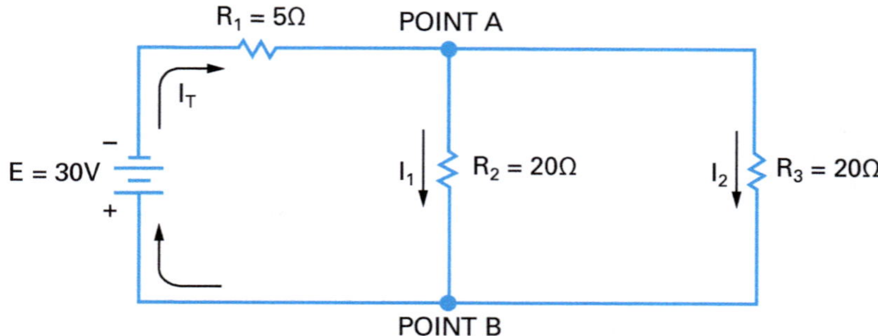

Figure 7 Reducing a series-parallel circuit.

In this circuit, resistors R_2 and R_3 are connected in parallel, but resistor R_1 is in series with both the battery (applied voltage) and the parallel combination of R_2 and R_3. The current I_T leaving the negative terminal of the voltage source travels through resistor R_1 before it is divided at the junction of resistors R_1, R_2, and R_3 (Point A). It then passes through the two branches formed by resistors R_2 and R_3.

Given the information in *Figure 7*, you can calculate the resistance of R_2 and R_3 in parallel, and then add it to the series resistance R_1 to find the total resistance of the circuit, R_T.

The total resistance of the circuit is the sum of R_1 and the equivalent resistance of R_2 and R_3 in parallel. To find R_T, first find the resistance of R_2 and R_3 in parallel. Because the two resistances have the same value of 20Ω, the resulting equivalent resistance is 10Ω. (As previously discussed, two equal resistances in parallel are equivalent to half of one of the resistances.) Therefore, the total resistance is 15Ω (5Ω + 10Ω).

For example, given the information in *Figure 8*, calculate the total resistance (R_T) and the total current (I_T).

Find R_T:

$$R_T = R_1 + R_2 + R_3$$
$$R_T = 20 + 50 + 120$$
$$R_T = 190\Omega$$

Find I_T using Ohm's law:

$$I_T = \frac{E_T}{R_T}$$

$$I_T = \frac{95}{190}$$

$$I_T = 0.5A$$

Now find the voltage drop across each individual resistor. In a series circuit, the current is the same through each resistor; that is, I = 0.5A through each resistor. Now that you know the current and resistance for each individual

1.4.0 Ohm's Law

Ohm's law is one of the primary tools used to find unknown values in resistive circuits. As previously discussed, unknown circuit parameters (voltage, current, and resistance) can be found by using Ohm's law and the techniques for determining equivalent resistance.

1.4.1 Applying Ohm's Law in Series Circuits

In resistive circuits, you can find unknown circuit parameters by using Ohm's law combined with the techniques for determining equivalent resistance. Ohm's law may be applied to an entire series circuit, and it may also be applied to the individual parts of the circuit. When it is used on a particular part of a circuit, the voltage across that part is equal to the current in that part multiplied by the resistance of that part.

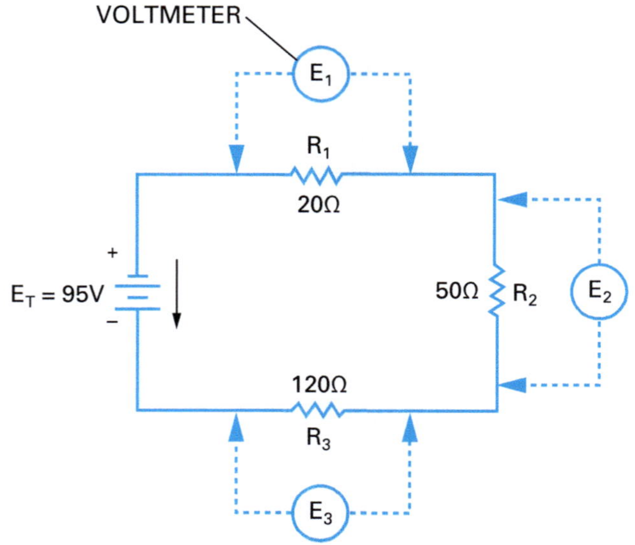

Figure 8 Calculating voltage drops.

resistor, you can use Ohm's law to calculate each resistor's voltage drop:

$$E_1 = I_1 \times R_1 = 0.5A \times 20\Omega = 10V$$
$$E_2 = I_2 \times R_2 = 0.5A \times 50\Omega = 25V$$
$$E_3 = I_3 \times R_3 = 0.5A \times 120\Omega = 60V$$

As we have just calculated, the voltage drops E_1, E_2, and E_3 in *Figure 8* are 10V, 25V, and 60V, respectively. Voltage drops are also known as *IR drops*. Their effect is to reduce the voltage that is available to be applied across the rest of the components in the circuit. The sum of the voltage drops in any series circuit is always equal to the voltage that is applied to the circuit. (The total voltage, E_T, is the same as the applied voltage.) In this example, $E_T = 10 + 25 + 60 = 95V$.

1.4.2 Applying Ohm's Law in Parallel Circuits

A parallel circuit is a circuit in which two or more components branch off of the same voltage source. The branches are like their own individual series circuits, but they all connected to the same battery (voltage source), as illustrated in *Figure 9*. The resistors R_1, R_2, and R_3 are in parallel with each other and with the battery. Each parallel branch has its own individual current. When the total current I_T leaves the voltage source E_T, part I_1 of the current the total current will flow through R_1, part I_2 will flow through R_2, and the remainder I_3 will flow through R_3. The branch currents (I_1, I_2, and I_3) can be different. However, if a voltmeter is connected across R_1, R_2, and R_3, the voltage drops E_1, E_2, and E_3 will each be equal to the source voltage E_T.

The total current I_T is equal to the sum of all branch currents. This formula applies for any number of parallel branches, whether the resistances are equal or unequal.

Using Ohm's law, the current for an individual branch equals the voltage for that branch divided by the total resistance of that branch. Since the voltage drop across each branch of a parallel circuit is equal to the applied voltage, you do not have to calculate individual voltage drops for each branch. The following equations illustrate this for the three branches in *Figure 9*:

$$\text{Branch 1: } I_1 = \frac{E_1}{R_1} = \frac{E_T}{R_1}$$

$$\text{Branch 2: } I_2 = \frac{E_2}{R_2} = \frac{E_T}{R_2}$$

$$\text{Branch 3: } I_3 = \frac{E_3}{R_3} = \frac{E_T}{R_3}$$

With the same applied voltage, any branch that has less resistance allows more current through it than a branch with higher resistance. The sum of the branch currents in a parallel circuit is equal to the total current (I_T).

Example 1:

The two branches R_1 and R_2, shown in *Figure 10 (A)*, draw a total line current of 20A across a 110V power line. (R_1 and R_2 are in parallel.) Branch R_1 takes 12A. What is the current I_2 in branch R_2?

Since we know that the total current (I_T) is equal to the sum of each branch current, we can rearrange the formula to find I_2, and then substitute given values, as follows:

$$I_T = I_1 + I_2$$
$$I_2 = I_T - I_1$$
$$I_2 = 20 - 12 = 8A$$

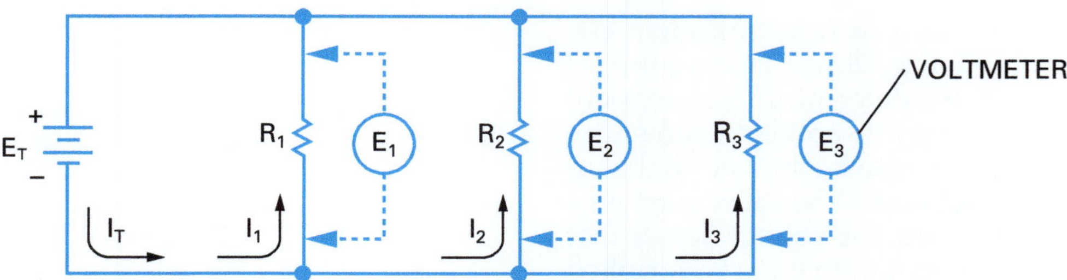

Figure 9 Parallel circuit.

NCCER – *Maritime Electrical*

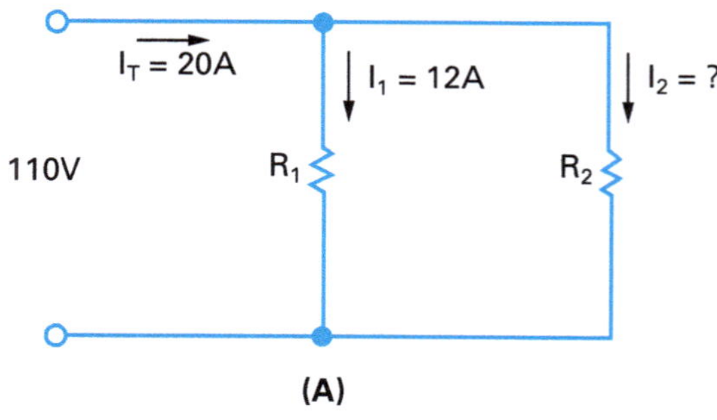

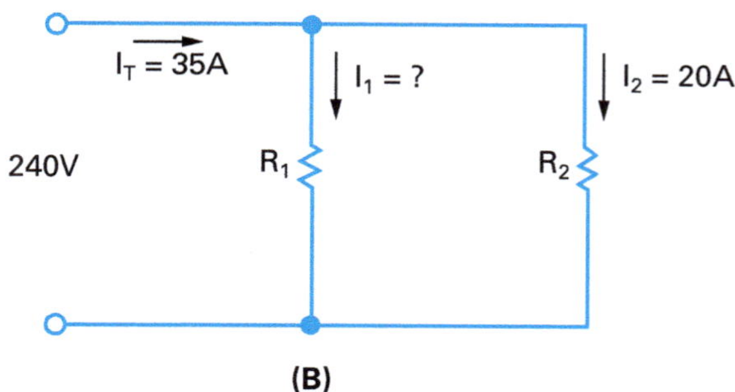

Figure 10 Solving for an unknown current.

Example 2:

As shown in *Figure 10 (B)*, the two branches R_1 and R_2 across a 240V power line draw a total line current of 35A. Branch R_2 takes 20A. What is the current I_1 in branch R_1?

Rearrange the formula to find I_1, and then substitute given values:

$$I_T = I_1 + I_2$$
$$I_1 = I_T - I_2$$
$$I_1 = 35 - 20 = 15A$$

1.4.3 Applying Ohm's Law in Series-Parallel Circuits

Series-parallel circuits combine the elements and characteristics of both the series and parallel configurations. By properly applying the equations and methods previously discussed, you can determine the values of individual components of the circuit. *Figure 11* shows a simple series-parallel circuit with a 1.5V battery.

The following steps outline the procedure for determining voltage drops across and current flows through all resistors in a series-parallel circuit:

> **NOTE**
>
> How these steps are applied to a particular circuit will depend strictly upon the arrangement of its components.

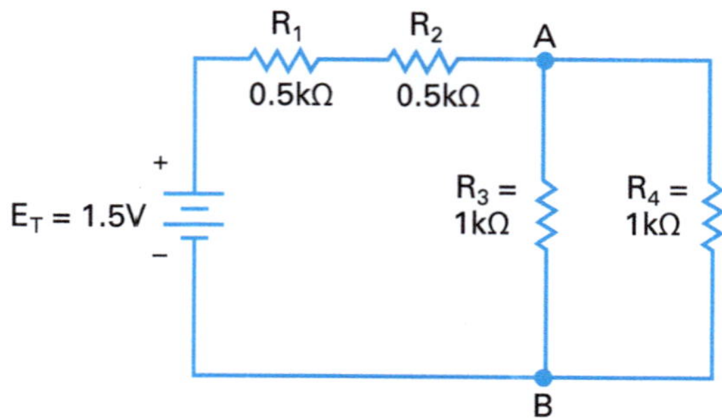

Figure 11 Series-parallel circuit.

Step 1 If necessary, redraw the circuit to make the series and parallel relationships of its components more obvious. Be sure to maintain the values and relationships of the original circuit. Unless careful consideration is taken during this step, it is easy to alter the circuit into a different one by accident. (Label this drawing "Diagram 1".)

Step 2 Reduce all groups of resistors in parallel to a single equivalent series resistor (using any of the previously discussed methods). Do this for each parallel group.

Step 3 Redraw the circuit using the equivalent series resistors in place of the parallel groups. The circuit should now be a simple series circuit. (Label this drawing "Diagram 2".)

Step 4 Determine the total circuit resistance (R_T) of Diagram 2 by adding all of the resistances.

Step 5 Determine the total circuit (I_T) by using the power supply voltage (E_T), the total circuit resistance (I_T), and the correct form of Ohm's law ($I_T = E_T \div R_T$).

Step 6 Determine the voltage drop (E_{Rn}) across each resistance in Diagram 2 by using the total current (I_T), the resistor's value (R_n), and the correct form of Ohm's law ($E_{Rn} = I_T \times R_n$).

Step 7 Referring back to Diagram 1, determine the branch current of each parallel resistor. The voltage drop across each parallel resistor group will be equal to the calculated voltage drop across its equivalent resistance (calculated in Step 6).

Step 8 Determine the branch current (I_{Rn}) of each individual parallel resistor using the voltage drop across the resistor (E_{Rn}), the branch resistance (R_n), and the correct form of Ohm's law ($I_{Rn} = E_{Rn} \div R_n$).

The circuit is now analyzed. After completing these steps, you will know the voltage drop across each resistor, the current flowing through it, the total circuit current, and the total circuit resistance.

The current and voltage associated with each component can be determined by first simplifying the circuit to find the total current, and then working across the individual components.

The circuit in *Figure 11* can be broken into two components: the series resistances R_1 and R_2, and the parallel resistances R_3 and R_4.

R_1 and R_2 can be added together to form the equivalent series resistance R_{1+2}:

$$R_{1+2} = R_1 + R_2$$
$$R_{1+2} = 0.5k\Omega + 0.5k\Omega$$
$$R_{1+2} = 1k\Omega$$

R_3 and R_4 can be totaled using either the general reciprocal formula we have been using, or, since there are two resistances in parallel, the product-over-sum method. Both methods are shown as follows:

Using the reciprocal formula:

$$R_{3+4} = \cfrac{1}{\cfrac{1}{R_3} + \cfrac{1}{R_4}}$$

$$R_{3+4} = \cfrac{1}{\cfrac{1}{1k\Omega} + \cfrac{1}{1k\Omega}}$$

$$R_{3+4} = \cfrac{1}{\cfrac{1}{1000\Omega} + \cfrac{1}{1000\Omega}}$$

$$R_{3+4} = \cfrac{1}{0.001 + 0.001} = \cfrac{1}{0.002}$$

$$R_{3+4} = 500\Omega \ \ or \ \ 0.5k\Omega$$

Using the product-over-sum method:

$$R_{3+4} = \cfrac{R_3 \times R_4}{R_3 + R_4}$$

$$R_{3+4} = \cfrac{1k\Omega \times 1k\Omega}{1k\Omega + 1k\Omega} = \cfrac{1k\Omega}{2k\Omega}$$

$$R_{3+4} = 500\Omega \ \ or \ \ 0.5k\Omega$$

The equivalent circuit containing the R_{1+2} resistance of 1kΩ and the R_{3+4} resistance of 0.5kΩ is shown in *Figure 12*.

Using the Ohm's law relationship that total current equals voltage divided by circuit resistance, the circuit current can be determined. First, however, total circuit resistance must be found. Because the simplified circuit consists of two resistances in series, they are simply added together to obtain total resistance.

$$R_T = R_{1+2} + R_{3+4}$$
$$R_T = 1k\Omega + 0.5k\Omega$$
$$R_T = 1.5k\Omega$$

Applying this to the current/voltage equation:

$$I_T = \cfrac{E_T}{R_T} = \cfrac{1.5kV}{1.5k\Omega} = 1mA \ \ or \ \ 0.001A$$

Now that the total current is known, voltage drops across individual components can be determined:

$$E = I \times R$$
$$E_1 = I_T \times R_1 = 1mA \times 0.5k\Omega = 0.5V$$
$$E_2 = I_T \times R_2 = 1mA \times 0.5k\Omega = 0.5V$$

Because the total voltage equals the sum of all voltage drops, the voltage drop from A to B (E_{AB}) can be determined by subtracting the E_{R1} and E_{R2} voltage drops from the total voltage, as follows:

$$E_T = E_{R1} + E_{R2} + E_{AB}$$
$$E_{AB} = E_T - E_{R1} - E_{R2}$$
$$E_{AB} = 1.5V - 0.5V - 0.5V$$
$$E_{AB} = 0.5V$$

Since R_3 and R_4 are in parallel, some of the total current must pass through each resistor. R_3 and R_4 are equal, so the same current should flow through each branch using the relationship:

$$I = \cfrac{E}{R}$$

$$I_{R3} = \cfrac{E_{R3}}{R_3}$$

$$I_{R3} = \cfrac{0.5V}{1k\Omega} \ \ or \ \ \cfrac{0.5V}{1000\Omega}$$

$$I_{R3} = 0.5mA \ \ or \ \ 0.0005A$$

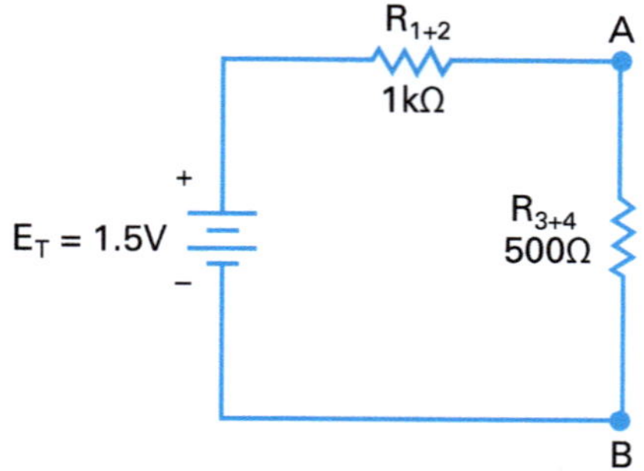

Figure 12 Simplified series-parallel circuit.

Therefore, the total current for the circuit passes through R_1 and R_2 and is evenly divided between R_3 and R_4.

Additional Resources

Electronics Fundamentals: Circuits, Devices, and Applications, Thomas L. Floyd. New York, NY: Pearson Eduction Inc.

Principles of Electric Circuits, Thomas L. Floyd. New York: Prentice Hall.

1.0.0 Section Review

1. Which of the following is true regarding a series circuit?

   a. It has more than one path for current flow.
   b. The current flow is the same through all resistors.
   c. The current flow varies depending on the strength of the resistor.
   d. The current flow through the resistors is always lower than the source current.

2. Which of the following is true regarding a parallel circuit?

   a. It has more than one path for current flow.
   b. The resistance is the same through all circuit components.
   c. The total resistance is equal to the sum of the individual resistances.
   d. The current flow through each resistor is always less than the total resistance.

3. Which of the following is true regarding a series-parallel circuit?

   a. Once reduced to an equivalent series circuit, the total resistance is found by adding the reciprocal of each resistance and then dividing by 1.
   b. The resistance is the same through all circuit components.
   c. If a circuit is reduced to an equivalent series circuit, the total resistance is found by adding the loads.
   d. The current flow through each resistor is always less than the total resistance.

4. Two branches of a circuit (R_1 and R_2) across a 120V power line draw a total line current of 30A. Branch R_1 takes 15A. What is the current I_2 in branch R_2?

   a. 12A
   b. 15A
   c. 17A
   d. 30A

2.0.0 KIRCHHOFF'S LAWS

Objective

Apply Kirchhoff's laws to various types of circuits.

a. Use Kirchhoff's current law.
b. Use Kirchhoff's voltage law.

Trade Terms

Kirchhoff's current law: The statement that the total amount of current flowing through a parallel circuit is equal to the sum of the amounts of current flowing through each current path.

Kirchhoff's voltage law: The statement that the sum of all the voltage drops in a circuit is equal to the source voltage of the circuit.

Kirchhoff's laws provide a simple, practical method of solving for unknown parameters in a circuit. Kirchhoff developed laws that applied to both voltage and current.

2.1.0 Kirchhoff's Current Law

In its most general form, **Kirchhoff's current law** states at any point in a circuit, the total current entering that point must equal the total current leaving that point. For parallel circuits, this implies that the current in a parallel circuit is equal to the sum of the currents in each branch.

When using Kirchhoff's laws to solve circuits, it is necessary to adopt conventions that determine the algebraic signs for current and voltage terms. A convenient system for current is to consider all current flowing into a branch point as positive, and all current directed away from that point as negative.

For example, in *Figure 13*, the currents can be written as:

$$I_A + I_B - I_C = 0 \quad or \quad 5A + 3A - 8A = 0$$

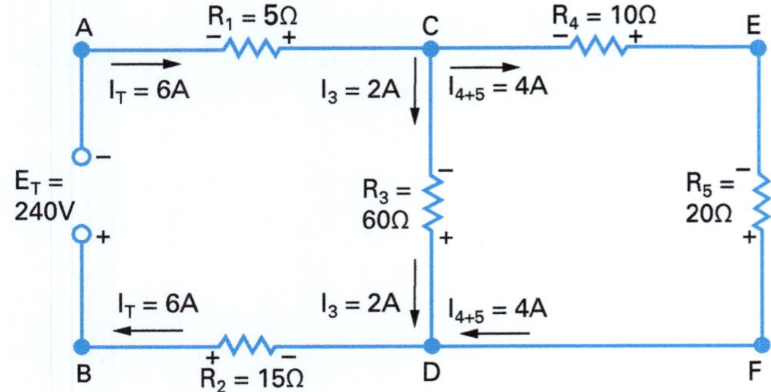

Figure 13 Kirchhoff's current law.

Currents I_A and I_B are positive terms because these currents flow into P, but I_C, directed out of P, is negative.

For a circuit application, refer to Point C at the top of the diagram in *Figure 14*. The total current I_T into Point C is 6A. It then divides into two currents, the I_3 (2A) and I_{4+5} (4A), which are both directed out of point C. Note that I_{4+5} is the current through R_4 and R_5. Write the algebraic equation again:

$$I_T - I_3 - I_{4+5} = 0$$

Then substitute the values for each current:

$$6A - 2A - 4A = 0$$

For the opposite direction, refer to Point D at the bottom of *Figure 14*. Here, the branch currents into Point D combine to equal the mainline current I_T returning to the voltage source. Now, I_T is directed out from Point D, with I_3 and I_{4+5} directed in. The algebraic equation is:

$$I_3 + I_{4+5} - I_T = 0$$
$$2A + 4A - 6A = 0$$

Note that at either Point C or Point D, the sum of the 2A and 4A branch currents must equal the 6A total line current. The amount of current coming into any point on a circuit must be equal to the amount of current leaving the circuit. Therefore, Kirchhoff's current law can also be stated as:

$$I_{IN} = I_{OUT}$$

For *Figure 14*, the equations for current can be written as follows.

At Point C:
$$6A = 2A + 4A$$

At Point D:
$$2A + 4A = 6A$$

Kirchhoff's current law is the basis for the practical rule in parallel circuits that the total line current must equal the sum of the branch currents.

Figure 14 Application of Kirchhoff's current law.

2.2.0 Kirchhoff's Voltage Law

Kirchhoff's voltage law states that the algebraic sum of the voltages around any closed path is zero.

An algebraic sum is the result of adding both positive (regular) numbers and negative numbers in the same addition problem. "Adding algebraically" is another way of saying that you used both addition and subtraction in the same problem. For example, adding –2 is the same as subtracting 2.

Referring to *Figure 15*, the sum of the voltage drops around the circuit must equal the voltage applied to the circuit:

$$E_A = E_1 + E_2 + E_3$$

Where:

E_A = voltage applied to the circuit
E_1, E_2, and E_3 = voltage drops in the circuit

Another way of stating this law is the algebraic sum of the voltage rises and voltage drops must be equal to zero. In the *Figure 15* example, the battery created the rise and the resistors created the drops. A voltage source is considered a voltage rise; a voltage across a resistor is a voltage drop. (For convenience in labeling, letter subscripts are shown for voltage sources and numerical subscripts are used for voltage drops.) This form of the law can be written by transposing the right members to the left side:

$$\text{Voltage applied} - \text{sum of voltage drops} = 0$$
$$E_A - E_1 - E_2 - E_3 = 0$$
$$E_A - (E_1 + E_2 + E_3) = 0$$

2.2.1 Loop Equations

Kirchhoff's voltage law may be applied more generally through something called a *loop equation*. Any closed circuit path for current flow is called a *loop*. A loop equation specifies the voltages around the loop.

Before writing loop equations, however, it is essential to develop a convention for deciding if a component in the loop is a voltage rise (+) or voltage drop (–). Do not assume that all sources (like batteries) are rises and all resistors are drops, as this convention can sometimes be faulty. Also, it doesn't work if you have a loop with nothing but resistors and no sources.

The following convention is simple and easy to apply. Work your way around the loop in whichever direction you wish. Whenever you encounter a component, note the polarity (positive or negative sign) that you encounter first. If the sign is positive, treat the component as a voltage rise. If the sign is negative, treat the component as a voltage drop. Examine *Figure 16*.

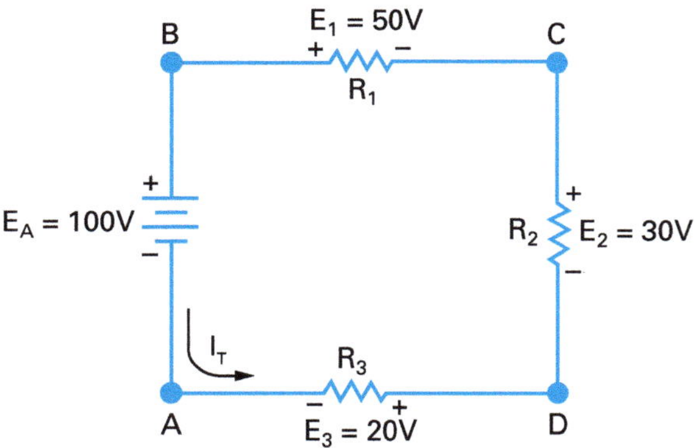

Figure 15 Kirchhoff's voltage law.

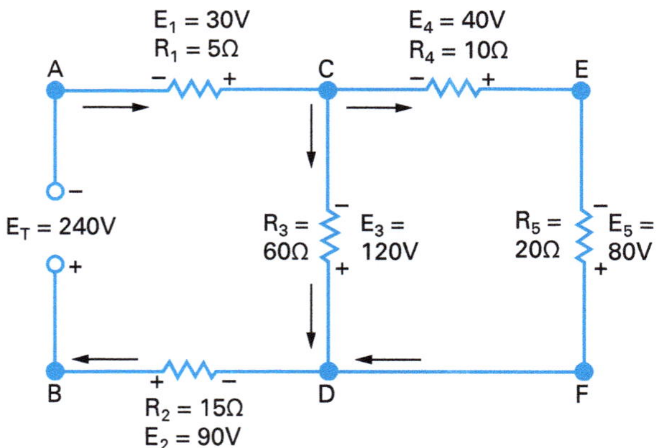

Figure 16 Loop equation.

Putting It All Together

Draw four 60W lamps in parallel with a 120V power source. What is the amperage in the circuit? What would happen to the amperage if you doubled the voltage?

Consider the inside loop through A, C, D, and B. This includes the voltage drops E_1, E_3, and E_2, and the source E_T. In a clockwise direction, starting at Point A, the algebraic sum of the voltages is:

$$- E_1 - E_3 - E_2 + E_T = 0$$

or

$$- 30V - 120V - 90V + 240V = 0$$

Voltages E_1, E_3, and E_2 have a negative value, because there is a decrease in voltage seen across each of the resistors in a clockwise direction. However, the source E_T is a positive term because an increase in voltage is seen in that same direction.

For the opposite direction, going counterclockwise in the same loop from Point B, E_T is negative while E_1, E_2, and E_3 have positive values. Therefore:

$$- E_T + E_2 + E_3 + E_1 = 0$$

or

$$- 240V + 90V + 120V + 30V = 0$$

When the negative term is transposed, the equation becomes:

$$240V = 90V + 120V + 30V$$

In this form, the loop equation shows that Kirchhoff's voltage law is really the basis for the practical rule in series circuits that the sum of the voltage drops must equal the applied voltage.

For example, determine the voltage E_B for the circuit shown in *Figure 17*. The direction of the current flow is shown by the arrow. First, mark the polarity of the voltage drops across the resistors and trace the circuit in the direction of the current flow starting at Point A. Then write the loop equation around the circuit:

$$- E_3 - E_B - E_2 - E_1 + E_A = 0$$

Solve for E_B:

$$E_B = E_A - E_3 - E_2 - E_1$$
$$E_B = 15V - 2V - 6V - 3V$$
$$E_B = 4V$$

Since E_B was found to be positive, the assumed direction of current is in fact the actual direction of current.

In its most general form, Kirchhoff's voltage law states that the algebraic sum of all the potential differences in a closed loop is equal to zero. A closed loop means any completely closed path consisting of wire, resistors, batteries, or other components. For series circuits, this implies that the sum of the voltage drops around the circuit is equal to the applied voltage. For parallel circuits, this implies that the voltage drops across all branches are equal.

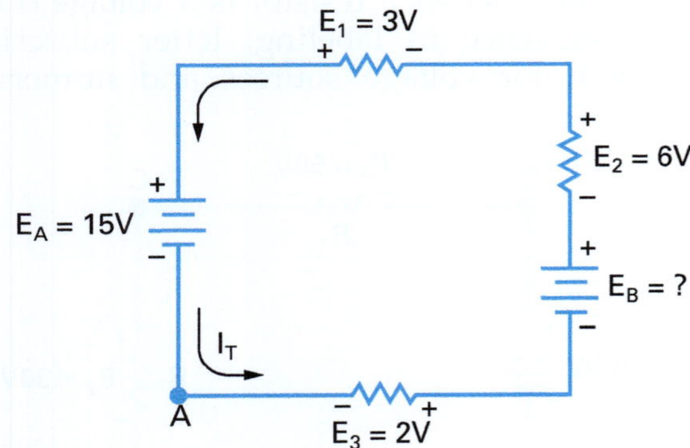

Figure 17 Applying Kirchhoff's voltage law.

2.0.0 Section Review

1. Who stated that the current entering a point must equal the total current leaving that point?

   a. Kirchhoff
   b. Joule
   c. Pascal
   d. Ohm

2. In a circuit with an applied voltage of 10V and voltage drops of $E_1 = 2V$, $E_2 = 5V$, and $E_3 = ?$, the value of E_3 must be _____.

   a. 3V
   b. 7V
   c. 12V
   d. 17V

SUMMARY

The relationships among current, voltage, resistance, and power in Ohm's law are the same for both DC series and DC parallel circuits. Understanding and being able to apply these concepts is necessary for effective circuit analysis and trouble-shooting. DC series-parallel circuits also have these fundamental relationships. Since DC series-parallel circuits are a combination of simple series and parallel circuits, Kirchhoff's voltage and current laws will apply. Calculating I, E, R, and P for series-parallel circuits is no more difficult than calculating these values for simple series or parallel circuits. However, for series-parallel circuits, these calculations require more careful circuit analysis in order to use Ohm's law correctly.

1. True or False? You can use Ohm's law to find the value of each branch current in a parallel circuit.

   a. True
   b. False

2. Find the total resistance in a series circuit with three resistances of 10Ω, 20Ω, and 30Ω.

   a. 1Ω
   b. 15Ω
   c. 20Ω
   d. 60Ω

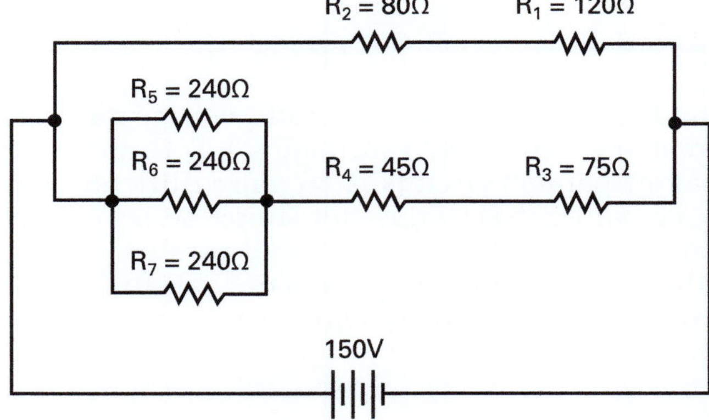

Figure RQ01

3. The total resistance in *Figure RQ01* is ______.

   a. 100Ω
   b. 129Ω
   c. 157Ω
   d. 1,040Ω

4. In a parallel circuit, the voltage across each path is equal to the ______.

   a. total circuit resistance times path current
   b. source voltage minus path voltage
   c. path resistance times total current
   d. applied voltage

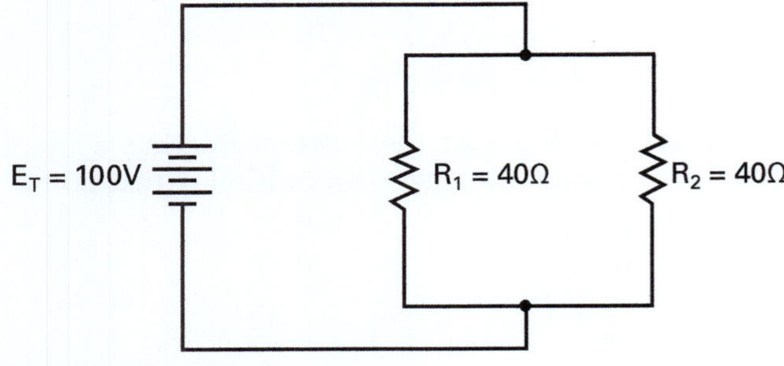

Figure RQ02

5. The value for total current in *Figure RQ02* is ______.

   a. 1.25A
   b. 2.50A
   c. 5A
   d. 10A

6. A resistor of 32Ω is in parallel with a resistor of 36Ω, and a 54Ω resistor is in series with the pair. When 350V is applied to the combination, the current through the 54Ω resistor is ______.

   a. 2.87A
   b. 3.26A
   c. 4.93A
   d. 5.86A

7. A 242Ω resistor is in parallel with a 180Ω resistor, and a 420Ω resistor is in series with the combination in a 27V circuit. A current of 22mA flows through the 242Ω resistor. The current through the 180Ω resistor is ______.

   a. 29.6mA
   b. 36.4mA
   c. 59.4mA
   d. 60.3mA

8. Which statement is *not* correct relative to Kirchhoff's current law?

   a. $I_A + I_B - I_C = 0$.
   b. The total current entering a point must equal total current leaving a point.
   c. The total line current must equal the sum of the branch currents.
   d. Kirchhoff's laws apply to current only.

9. Kirchhoff's voltage law states that the algebraic sum of the voltages around any closed current path is ______.

   a. infinity
   b. zero
   c. twice the current
   d. always less than the individual voltages due to voltage drop

10. Two 24Ω resistors are in parallel, and a 42Ω resistor is in series with the combination. When 78V is applied to the three resistors, the voltage drop across the 42Ω resistor is about ______.

   a. 49.8V
   b. 55.8V
   c. 60.5V
   d. 65.3V

Fill in the blank with the correct term that you learned from your study of this module.

1. __________ states that the total amount of current flowing through a parallel circuit is equal to the sum of the amounts of current flowing through each current path.

2. __________ states that the sum of all the voltage drops in a circuit is equal to the source voltage of the circuit.

3. __________ contain both series and parallel current paths.

4. __________ contain two or more parallel paths through which current can flow.

5. A __________ contains only one path for current flow.

Trade Terms

Kirchhoff's current law
Kirchhoff's voltage law
Parallel circuits

Series circuit
Series-parallel circuits

1. Which formula is used for series circuits?

   a. $R_T = \dfrac{R}{N}$

   b. $R_T = \dfrac{1}{\dfrac{1}{R_1} + \dfrac{1}{R_2} + \dfrac{1}{R_3}}$

   c. $R_T = R_1 + R_2 + R_3$

   d. $R_T = \dfrac{R_1 \times R_2}{R_1 + R_2}$

2. If two or more resistors are connected in parallel, the ______.

   a. total resistance is higher than any single resistor

   b. total resistance is lower than any single resistor

   c. current flow varies based on voltage fluctuation

   d. resistance depends on the power rating of the circuit

3. The formula for calculating the total resistance in a series circuit with three resistors is ______.

   a. $R_T = R_1 + R_2 + R_3$
   b. $R_T = R_1 - R_2 - R_3$
   c. $R_T = R_1 \times R_2 \times R_3$
   d. $R_T = \dfrac{1}{\dfrac{1}{R_1} + \dfrac{1}{R_2} + \dfrac{1}{R_3}}$

4. True or False? Parallel circuits are current dividers.

5. The formula for calculating the total resistance in a parallel circuit with three resistors is ______.

   a. $R_T = R_1 + R_2 + R_3$
   b. $R_T = R_1 - R_2 - R_3$
   c. $R_T = R_1 \times R_2 \times R_3$
   d. $R_T = \dfrac{1}{\dfrac{1}{R_1} + \dfrac{1}{R_2} + \dfrac{1}{R_3}}$

6. True or False? In a series circuit, E_T must equal the sum of the individual voltage drops.

7. When calculating series-parallel combination circuits, it is important to ______.

   a. reduce to series loads, then add the loads

   b. replace parallel combinations with one value

   c. reduce to simpler circuits where possible

   d. All of the above.

8. True or False? Kirchhoff's voltage law can be used to determine voltage drops.

9. Two resistors are connected in parallel; R_1 is 90 ohms, R_2 is 45 ohms. What is the total resistance of R_1 and R_2 in parallel?

   ____

10. Two resistors are connected in parallel; R_2 is 90 ohms, R_3 is 45 ohms. A third resistor (R_1, which is 20 ohms) is connected in series with R_2 and R_3.

    a. What is the total resistance of this circuit?

    ____

    b. If the voltage source of this circuit is 150V, what is the total current flow?

    ____

    (Hint: sketch the circuit before solving.)

James Mitchem

JEM Electrical Consulting Services

Jim Mitchem owns his own electrical consulting firm. During his career in the electrical industry, he worked his way up from apprentice to technical services manager, and is now a consultant on large commercial and industrial installations.

How did you become an electrician?

Quite by accident. A couple of years after college, I was working as a relief operator in a plant when the lead electrician retired, creating a vacancy. I liked the idea that electricians were expected to use their knowledge and initiative to keep the place running. I applied and was accepted as a trainee.

How did you get your training?

I took an electrical apprenticeship course by correspondence, and I was fortunate enough to work with good people who helped me along. I worked in an environment that exposed me to a variety of equipment and applications, and just about everyone I've ever worked with has taught me something. Now I'm passing my knowledge on to others.

What kinds of work have you done in your career?

I've worked as an apprentice, journeyman, instrument and controls technician, instrument fitter, foreman, general foreman, superintendent, and startup engineer. Each of these positions required that I learn new skills, both technical and managerial. My experience in many disciplines and types of projects has given me a high level of credibility with my clients.

Now I act as a technical resource and troubleshooter in functions such as safety, quality assurance, and training. I visit job sites to help solve problems and help out with commissioning and startup.

What factor or factors have contributed the most to your success?

There are several factors. Two very important ones have been a desire to learn and a willingness to do whatever is asked of me. I also keep an eye on the big picture. When I'm on a job, I'm not just pulling wire, I'm building a power plant or whatever the project is.

Any advice for apprentices just beginning their careers?

Keep learning! And don't depend on others to train you. Take the initiative to buy or borrow books and trade journals. Take licensing tests and do whatever is necessary to keep your licenses current. Finally, make sure you know your own personal and professional values and work with a company that shares those values.

Trade Terms Introduced in This Module

Kirchhoff's current law: The statement that the total amount of current flowing through a parallel circuit is equal to the sum of the amounts of current flowing through each current path.

Kirchhoff's voltage law: The statement that the sum of all the voltage drops in a circuit is equal to the source voltage of the circuit.

Parallel circuits: Circuits containing two or more parallel paths through which current can flow.

Series circuit: A circuit that has only one path for current flow.

Series-parallel circuits: Circuits that contain both series and parallel current paths.

Additional Resources

This module presents thorough resources for task training. The following resource material is recommended for further study.

Electronics Fundamentals: Circuits, Devices, and Applications, Thomas L. Floyd. New York, NY: Pearson Education, Inc.

Principles of Electric Circuits, Thomas L. Floyd. New York, NY: Pearson Ecustion, Inc.

Figure Credit

Greenlee / A Textron Company, Module Opener

Section Review Answer Key

Answer	Section Reference	Objective
Section One		
1. b	1.1.0	1a
2. a	1.2.0	1b
3. c	1.3.1	1c
4. b	1.4.2	1d
Section Two		
1. a	2.1.0	2a
2. a	2.2.0	2b

1.0.0 SECTION REVIEW

Question 4

Since the total current (I_T) is equal to the sum of the branch currents, you can use this equation to find the unknown branch current (I_2), as follows:

$$I_T = I_1 + I_2$$
$$I_2 = I_T - I_1$$
$$I_2 = 30 - 15 = 15A$$

The current I_2 in branch R_2 is **15A**.

2.0.0 SECTION REVIEW

Question 2

Use Kirchhoff's voltage law, and use the principles of algebra (rearrange the equation) to solve for E_3:

$$E_A = E_1 + E_2 + E_3$$
$$E_3 = E_A - (E_1 + E_2)$$
$$E_3 = 10V - (2V + 5V)$$
$$E_3 = 10V - 7V$$
$$E_3 = 3V$$

The value of E_3 is **3V**.

NCCER CURRICULA — USER UPDATE

NCCER makes every effort to keep its textbooks up-to-date and free of technical errors. We appreciate your help in this process. If you find an error, a typographical mistake, or an inaccuracy in NCCER's curricula, please fill out this form (or a photocopy), or complete the online form at **www.nccer.org/olf**. Be sure to include the exact module ID number, page number, a detailed description, and your recommended correction. Your input will be brought to the attention of the Authoring Team. Thank you for your assistance.

Instructors – If you have an idea for improving this textbook, or have found that additional materials were necessary to teach this module effectively, please let us know so that we may present your suggestions to the Authoring Team.

NCCER Product Development and Revision

13614 Progress Blvd., Alachua, FL 32615

Email: curriculum@nccer.org
Online: www.nccer.org/olf

❏ Trainee Guide　　❏ Lesson Plans　　❏ Exam　　❏ PowerPoints　　Other ___________________

Craft / Level: ____　Copyright Date: ______________

Module ID Number / Title: ____

Section Number(s): ____

Description: ____

____

____

____

____

Recommended Correction: ____

____

____

____

____

Your Name: ____

Address: ____

____

Email: ____　Phone: ______________________

Introduction to Electrical Circuits

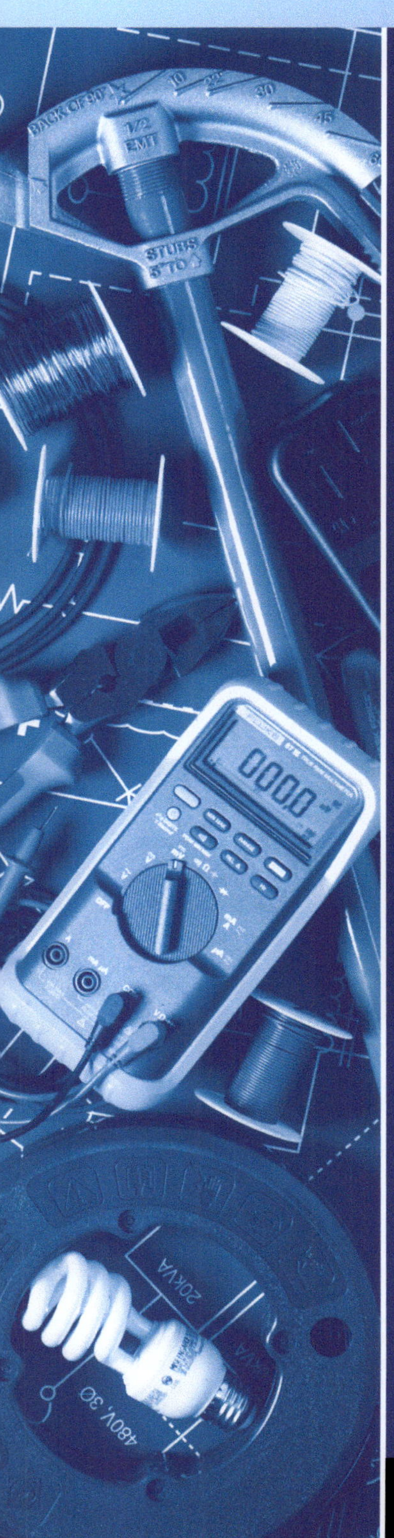

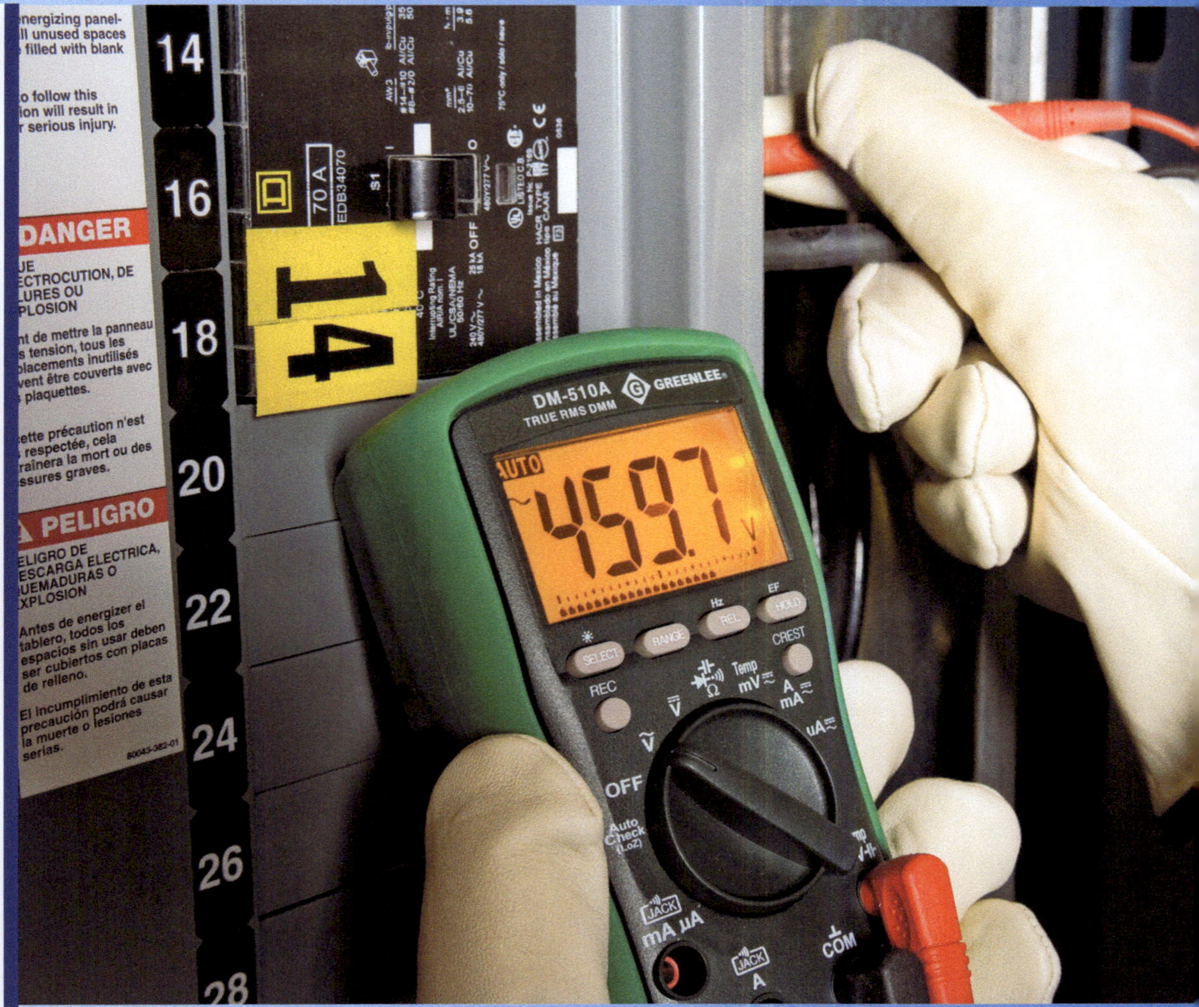

OVERVIEW

All kinds of instruments use electrical circuitry to function. This module discusses basic atomic theory and electrical theory, which are the fundamental concepts behind electricity in every setting. It also covers electrical units of measurement and explains how Ohm's law and the power equation can be used to determine unknown values. This module also includes electrical schematic diagrams.

Module 26103-17

Trainees with successful module completions may be eligible for credentialing through the NCCER Registry. To learn more, go to **www.nccer.org** or contact us at 1.888.622.3720. Our website has information on the latest product releases and training, as well as online versions of our *Cornerstone* magazine and Pearson's product catalog.

Your feedback is welcome. You may email your comments to **curriculum@nccer.org**, send general comments and inquiries to **info@nccer.org**, or fill in the User Update form at the back of this module.

This information is general in nature and intended for training purposes only. Actual performance of activities described in this manual requires compliance with all applicable operating, service, maintenance, and safety procedures under the direction of qualified personnel. References in this manual to patented or proprietary devices do not constitute a recommendation of their use.

Introduction to Electrical Circuits

Objectives

When you have completed this module, you will be able to do the following:

1. Describe atomic structure as it relates to electricity.
   a. Identify the components of an atom.
   b. Compare the atomic structures of conductors and insulators.
   c. Identify the role of magnetism in electrical devices.
   d. Identify the basic components in a power distribution system.
2. Identify electrical units of measurement.
   a. Define current.
   b. Define voltage.
   c. Define resistance.
   d. Use Ohm's law to solve for unknown circuit values.
3. Read schematic diagrams.
   a. Identify the symbol for a resistor and determine its value based on color codes.
   b. Distinguish between series and parallel circuits.
   c. Identify the instruments used to measure circuit values.
   d. Calculate electrical power.

Performance Tasks

This is a knowledge-based module. There are no performance tasks.

Trade Terms

Ammeter	Electrons	Ohmmeter	Solenoids
Amperes (A)	Insulator	Ohm's law	Transformers
Atoms	Joule (J)	Power	Valence shell
Battery	Kilo	Proton	Volts (V)
Charge	Matter	Relays	Voltage
Circuit	Mega	Resistance	Voltage drop
Conductors	Neutrons	Resistors	Voltmeter
Coulomb	Nucleus	Schematic	Watts (W)
Current	Ohms (Ω)	Series circuit	

Industry Recognized Credentials

If you are training through an NCCER-accredited sponsor, you may be eligible for credentials from NCCER's Registry. The ID number for this module is 26103-17. Note that this module may have been used in other NCCER curricula and may apply to other level completions. Contact NCCER's Registry at 888.622.3720 or go to **www.nccer.org** for more information.

Note

NFPA 70®, *National Electrical Code*® and *NEC*® are registered trademarks of the National Fire Protection Association, Quincy, MA.

Contents

Figures and Tables

1.0.0 ATOMIC STRUCTURE AND ELECTRICITY

Objective

Describe atomic structure as it relates to electricity.

a. Identify the components of an atom.
b. Compare the atomic structures of conductors and insulators.
c. Identify the role of magnetism in electrical devices.
d. Identify the basic components in a power distribution system.

Trade Terms

Amperes (A): The basic unit of measurement for electrical current, represented by the letter A.

Atoms: The smallest particles to which an element may be divided and still retain the properties of the element.

Battery: A DC voltage source consisting of two or more cells that convert chemical energy into electrical energy.

Charge: A quantity of electricity that is either positive or negative.

Circuit: A complete path for current flow.

Conductors: Materials through which it is relatively easy to maintain an electric current.

Current: The movement, or flow, of electrons in a circuit. Current (I) is measured in amperes.

Electrons: Negatively charged particles that orbit the nucleus of an atom.

Insulator: A material through which it is difficult to conduct an electric current.

Matter: Any substance that has mass and occupies space.

Neutrons: Electrically neutral particles (neither positive nor negative) that have the same mass as a proton and are found in the nucleus of an atom.

Nucleus: The center of an atom. It contains the protons and neutrons of the atom.

Ohms (Ω): The basic unit of measurement for resistance, represented by the symbol Ω.

Power: The rate of doing work, or the rate at which energy is used or dissipated. Electrical power is measured in watts.

Proton: The smallest positively charged particle of an atom. Protons are contained in the nucleus of an atom.

Relays: Electromechanical devices consisting of a coil and one or more sets of contacts. Used as a switching device.

Resistance: An electrical property that opposes the flow of current through a circuit. Resistance (R) is measured in ohms.

Solenoids: Electromagnetic coils used to control a mechanical device such as a valve.

Transformers: Devices consisting of one or more coils of wire wrapped around a common core. Transformers are commonly used to step voltage up or down.

Valence shell: The outermost ring of electrons that orbit about the nucleus of an atom.

Volts (V): The unit of measurement for voltage, represented by the letter V. One volt is equivalent to the force required to produce a current of one ampere through a resistance of one ohm.

Voltage: The driving force that makes current flow in a circuit. Voltage, often represented by the letter E, is also referred to as *electromotive force (emf)*, *difference of potential*, or *electrical pressure*.

Watts (W): The basic unit of measurement for electrical power, represented by the letter W.

Electricity is a form of energy that can be used by electrical devices such as motors, lights, TVs, heaters, and numerous other devices to perform work. Electricity is also used to control non-electrical devices that perform work. For example, although your car's engine runs on gasoline (rather than electricity), you wouldn't be able to start it or turn it off without the electrical system and battery. In order to work with electricity, you need to know how it is produced and how it acts in electrical circuits.

An electrical circuit contains, at minimum, a voltage source, a load, and conductors (wires) to carry the electrical current (*Figure 1*). The circuit should also have a means to stop and start the current, such as a switch.

Electricity is all about cause and effect. The presence of voltage (measured in volts) in a closed circuit will cause current (measured in amperes, or amps) to flow. Voltage can also be described as electrical pressure; the more pressure you apply, the more current will flow. However, the amount of current flow is also determined by how much resistance (measured in ohms) the load offers to

---

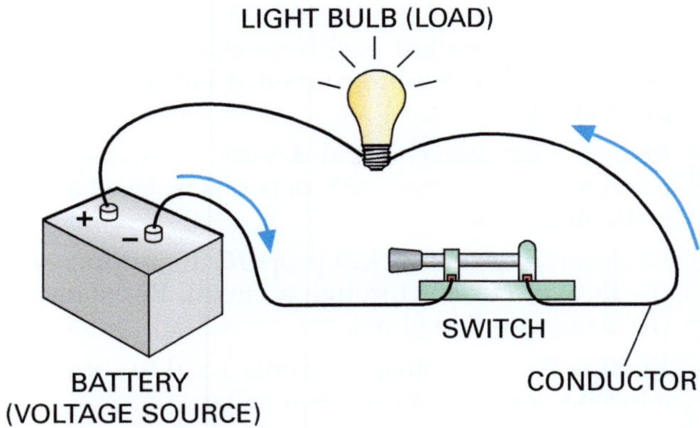

Figure 1 Basic electrical circuit.

the flow of current. In order to convert electrical energy into work, the load consumes energy. The amount of energy a device consumes is called **power**, and is expressed in watts. **Volts (V)**, **amperes (A)**, **ohms (Ω)**, and **watts (W)** are related in such a way that if any one of them changes, the others are proportionally affected. This relationship can be seen using basic math principles that you will learn in this module. You will also learn how electricity is produced and how test instruments are used to measure electricity.

In order to understand electrical theory, you must first understand the basic concepts of atomic theory. Atomic theory explains the construction and behavior of **atoms**, including the transfer of **electrons** that results in current flow.

1.1.0 Components of an Atom

The atom is the smallest part of an element that enters into a chemical change, but it does so in the form of a charged particle. These charged particles are called ions and are of two types—positive and negative. A positive ion may be defined as an atom that has become positively charged. A negative ion may be defined as an atom that has become negatively charged. One of the properties of charged ions is that ions of the same charge tend to repel one another, whereas ions of unlike charge will attract one another. The term **charge** can be taken to mean a quantity of electricity that is either positive or negative.

The structure of an atom is best explained by a detailed analysis of the simplest of all atoms: the hydrogen atom. A hydrogen atom, shown in *Figure 2*, is composed of a **nucleus** containing one **proton** and a single orbiting electron. As the electron revolves around the nucleus, it is held in this orbit by two counteracting forces. One of these forces is called centrifugal force, which is the force that tends to cause the electron to fly outward as it travels around its circular orbit. The second force acting on the electron is electrostatic force, which pulls the electron toward the nucleus. This is caused by the mutual attraction between the positively charged nucleus and the negatively charged electron. At some given radius (distance from the nucleus), the two forces will balance each other, providing a stable path for the electron.

Electrostatic force can be either a repelling force or an attracting force, depending on the polarity (positive or negative charge) of the components involved. The following rules apply:

- A proton (+) repels another proton (+).
- An electron (–) repels another electron (–).
- A proton (+) attracts an electron (–).

An atom is made up of three types of subatomic particles that each have unique electrical characteristics: electrons, protons, and **neutrons**. The protons and neutrons are located in the center, or nucleus, of the atom, and the electrons travel around the nucleus in orbits.

Because protons are relatively heavy, the repelling force they exert on one another in the nucleus of an atom has little effect.

The attracting and repelling forces on charged materials occur because of the electrostatic lines of force that exist around the charged materials. In a negatively charged object, the lines of force of the excess electrons combine to produce an electrostatic field that has lines of force coming into the object from all directions. In a positively charged object, the lines of force of the excess protons combine to produce an electrostatic field that has lines of force going out of the object in all directions. The electrostatic fields either aid or oppose each other to attract or repel.

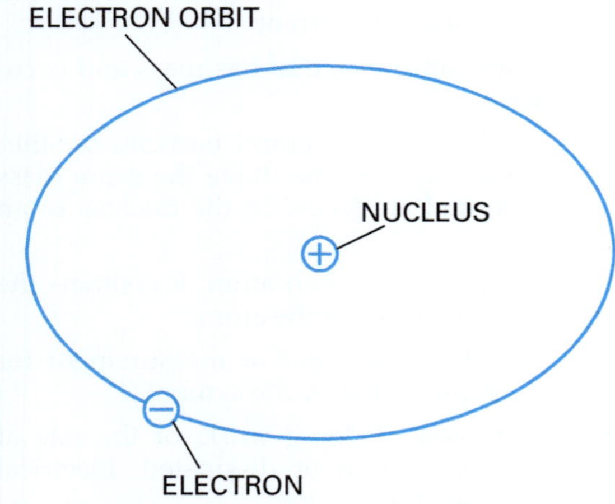

Figure 2 Hydrogen atom.

Why Bother Learning Theory?

Many trainees wonder why they need to bother learning the theory behind how things operate. They figure, why should I learn how it works as long as I know how to install it? The answer is, if you only know how to install something (e.g., run wire, connect switches, etc.), that's all you are ever going to be able to do. For example, if you don't know how your car operates, how can you troubleshoot it? The answer is, you can't. You can only keep changing out the parts until you finally hit on what is causing the problem. (How many times have you seen people do this?) Remember, unless you understand not only how things work but why they work, you'll only be a parts changer. With theory behind you, there is no limit to what you can do.

1.1.1 The Nucleus

The nucleus is the central part of the atom. It is made up of heavy particles called protons and neutrons. The proton is a charged particle containing the smallest known unit of positive electricity. The neutron has no electrical charge. The number of protons in the nucleus determines how the atom of one element differs from the atom of another element.

Although a neutron is actually a unique particle, it is generally thought of as an electron and proton combined, and is electrically neutral. Because they are electrically neutral, neutrons are not considered relevant to the electrical nature of atoms.

1.1.2 Electrical Charges

The negative charge of an electron is equal to the positive charge of a proton. The charges of an electron and a proton are called *electrostatic charges*. The lines of force associated with each particle produce electrostatic fields. Because of the way these fields act together, charged particles can attract or repel one another. The Law of Electrical Charges states that particles with like charges repel each other and those with unlike charges attract each other. The Law of Electrical charges is illustrated in *Figure 3*.

1.2.0 Atomic Structures of Conductors and Insulators

The difference between atoms, with respect to chemical activity and stability, depends on the number and position of the electrons included within the atom. In general, the electrons reside in groups of orbits called *shells*. The shells are arranged in steps that correspond to fixed energy levels.

The outer shell of an atom is called the **valence shell**, and the electrons contained in this shell are called *valence electrons* (*Figure 4*). The number

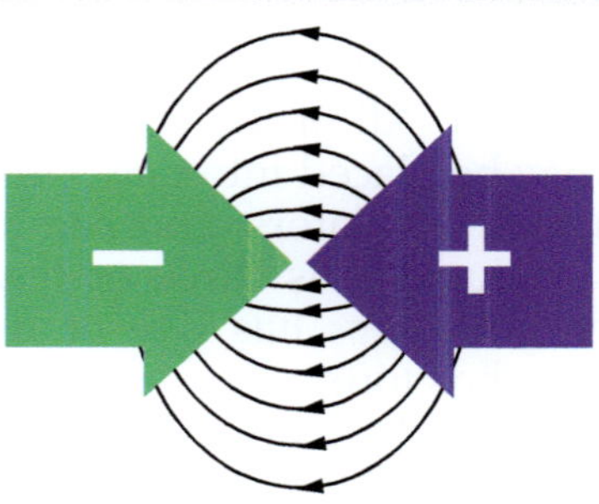

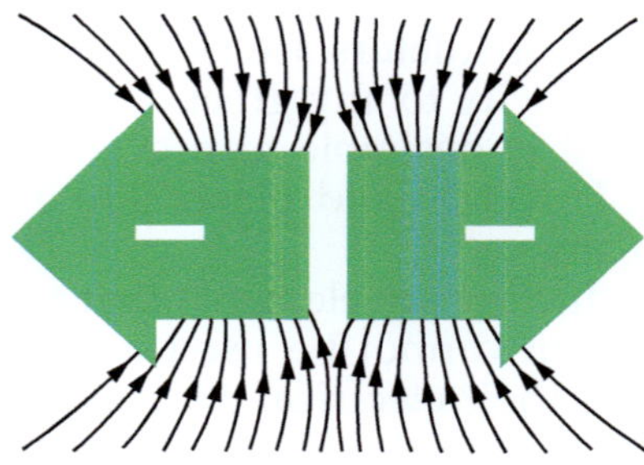

Figure 3 Law of Electrical Charges.

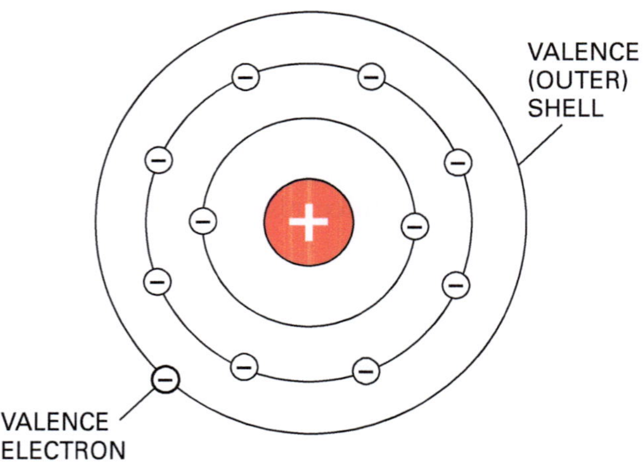

Figure 4 Valence shell and electrons.

of valence electrons determines an atom's ability to gain or lose an electron, which in turn determines the chemical and electrical properties of the atom. An atom that is lacking only one or two electrons from its outer shell will easily gain electrons to complete its shell, but a large amount of energy is required to free any of its electrons. An atom having a relatively small number of electrons in its outer shell in comparison to the number of electrons required to fill the shell will easily lose these valence electrons.

When it comes to electricity, the valence electrons are of the most concern. That is because they are easiest electrons to break loose from their parent atom. All of the elements that make up **matter** may be placed into one of three categories: conductors, insulators, and semiconductors. Usually, a conductor has three or fewer valence electrons, an **insulator** has five or more, and a semiconductor has four.

Conductors are elements that will readily conduct a flow of electricity. Because of their strong conducting abilities, they are formed into wire and used whenever it is desired to transfer electrical energy from one point to another. Copper and silver are examples of conductors.

In contrast, insulators are elements that do not conduct electricity to any great degree. These are used when it is desirable to prevent the flow of electricity. Porcelain and plastic are examples of good insulators.

Semiconductors are elements that are not good conductors but cannot be used as insulators either because their electrical characteristics fall in between those of conductors and those of insulators. Germanium and silicon are examples of semiconductors. As you will learn later in your training, semiconductors play a crucial role in electronic circuits.

1.3.0 Magnetism in Electrical Devices

The operation of many electrical components relies on the power of magnetism. Motors, **relays**, **transformers**, and **solenoids** are examples. Magnetized iron generates a magnetic field consisting of magnetic lines of force, also known as magnetic

flux lines (*Figure 5*). Magnetic objects within the field will be attracted or repelled by the magnetic field. The more powerful the magnet, the more powerful the magnetic field around it. Each magnet has a north pole and a south pole. Opposing poles attract each other, and like poles repel each other.

Electricity also produces magnetism. Current flowing through a conductor produces a small magnetic field around the conductor. If the conductor is coiled around an iron bar, the result is an electromagnet (*Figure 6*) that attracts and repels other magnetic objects just like an iron magnet. This is the basis on which electric motors and other components operate.

1.4.0 Power Distribution

Electricity comes from electrical generating plants operated by utilities like your local power company. Steam from coal-burning or nuclear power plants is used to power huge generators called turbines, which generate electricity. There are also hydroelectric power plants, solar power generating plants, and wind-driven turbines. *Figure 7* illustrates how electricity is safely distributed from generating stations to industrial facilities, businesses, and homes.

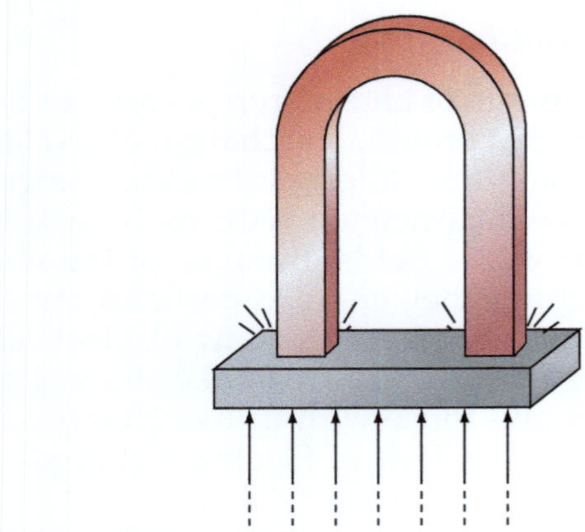

Figure 5 Magnetism.

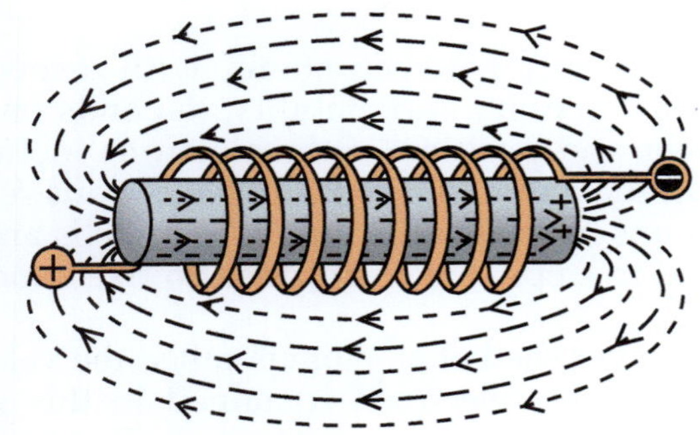

Figure 6 Electromagnet.

The electrical power that travels through long-distance transmission lines may be as high as 750,000V. Devices known as *transformers* are used to step the voltage down to lower levels as it reaches electrical substations and eventually homes, offices, and factories. The voltage you receive at home is usually about 240V. At the wall outlet where you plug in small appliances such as televisions and toasters, the voltage is about 120V (*Figure 8*). Electric stoves, clothes dryers, water heaters, and central air conditioning systems usually require the full 240V. Commercial buildings and factories may receive anywhere from 208V to 575V, depending on the amount of power their machines consume.

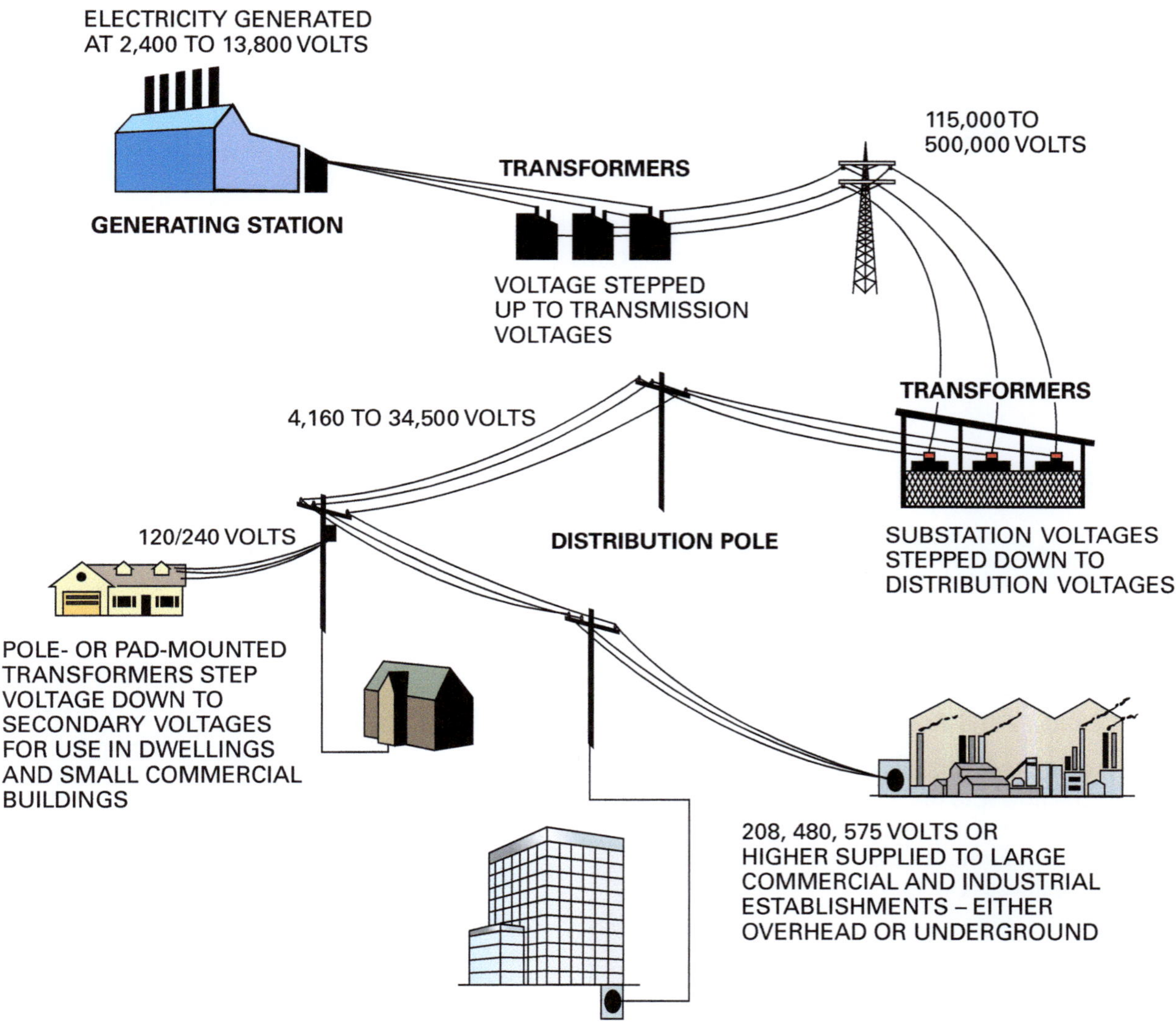

Figure 7 Electrical power distribution.

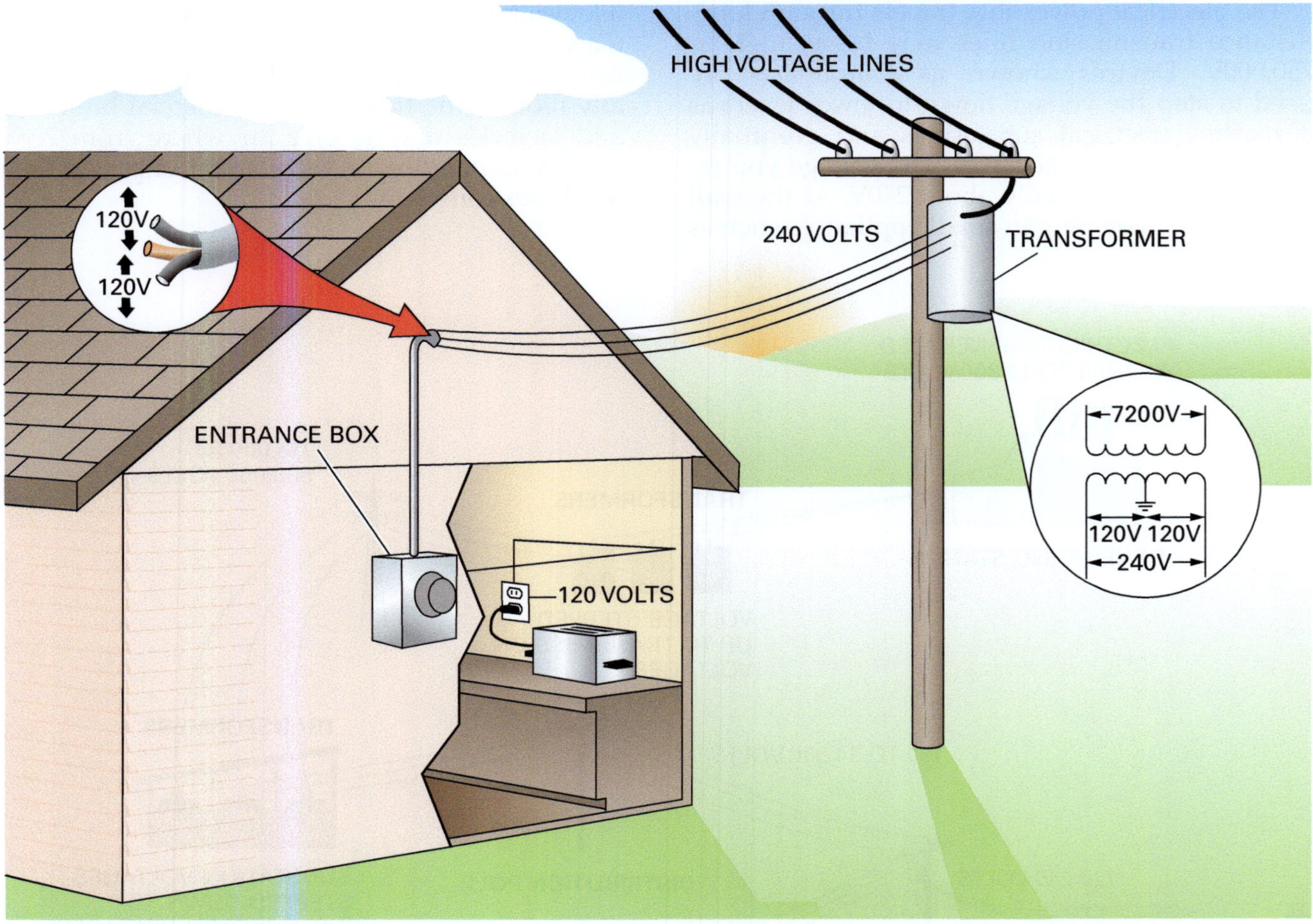

Figure 8 Power distribution within a home.

Transformers

Large distribution transformers at power substations step down the power to the level required for local distribution. Pole transformers like the one shown here step it down further to the voltages needed for homes and businesses.

Hydroelectric Plants

Hydroelectric plants use the power generated by water to drive turbines that produce electricity.

Figure Credit: U.S. Army Corps of Engineers

Additional Resources

Electronics Fundamentals: Circuits, Devices, and Applications, Thomas L. Floyd. New York, NY: Pearson Education, Inc.

Principles of Electric Circuits, Thomas L. Floyd. New York, NY: Pearson Education, Inc.

1.0.0 Section Review

1. A proton repels a(n) ______.

   a. electron
   b. proton
   c. neutron
   d. negative ion

2. An atom with seven valence electrons is most likely a(n)______.

   a. insulator
   b. conductor
   c. capacitor
   d. semiconductor

3. Current flowing through a conductor coiled around an iron bar produces a(n)______.

   a. insulator
   b. capacitor
   c. electromagnet
   d. turbine

4. The voltage used by a typical television or toaster is ______.

   a. 60V
   b. 120V
   c. 180V
   d. 240V

2.0.0 ELECTRICAL UNITS OF MEASUREMENT

Objective

Identify electrical units of measurement.

a. Define current.
b. Define voltage.
c. Define resistance.
d. Use Ohm's law to solve for unknown circuit values.

Trade Terms

Coulomb: A unit of electrical charge equal to 6.25×1018 electrons (or 6.25 quintillion electrons). A coulomb is the common unit of quantity used for specifying the size of a given charge.

Joule (J): A unit of measurement for doing work, represented by the letter J. One joule is equal to one newton-meter (Nm).

Ohm's law: A statement of the relationships among current, voltage, and resistance in an electrical circuit: current (I) equals voltage (E) divided by resistance (R). Generally expressed as a mathematical formula: $I = E/R$.

Resistors: Any devices in a circuit that resist the flow of electrons.

Various circuit values, such as voltage, current, and resistance, can be measured to determine circuit characteristics. Even if only two of these values are known, the remaining unknown value can be calculated using Ohm's law.

2.1.0 Defining Current

The movement of the flow of electrons is called *current*. Electrical current is often represented by the letter I. The basic unit in which current is measured is the ampere (A), also called an amp. One ampere of current is defined as the movement of 1 **coulomb** of charge past any point of a conductor in a second. An electron has 1.6×10^{-19} coulombs of charge. Therefore, it takes 6.25×10^{18} electrons to make up one coulomb of charge, as shown in the following equation:

$$\frac{1}{1.6 \times 10^{-19}} = 6.25 \times 10^{18} \text{ electrons}$$

If two particles, one having charge Q_1 and the other charge Q_2, are a distance (d) apart, then the force between them is given by Coulomb's law, which states that the force is directly proportional to the product of the two charges and inversely proportional to the square of the distance between them:

$$\text{Force} = \frac{k \times Q_1 \times Q_2}{d^2}$$

If Q_1 and Q_2 are both positive or both negative, then the force is positive; it is repulsive (a repelling force). If Q_1 and Q_2 are of opposite charges, then the force is negative; it is attractive (an attracting force). The letter k equals a constant with a value of 10^9.

Electrical current is a rate, which can be defined as an equation:

$$I = \frac{Q}{T}$$

Where:

I = current (amperes)
Q = charge (coulombs)
T = time (seconds)

Charge differs from current in that charge (Q) is an accumulation of charge, whereas current (I) measures the intensity of moving charges.

In a conductor, such as copper wire, the free electrons are charges that can be forced to move with relative ease by a potential difference. If a potential difference is connected across two ends of a copper wire, as shown in *Figure 9*, the applied voltage forces the free electrons to move. This current is a flow of electrons from the point of negative charge (–) at one end of the wire, moving through the wire to the positive charge (+) at the other end. The direction of the electron flow is from the negative side of the battery, through the wire, and back to the positive side of the battery. The direction of current flow is therefore from a point of negative potential to a point of positive potential.

Think About It

Current Flow

Why do you need two wires to use electrical devices? Why can't current simply move to a lamp and be released as light energy?

Law of Electrical Force

In the 18th century, a French physicist named Charles de Coulomb was concerned with how electric charges behaved. He watched the repelling forces exerted by opposite electric charges measuring the twist in a wire. An object's weight acted as a turning force to twist the wire, and the amount of twist was proportional to the object's weight. After many experiments with opposing forces, de Coulomb proposed the Inverse Square Law, later known as the Law of Electrical Force.

2.2.0 Defining Voltage

An electric charge has the ability to do the work of moving another charge by attraction or repulsion. The ability of a charge to do work is called its *potential*. When one charge is different from another, there is a difference in potential between them. The sum of the difference of potential of all the charges in the electrostatic field is referred to as the *potential difference, electromotive force (emf)*, or *voltage*. Voltage is often represented by the letter E.

One volt is the potential difference between two points for which one coulomb of electricity will do one **joule (J)** of work. A battery is one of several means of creating voltage. It chemically creates a large reserve of free electrons at the negative (–) terminal. The positive (+) terminal has electrons chemically removed and will therefore accept electrons if an external path is provided from the negative (–) terminal. When a battery is no longer able to chemically deposit electrons at the negative (–) terminal, it is said to be dead, or in need of recharging. Batteries are usually rated in volts. Large batteries are also rated in ampere-hours, where one ampere-hour is a current of one amp supplied for one hour.

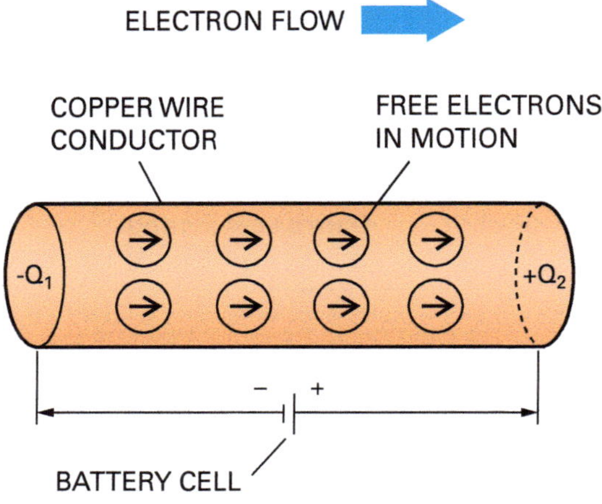

Figure 9 Potential difference causing electric current.

2.3.0 Defining Resistance

Resistance is directly related to the ability of a material to conduct electricity. All conductors have very low resistance; insulators have very high resistance. Resistance can be defined as the opposition to current flow. To add resistance to a circuit, electrical components called **resistors** are used. A resistor is a device whose resistance to current flow is a known, specified value. Resistance is measured in ohms and is represented by the symbol R in equations. One ohm is defined as the amount of resistance that will limit the current in a conductor to one ampere when the voltage applied to the conductor is one volt. The symbol for an ohm is Ω.

The resistance of a wire is proportional to the length of the wire, inversely proportional to the cross-sectional area of the wire, and dependent upon the kind of material of which the wire is made. The relationship for finding the resistance of a wire is:

$$R = \rho \frac{L}{A}$$

Where:

R = resistance (ohms)
L = length of wire (feet)
A = area of wire (circular mils, CM, or cm^2)
ρ = specific resistance (ohm-CM/ft or microhm-CM)

A mil equals 0.001 inch; a circular mil is the cross-sectional area of a wire one mil in diameter.

The specific resistance is a constant that depends on the material of which the wire is made. (The specific resistance of a material is also called its *resistivity*.) The resistance of copper building wire is 12.9 ohm-CM/ft, whereas that of aluminum is 21.3 ohm-CM/ft. Note that the resistance also varies by ambient temperature.

2.4.0 Calculating Circuit Values

Ohm's law defines the relationship between current, voltage, and resistance. If any two of these quantities are known, the third can be calculated using Ohm's law. (For example, if you know the current flowing to a lamp and how much voltage is applied, you can determine the resistance of the lamp.) There are three ways to express Ohm's law mathematically, each of which solves for a different circuit value:

- *Current* – The current in a circuit is equal to the voltage applied to the circuit divided by the resistance of the circuit:

$$I = \frac{E}{R}$$

- *Resistance* – The resistance of a circuit is equal to the voltage applied to the circuit divided by the current in the circuit:

$$R = \frac{E}{I}$$

- *Voltage* – The applied voltage to a circuit is equal to the product of the current and the resistance of the circuit:

$$E = I \times R \quad or \quad E = IR$$

Where:

I = current (amperes)
R = resistance (ohms)
E = voltage, or emf (volts)

If any two of the quantities E, I, or R are known, the third can be calculated. The Ohm's law equations can be memorized and practiced effectively by using an Ohm's law circle, as shown in *Figure 10*. To find the equation for E, I, or R when two quantities are known, cover the unknown third quantity. The other two quantities in the circle will indicate how the covered quantity may be found.

Example 1:
Find I when E = 120V and R = 30Ω.

$$I = \frac{E}{R}$$

$$I = \frac{120V}{30Ω}$$

$$I = 4A$$

This formula shows that in a DC circuit, current (I) is directly proportional to voltage (E) and inversely proportional to resistance (R).

What's Measured	Unit of Measurement and Symbol		Ohm's Law Symbol
Amount of current	Amp	A	I
Electrical power	Watt	W	P
Force of current	Volt	V	E
Resistance to current	Ohm	Ω	R

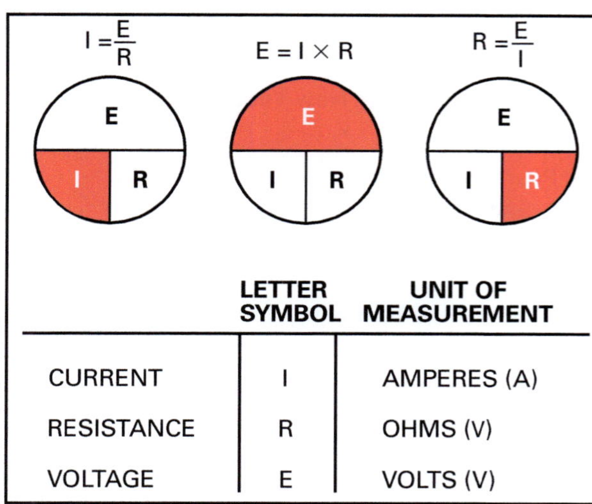

Figure 10 Ohm's law circle.

Example 2:
Find R when E = 240V and I = 20A.

$$R = \frac{E}{I}$$

$$R = \frac{240V}{20\Omega}$$

$$R = 12\Omega$$

Example 3:
Find E when I = 15A and R = 8Ω.

$$E = I \times R$$
$$E = 15A \times 8\Omega$$
$$E = 120V$$

Joule's Law

While other scientists of the 19th century were experimenting with batteries, cells, and circuits, James Joule was theorizing about the relationship between heat and energy. He discovered, contrary to popular belief, that work did not just move heat from one place to another; work, in fact, generated heat. Furthermore, he demonstrated that over time a relationship existed between the temperature of water and electric current. These ideas formed the basis for the concept of energy. In his honor, the modern unit of energy was named the joule.

Think About It

Voltage Matters

Standard household voltage is different around the world, from 100V in Japan to 600V in Bombay, India. Many countries have no standard voltage; for example, France varies from 110V to 360V. If you were to plug a 120V hair dryer into England's 240V, you would burn out the dryer. Use basic electric theory to explain exactly what would happen to destroy the hair dryer.

Units of Electricity and Volta

A disagreement with a fellow scientist over the twitching of a frog's leg eventually led 18th-century physicist Alessandro Volta to theorize that when certain objects and chemicals come into contact with each other, they produce an electric current. Believing that electricity came from contact between metals only, Volta coined the term *metallic electricity*. To demonstrate his theory, Volta placed two discs, one of silver and the other of zinc, into a weak acidic solution. When he linked the discs together with wire, electricity flowed through the wire. Thus, Volta introduced the world to the battery, also known as the Voltaic pile. Volta needed a term to measure the strength of the electric push or the flowing charge, which is now called a *volt*.

Additional Resources

Electronics Fundamentals: Circuits, Devices, and Applications, Thomas L. Floyd. New York, NY: Pearson Education, Inc.

Principles of Electric Circuits, Thomas L. Floyd. New York, NY: Pearson Education, Inc.

2.0.0 Section Review

1. Coulomb's law can be used to calculate the _____.

    a. direction of current flow
    b. force between two charges
    c. speed of electron movement
    d. loss of current

2. The negative terminal of a battery contains a large reserve of free _____.

    a. electrons
    b. protons
    c. neutrons
    d. acid

3. Which of the following does *not* have an effect on wire resistance?

    a. temperature
    b. type of power
    c. wire length
    d. material

4. Find the applied voltage when the current is 30A and the resistance is 4Ω.

    a. 30V
    b. 60V
    c. 120V
    d. 180V

3.0.0 READING SCHEMATIC DIAGRAMS

Objective

Read schematic diagrams.

a. Identify the symbol for a resistor and determine its value based on color codes.
b. Distinguish between series and parallel circuits.
c. Identify the instruments used to measure circuit values.
d. Calculate electrical power.

Trade Terms

Ammeter: An instrument for measuring electrical current.

Kilo: A prefix used to indicate one thousand (for example, one kilowatt is equal to one thousand watts).

Mega: A prefix used to indicate one million; for example, one megawatt is equal to one million watts.

Ohmmeter: An instrument used for measuring resistance.

Schematic: A type of drawing in which symbols are used to represent the components in a system.

Series circuit: A circuit with only one path for current flow.

Voltage drop: The change in voltage across a component that is caused by the current flowing through it and the amount of resistance opposing it.

Voltmeter: An instrument for measuring voltage. The resistance of the voltmeter is fixed. When the voltmeter is connected to a circuit, the current passing through the meter will be directly proportional to the voltage at the connection points.

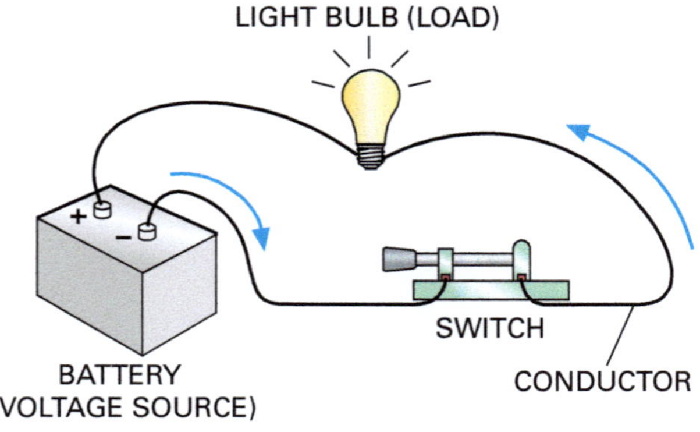

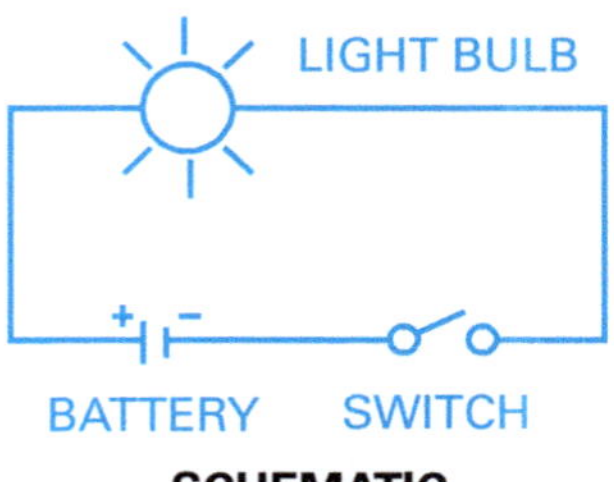

Figure 11 Electrical circuit.

The simple electric circuit shown in *Figure 1* appears again in *Figure 11* in both pictorial and **schematic** forms. The schematic diagram is a shorthand way to draw an electric circuit, and circuits are usually represented in this way. In addition to the connecting wire, three components are shown symbolically: the battery, the switch, and the lamp. Note the positive (+) and negative (–) markings in both the pictorial and schematic representations of the battery. The schematic components represent the pictorial components in a simplified manner. A schematic diagram is one that shows, by means of graphic symbols, the electrical connections and functions of the different parts of a circuit.

The standard graphic symbols for commonly used electrical and electronic components are shown in *Figure 12*.

3.1.0 Resistors

The function of a resistor is to offer a particular resistance to current flow. For a given current and known resistance, the change in voltage across the component, or **voltage drop**, can be predicted using Ohm's law. Voltage drop refers to a specific amount of voltage used, or developed, by that component. An example is a very basic circuit of a 10V battery and a single resistor in a **series circuit**. The voltage drop across that resistor is 10V because it is the only component in the circuit and all voltage must be dropped across that resistor. Similarly, for a given applied voltage, the current that flows may be predetermined by selection of the resistor value. The required power dissipation largely dictates the construction and physical size of a resistor.

The two most common types of electronic resistors are wire-wound and carbon composition construction. A typical wire-wound resistor consists of a length of nickel wire wound on a

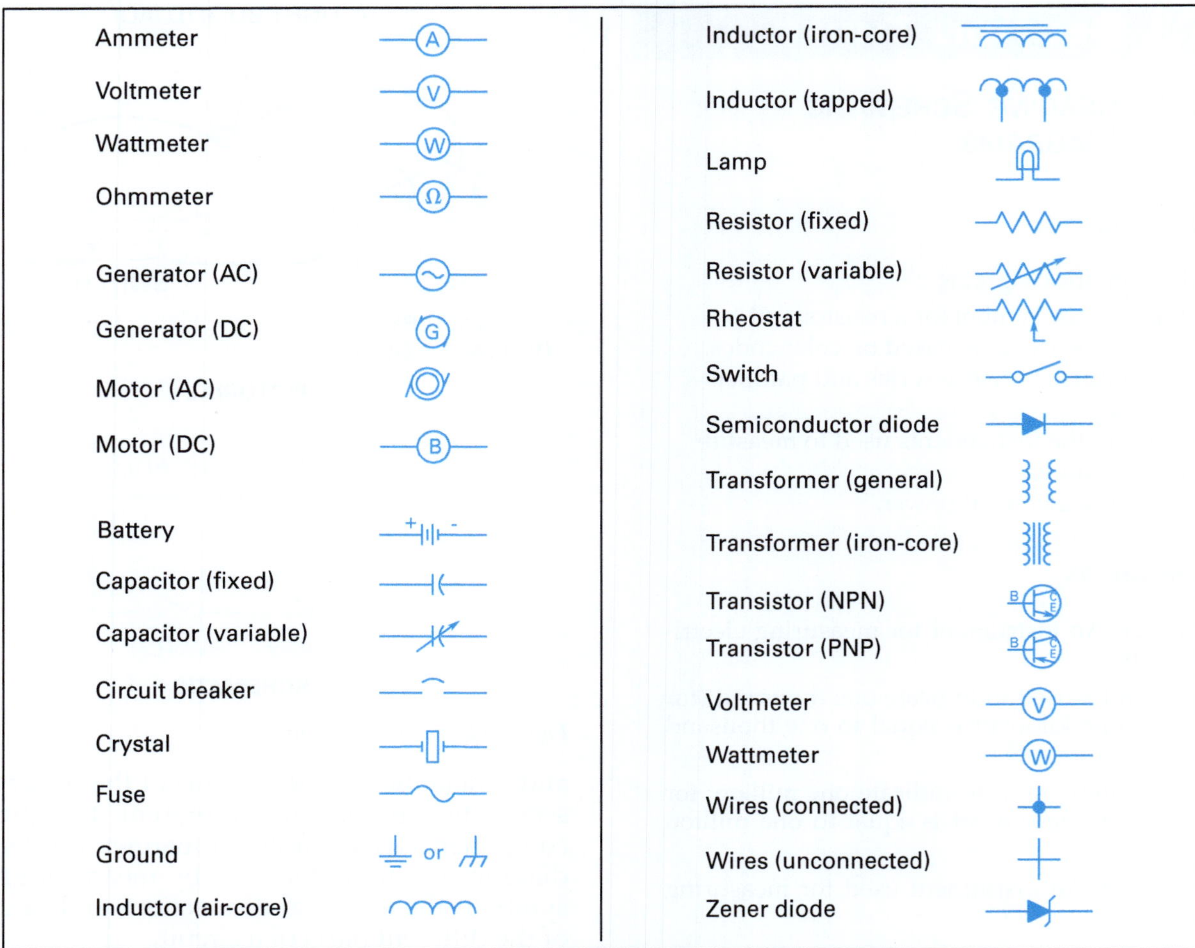

Figure 12 Standard schematic symbols.

ceramic tube and covered with porcelain. Low-resistance connecting wires are provided, and the resistance value is usually printed on the side of the component. *Figure 13* illustrates the construction of typical resistors. Carbon composition resistors are constructed by molding mixtures of powdered carbon and insulating materials into a cylindrical shape. An outer sheath of insulating material affords mechanical and electrical protection, and copper connecting wires are provided at each end. Carbon composition resistors are smaller and less expensive than the wire-wound type. However, the wire-wound type is the more rugged of the two and is able to survive much larger power dissipations than the carbon composition type.

Most resistors have standard fixed values, so they can be termed fixed resistors. Variable

Using Your Intuition

Learning the meanings of various electrical symbols may seem overwhelming, but if you take a moment to study *Figure 12*, you will see that most of them are intuitive—that is, they are shaped (in a symbolic way) to represent the actual object. For example, the battery shows + and −, just like an actual battery. The motor has two arms that suggest a spinning rotor. The transformer shows two coils. The resistor has a jagged edge to suggest pulling or resistance. Connected wires have a black dot that reminds you of solder. Unconnected wires simply cross. The fuse stretches out in both directions as though to provide extra slack in the line. The circuit breaker shows a line with a break in it. The capacitor shows a gap. The variable resistor has an arrow like a swinging compass needle. As you learn to read schematics, take the time to make mental connections between the symbol and the object it represents.

NCCER – *Maritime Electrical*

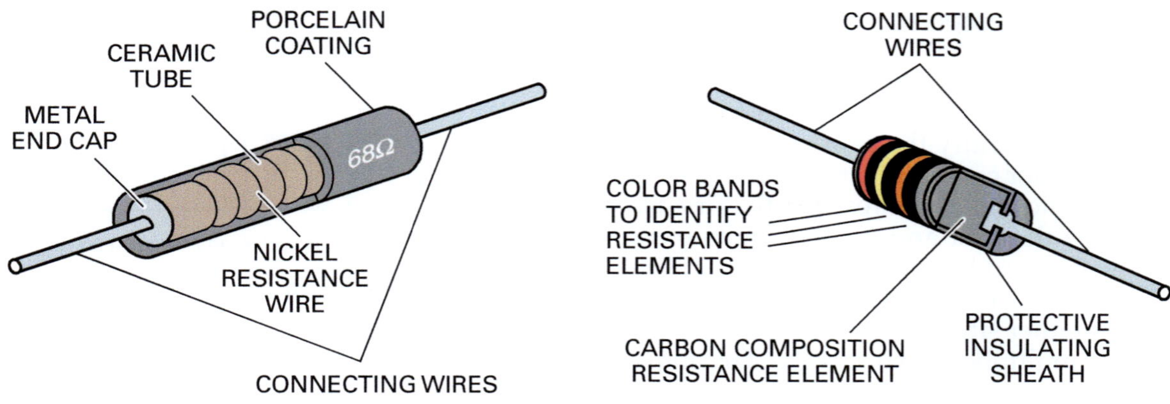

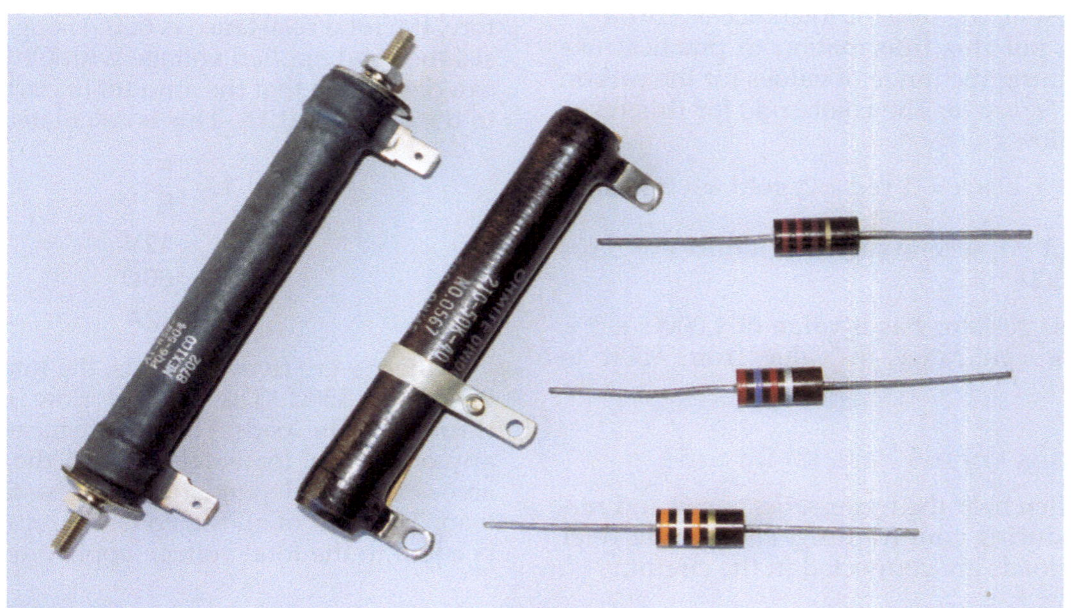

Figure 13 Common resistors.

resistors, also known as adjustable resistors, are used a great deal in electronics. Two common symbols for a variable resistor are shown in *Figure 14*.

A variable resistor consists of a coil of closely wound insulated resistance wire formed into a partial circle. The coil has a low-resistance terminal at each end, and a third terminal is connected to a movable contact with a shaft adjustment facility. The movable contact can be set to any point on a connecting track that extends over one (uninsulated) edge of the coil. Using the adjustable contact, the resistance from either end terminal to the center terminal can be adjusted from zero to the maximum coil resistance. Another type of variable resistor is known as a decade resistance box—a laboratory component that contains precise values of switched series-connected resistors.

Think About It

Drawing a Schematic

Draw a schematic diagram showing a voltage source, switch, motor, and fuse.

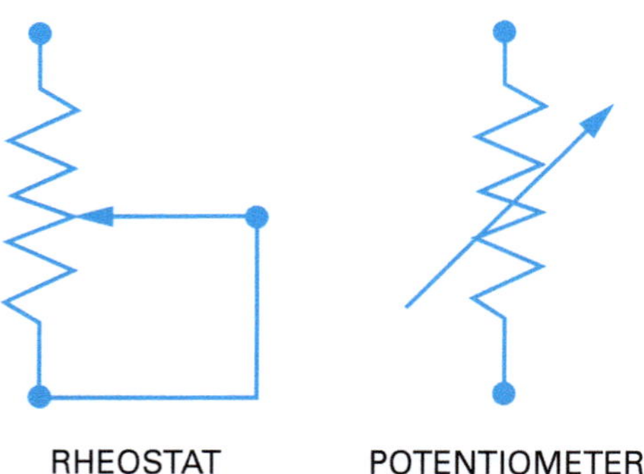

Figure 14 Symbols used for variable resistors.

Because carbon composition resistors are small in size—some less than 0.4 inch (1 cm) long—it is not convenient to print the resistance value on the side. Instead, a color code in the form of colored bands is used to identify the resistance value and tolerance. The color code is illustrated in *Figure 15*. Starting from one end of the resistor, the first two bands identify the first and second digits of the resistance value, and the third band indicates the number of zeros. An exception to this is when the third band is either silver or gold, which indicates a 0.01 or 0.1 multiplier, respectively. The fourth band is always either silver or gold, and in this position, silver indicates a ± 10% tolerance and gold indicates a ± 5% tolerance. Where no fourth band is present, the resistor tolerance is ± 20%.

You can put this information to practical use by determining the range of values for the carbon resistor in *Figure 16*. The color code for this resistor is as follows:

- Brown = 1, black = 0, red = 2, gold = a tolerance of ± 5%
- First digit = 1, second digit = 0, number of zeros (2) = 1,000Ω

Since this resistor has a value of 1,000Ω ± 5%, the resistor can range in value from 950Ω to 1,050Ω.

3.2.0 Series versus Parallel Circuits

You will often hear the terms *series circuit* and *parallel circuit* during your training. These terms refer to the way loads are connected in the circuit.

Figure 16 Sample color codes on a fixed resistor.

3.2.1 Series Circuits

A series circuit provides only one path for current flow and is a voltage divider. The total resistance (R_T) of a series circuit is equal to the sum of the individual resistances in the circuit. The 12V series circuit in *Figure 17* (A) has two 30Ω loads. Therefore, the total resistance is 60Ω. Using the total resistance and applied voltage with Ohm's law, you can determine that the amount of current flowing in the circuit is 0.2A. This is calculated as follows:

$$I = \frac{E}{R}$$

$$I = \frac{12V}{60\Omega}$$

$$I = 0.2A$$

If there were five 30Ω loads, the total resistance would be 150Ω. The current flow is the same through all the loads. The voltage measured across any one of the loads (also called the *voltage drop across a load*) depends on the resistance of that load. The sum of all the voltage drops in a circuit is equal to the total voltage applied to the circuit.

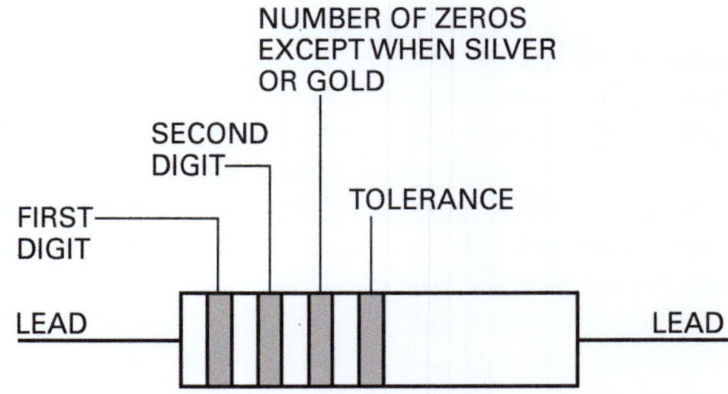

0	BLACK	7		VIOLET
1	BROWN	8		GRAY
2	RED	9		WHITE
3	ORANGE	0.1		GOLD
4	YELLOW	0.01		SILVER
5	GREEN	5%		GOLD – TOLERANCE
6	BLUE	10%		SILVER – TOLERANCE

Figure 15 Resistor color codes.

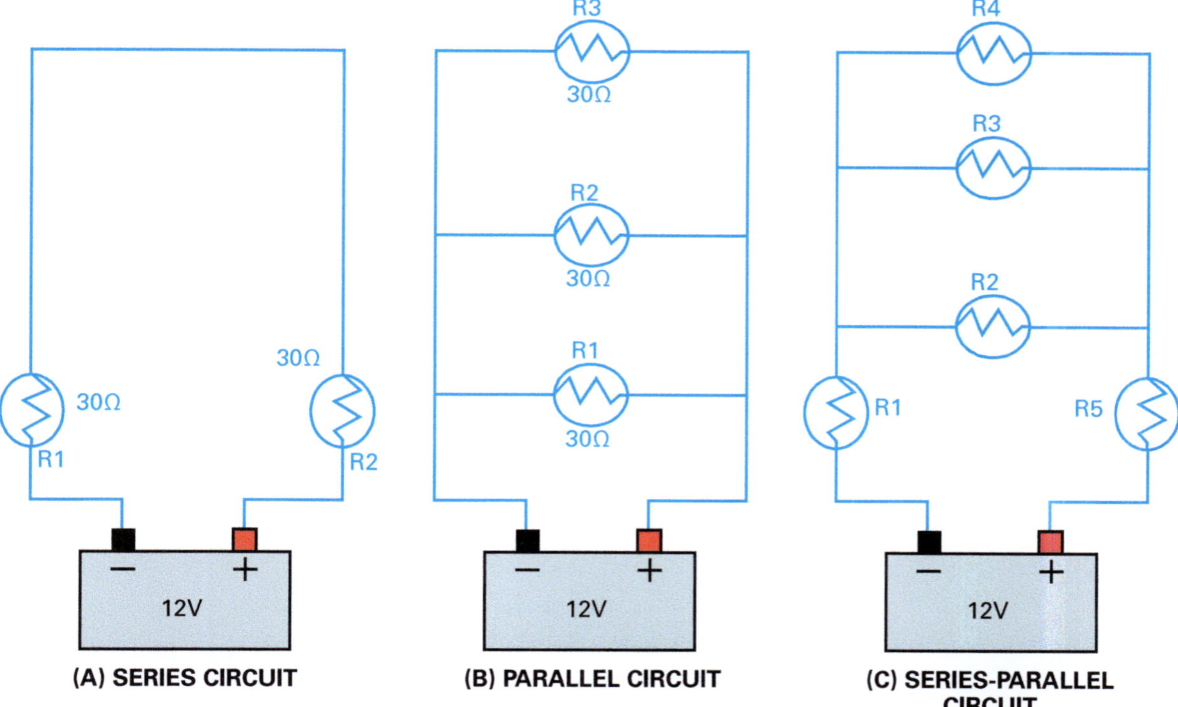

Figure 17 Types of circuits.

Circuits containing loads in series are uncommon. An important trait of a series circuit is that if the circuit is open at any point, no current will flow. For example, if you have five light bulbs connected in series and one of them blows, all five lights will go off.

3.2.2 Parallel Circuits

In a parallel circuit, each load is connected directly to the voltage source; therefore, the voltage drop through each of the loads is the same, and the current is divided between the loads. (The amount of current across each load depends on its resistance.) The source sees the circuit as two or more individual circuits containing one load each. In the parallel circuit in *Figure 17* (B), the source sees three circuits, each containing a 30Ω load. The current flow through any load is determined by the resistance of that load. Thus, the total current drawn by the circuit is the sum of the individual currents. The total resistance of a parallel circuit is calculated differently from that of a series circuit. In a parallel circuit, the total resistance is less than the smallest of the individual resistances.

For example, each of the 30Ω loads in *Figure 17* (B) draws 0.4A at 12V; therefore, the total current is 1.2A. This is calculated as follows:

$$I = \frac{E}{R} = \frac{12V}{30\Omega} = 0.4A \text{ per circuit}$$

0.4A per circuit $\times$ 3 circuits = 1.2A

Now, Ohm's law can be used again to calculate the total resistance:

$$R = \frac{E}{I} = \frac{12V}{1.2\Omega} = 10\Omega$$

This example was simple because all the resistances were the same value. The process is the same when the resistances are different, but the current calculation has to be done for each load. The individual currents are added to get the total current.

Unlike series circuits, parallel circuits continue working even if one circuit opens. Household circuits are wired in parallel. In fact, almost all the load circuits you encounter will be parallel circuits.

The following formulas can be used to convert parallel resistances to a single resistance value:

- For two resistances connected in parallel:

$$R_T = \frac{R1 \times R2}{R1 + R2}$$

- For three or more resistances connected in parallel:

$$R_T = \frac{1}{\dfrac{1}{R1} + \dfrac{1}{R2} + \dfrac{1}{R3}}$$

Example:

1. The total resistance of the parallel circuit below is 6Ω.

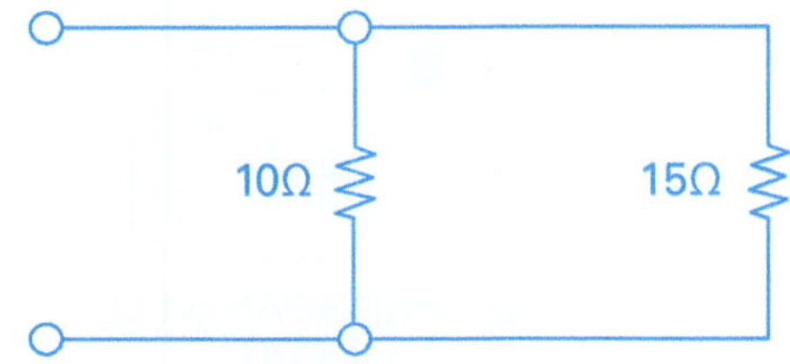

Total resistance (R_T) =

$$\frac{R1 \times R2}{R1 + R2} = \frac{10 \times 15}{10 + 15} = \frac{50}{25} = 6\Omega$$

2. The total resistance of the parallel circuit below is 4.76Ω.

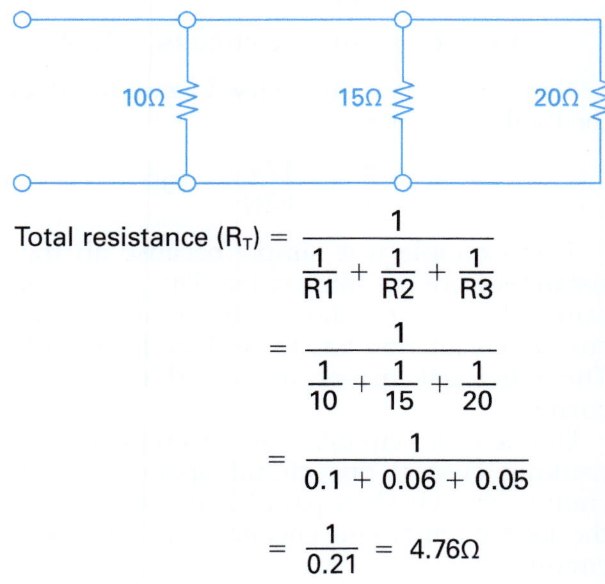

Total resistance (R_T) =

$$\frac{1}{\dfrac{1}{R1} + \dfrac{1}{R2} + \dfrac{1}{R3}}$$

$$= \frac{1}{\dfrac{1}{10} + \dfrac{1}{15} + \dfrac{1}{20}}$$

$$= \frac{1}{0.1 + 0.06 + 0.05}$$

$$= \frac{1}{0.21} = 4.76\Omega$$

3.2.3 Series-Parallel Circuits

Electronic circuits sometimes contain a hybrid arrangement known as a *series-parallel circuit*, shown in *Figure 17* (C). However, you will rarely find loads connected in this arrangement.

To determine the total resistance of a series-parallel circuit, the parallel loads must be converted to their equivalent series resistance. The load resistances are then added to determine total circuit resistance.

> **NOTE**
>
> Because loads are rarely connected in a series-parallel arrangement, you will not have to determine characteristics of series-parallel circuits very often. However, these calculations are presented in detail in Module 26104-17 of this curriculum, "Electrical Theory".

3.3.0 Electrical Meters

Electricians frequently use test meters to measure voltage, current, and resistance. The most common test meter is the volt-ohm-milliammeter (VOM), also called a *multimeter*. (Multimeters usually measure voltage, resistance, and current.) *Figure 18* shows both digital and analog multimeters.

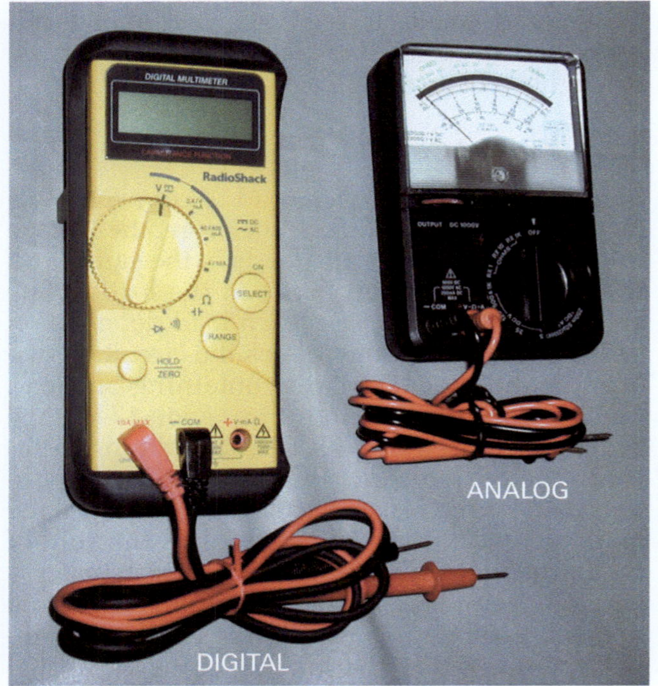

Figure 18 Digital and analog meters.

An analog meter is described as *analog* because the pointer moves in proportion to the value being measured. The person using the meter must then interpret the scale to determine the measured value, although digital meters display the result numerically on the screen.

Multimeters are commonly used to measure AC and DC voltage, DC current, and resistance. They can also be used to measure AC current in the milliamp range. For larger current values, it is usually necessary to use a clamp-on **ammeter** (*Figure 19*).

Only qualified individuals may use these meters. Consult your company's safety policy for applicable rules.

3.3.1 Measuring Current

A clamp-on ammeter is used to measure current. The jaws of the ammeter are placed around a single conductor (*Figure 20*), and current flowing through the wire creates a magnetic field, inducing a proportional current in the ammeter jaws. This current is read by the meter movement and appears as a direct readout or, on an analog meter, as a deflection of the meter needle.

In-line ammeters (*Figure 21*) are less common and must be connected in series with the circuit, which means that the circuit must be opened.

Aside from following good safety practices, there are a few things to remember when measuring current:

- If the ammeter jaws are dirty or misaligned, the meter will not read correctly.
- Meters are precision instruments and should be handled carefully to avoid damage.
- An analog meter can be damaged when the power of the current far exceeds the selected scale. When using an analog meter, always start at the highest range on the meter and work down.
- Do not clamp the meter jaws around two different conductors at the same time, or an inaccurate reading will result.

3.3.2 Measuring Voltage

A **voltmeter** must be connected in parallel with (across) the component or circuit to be tested (*Figure 22*). If a circuit function is not operating, the voltmeter can be used to determine if the correct voltage is available to the circuit. Voltage must be checked with power applied.

Is It a Series Circuit?

When the term *series circuit* is used, it refers to the way the loads are connected. The same is true for parallel and series-parallel circuits. You will rarely, if ever, find loads connected in series, or in a series-parallel arrangement. The simple circuit shown here illustrates this point. At first glance, you might think it is a series-parallel circuit. On closer examination, you can see that there are only two loads—the relay and the contactor—and they are connected in parallel. Therefore, it is a parallel circuit. The control devices are wired in series with the loads, but only the loads are considered in determining the type of circuit.

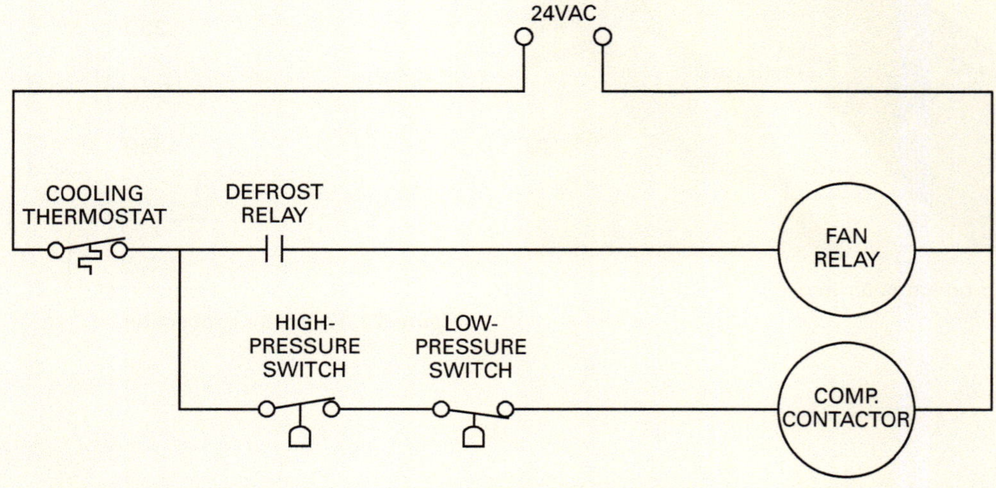

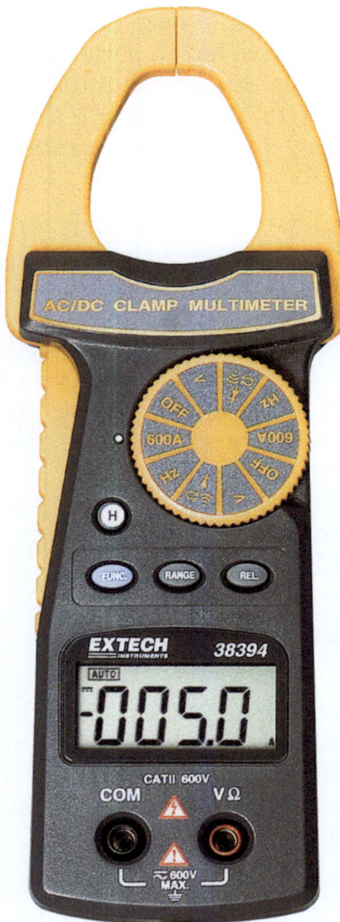

Figure 19 Clamp-on ammeter.

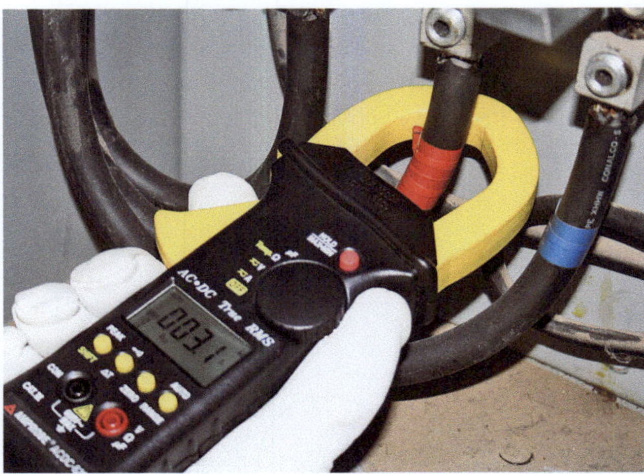

Figure 20 Clamp-on ammeter in use.

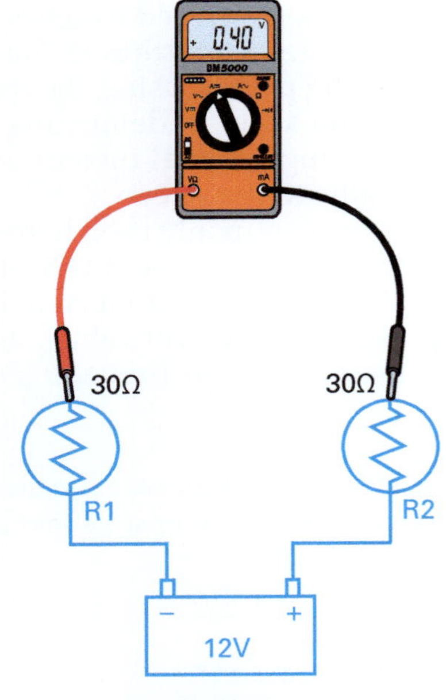

Figure 21 In-line ammeter test setup.

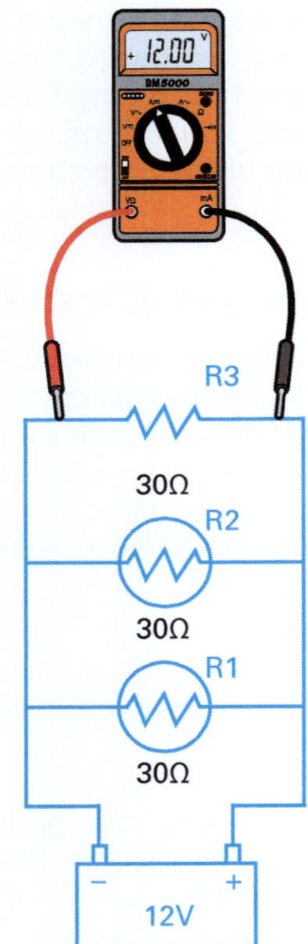

Figure 22 Voltmeter connection.

NCCER – *Maritime Electrical*

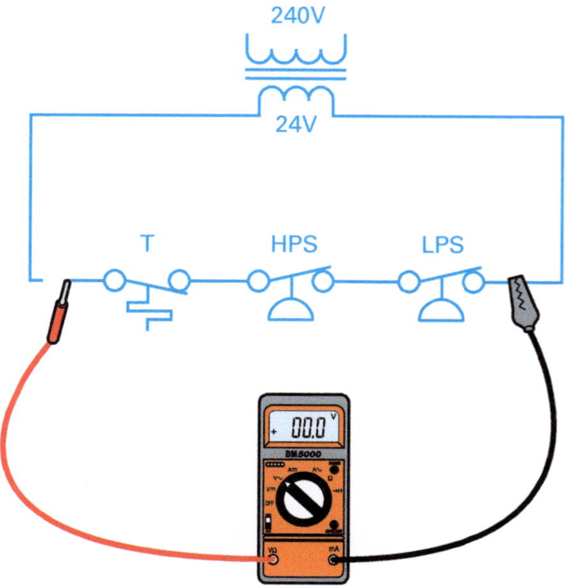

Figure 23 Ohmmeter connection for continuity testing.

3.3.3 Measuring Resistance

An **ohmmeter** contains an internal battery that acts as a voltage source. Therefore, resistance measurements are always made with the system power shut off. Sometimes, an ohmmeter is used to measure resistance in a load; motor windings are a good example. More often, an ohmmeter is used to check continuity in a circuit. A wire or closed switch offers negligible resistance. With the ohmmeter connected as in *Figure 23* and the three switches closed, the current produced by the ohmmeter battery will flow unopposed and the meter will show zero resistance. The circuit has continuity; that is, it is continuous. If a switch is open, however, there is no path for current and the meter will see infinite resistance (lack of continuity).

A continuity tester (*Figure 24*) is a simple device consisting primarily of a battery and either an audible or visual indicator. It can be used in place of an ohmmeter to test the continuity of a wire and to identify individual wires contained in a conduit or other raceway. To test the continuity of a wire, strip the insulation off the end of the wire to be tested at one end of the conduit run, then connect (short) the wire to the metal conduit. At the other end of the conduit run, clip the alligator clip lead of the tester to the conduit and touch the probe to the end of the wire under test. If the tester audible alarm sounds or the indicator light comes on, there is continuity. Note that this only indicates there is continuity between the two points being tested; it does not indicate the actual value of the resistance. If there is no indication, the wire is open.

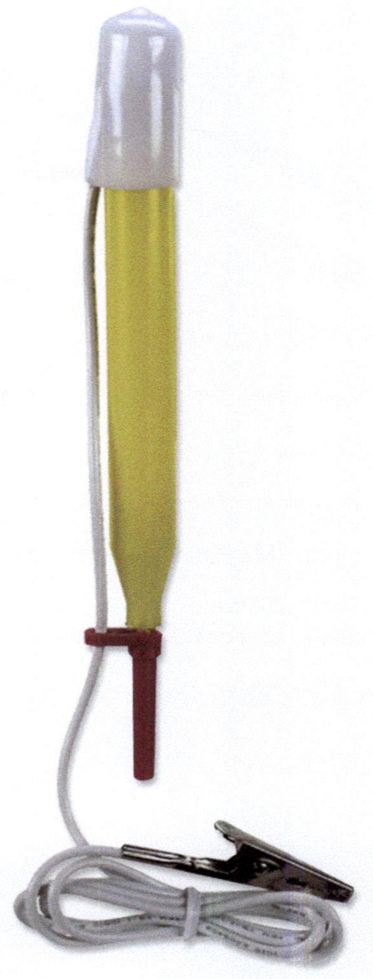

Figure 24 Continuity tester.

To identify individual wires in a conduit run, touch the tester probe to the wires in the conduit one at a time until the tester audible alarm sounds or the indicator lights. Then, put matching identification tags on both ends of the wire. Continue this procedure until all the wires have been identified.

3.3.4 Voltage Testers

Figure 25 shows one of the wide varieties of devices available for checking for the presence of voltage. It can be used as a troubleshooting tool and as a safety device to make sure the voltage is turned off before touching any terminals or conductors. When the probes are touched to the circuit, the light on the instrument will turn on if a voltage is present. Instruments like these are available in several voltage ranges, so it is important to know something about the circuit you are checking.

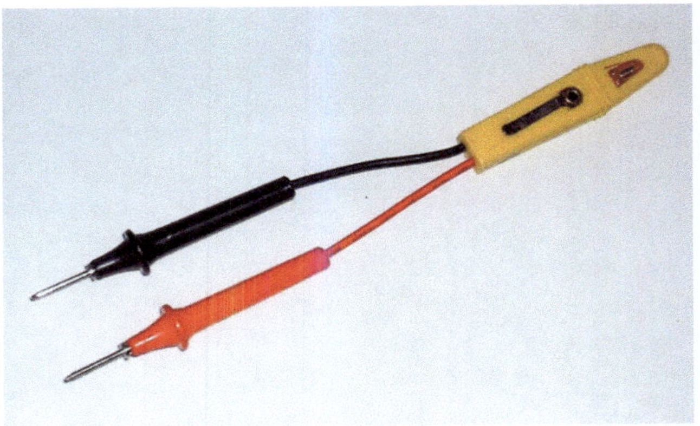

Figure 25 Voltage tester.

3.4.0 Calculating Electrical Power

Power is defined as the rate of doing work, which is equivalent to the rate at which energy is used or dissipated. Electrons passing through a resistance dissipate energy in the form of heat. In electrical circuits, power is measured in units called watts (W). The power in watts equals the rate of energy conversion. One watt of power is equal to the energy used by 1 volt to move electrical charge at a rate of 1 coulomb per second. Since 1 ampere is equal to 1 coulomb per second, power in watts is equal to the product of amperes and volts.

The work done in an electrical circuit can be useful work or wasted work. In both cases, the

Old and New Test Instruments

Early electricians used individual meters to test circuit parameters. Today, those instruments seem primitive given the availability of all-purpose instruments like the clamp-on ammeter with remote read capability shown here.

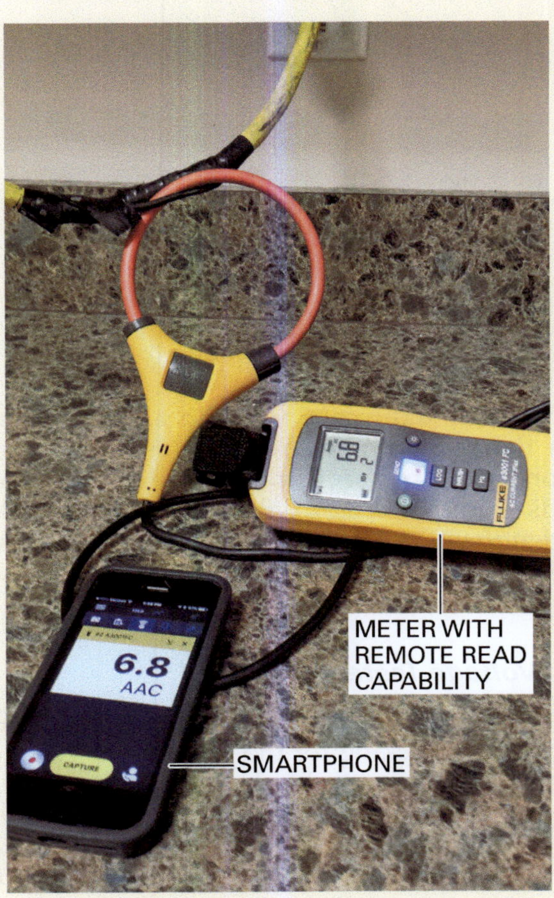

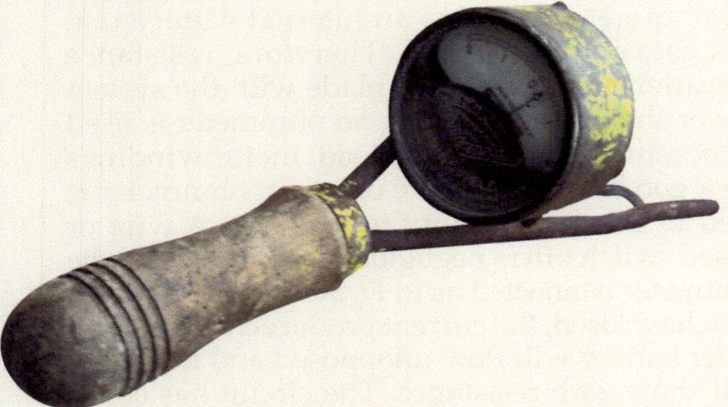

NCCER – *Maritime Electrical*

rate at which the work is done is still measured in power. The turning of an electric motor is useful work. The heating of wires or resistors in a circuit is wasted work, since no useful function is performed by the heat.

The unit of electrical work is the joule. This is the amount of work done by one coulomb flowing through a potential difference of one volt. If five coulombs flow through a potential difference of one volt, five joules of work are done. The time it takes these coulombs to flow through the potential difference has no bearing on the amount of work done.

Amperes are a measurement of current, which is a rate of current flow. This rate measures how much electrical charge passes a point in a certain amount of time. As previously discussed, one ampere is equal to a rate of 1 coulomb per second. A joule is a measurement of work (or dissipated energy); one joule of work is done when 1 ampere moves through 1 volt in a second. This rate of one joule per second is the basic unit of power, and is called a *watt*. Therefore, a watt is the power used when one ampere of current flows through a potential difference of 1 volt, as shown in *Figure 26*.

Mechanical power is usually measured in units of horsepower (hp). To convert from horsepower to watts, multiply the number of horsepower by 746. To convert from watts to horsepower, divide the number of watts by 746. *Table 1* lists conversions for common units of power.

The kilowatt-hour (kWh) is commonly used for large amounts of electrical work or energy. (The prefix **kilo** means one thousand, so 1 kilowatt is equal to 1,000 watts.) The amount is calculated as the product of the power in kilowatts multiplied by the time in hours during which the power

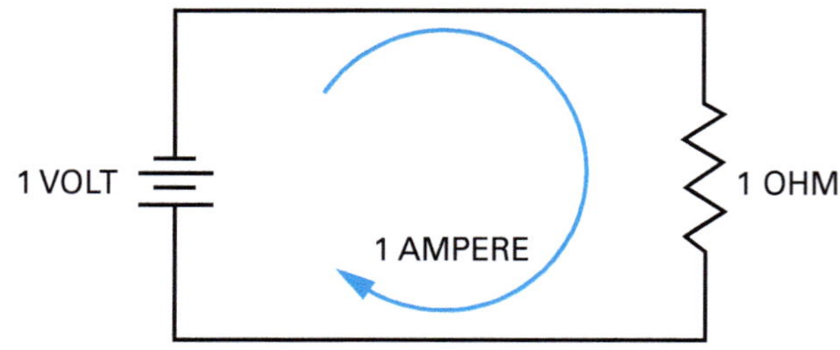

Figure 26 One watt.

Table 1 Conversion Table

1,000 watts (W)	= 1 kilowatt (kW)
1,000,000 watts (W)	= 1 megawatt (MW)
1,000 kilowatts (kW)	= 1 megawatt (MW)
1 watt (W)	= 0.00134 horsepower (hp)
1 horsepower (hp)	= 746 watts (W)

is used. If a light bulb uses 300W or 0.3kW for 4 hours, the amount of energy is 0.3 × 4, which equals 1.2kWh.

Very large amounts of electrical work or energy are measured in megawatts (MW). (The prefix **mega** means one million, so 1 megawatt is equal to 1 million watts.)

3.4.1 Power Equation

When one ampere flows through a difference of two volts, two watts of power must be used. In other words, the number of watts used is equal to the number of amperes of current times the potential difference. This is called *the power equation*, and it is expressed as follows:

$$P = I \times E \quad or \quad P = IE$$

Where:

  P = power used in watts
  I = current in amperes
  E = potential difference in volts

The equation is sometimes called Ohm's law for power, because it is similar to Ohm's law. This equation is used to find the power consumed in a circuit or load when the values of current and voltage are known.

As an example, suppose the total load current in the main line equals 20A. Then the power in watts from the 120V line is:

$$P = I \times E$$
$$P = 20A \times 120V$$
$$P = 2400W \quad or \quad 2.4kW$$

If this power is used for five hours, then the energy of work supplied in kilowatt-hours (kWh) equals:

$$2.4kW \times 5 \text{ hours} = 12kWh$$

The second form of the power equation is used to find the voltage when the power and current are known:

$$E = \frac{P}{I}$$

Power

We take electrical power for granted, never stopping to think how surprising it is that a flow of submicroscopic electrons can pump thousands of gallons of water or illuminate a skyscraper. Our lives now constantly rely on the ability of the electron to do work. Think about your day up to this moment. How has electrical power shaped your experience?

The third form of the power equation is used to find the current when the power and voltage are known:

$$I = \frac{P}{E}$$

Using these three equations, the power, voltage, or current in a circuit can be calculated whenever any two of the values are already known.

Example 1:
Calculate the power in a circuit where the source of 100V produces 2A in a 50Ω resistance.

$$P = IE$$
$$P = 2A \times 100V$$
$$P = 200W$$

This means the source generates 200W of power while the resistance dissipates 200W in the form of heat.

Example 2:
Calculate the source voltage in a circuit that consumes 1,200W at a current of 5A.

$$E = \frac{P}{I}$$
$$E = \frac{1200W}{5A}$$
$$E = 240V$$

Example 3:
Calculate the current in a circuit that consumes 600W with a source voltage of 120V.

$$I = \frac{P}{E}$$
$$I = \frac{600W}{120V}$$
$$I = 5A$$

Components that use the power dissipated in their resistance are generally rated in terms of power. The power is rated at normal operating voltage, which is usually 120V. For instance, an appliance that draws 5A at 120V would dissipate 600W. The rating for the appliance would then be 600W/120V.

To calculate I or R for components rated in terms of power at a specified voltage, it may be convenient to use the power formula in different forms. There are three basic power formulas, but each can be rearranged into two other forms for a total of nine combinations, as shown in *Figure 27*. Note that all of these formulas are based on Ohm's law (E = IR) and the power formula (P = IE).

3.4.2 Power Rating of Resistors

If too much current flows through a resistor, the heat caused by the current will damage or destroy the resistor. This heat is caused by I^2R heating, which is power loss expressed in watts. This is expressed as the following equation:

$$P = I^2R$$

Every resistor is given a wattage, or power rating, to show how much I^2R heating it can take before it burns out. This means that a resistor with a power rating of 1W will burn out if it is used in a circuit where the current causes it to dissipate heat at a rate greater than 1W.

If the power rating of a resistor is known, the maximum current it can carry is found by using an equation derived from $P = I^2R$:

$$P = I^2R$$

Divide both sides of the equation by R:

$$\frac{P}{R} = \frac{I^2R}{R}$$
$$P/R = I^2$$

Take the square root of each side:

$$\sqrt{P/R} = \sqrt{I^2}$$

Transpose the equation; this is the equation used to find the current.

$$I = \sqrt{P/R}$$

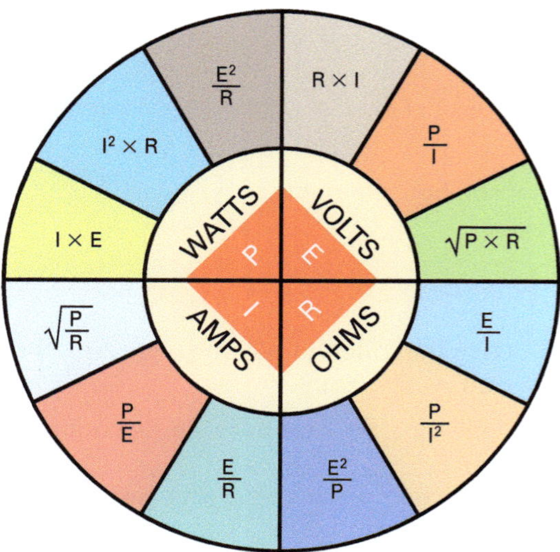

Figure 27 Expanded Ohm's law circle.

Using this equation, find the maximum current that can be carried by a 1Ω resistor with a power rating of 4W:

$$I = \sqrt{P/R} = \sqrt{4/1} = \sqrt{4} = 2A$$

If such a resistor conducts more than 2A, it will dissipate more than its rated power, causing it to burn out.

Power ratings assigned by resistor manufacturers are usually based on the resistors being mounted in an open location where there is free air circulation, and where the temperature is not higher than 104°F (40°C). If a resistor is mounted in a small, crowded, enclosed space, or where the temperature is higher than 104°F (40°C), there is a good chance it will burn out even before its power rating is exceeded. Also, some resistors are designed to be attached to a chassis or frame that will carry away the heat.

Think About It

Putting It All Together

Notice the common electrical devices in the building you're in. What is their wattage rating? How much current do they draw? How would you test their voltage or amperage?

3.0.0 Section Review

1. A resistor with a gold band in the fourth position has a tolerance of ______.

   a. ± 1%
   b. ± 5%
   c. ± 10%
   d. ± 20%

2. The total resistance in a 120V series circuit with three 10Ω resistors is ______.

   a. 1.2Ω
   b. 12Ω
   c. 30Ω
   d. 120Ω

3. Which of the following is true regarding the use of a clamp-on ammeter?

   a. Place the jaws around a single conductor at a time.
   b. Place the jaws around an even number of conductors only.
   c. Start at the lowest meter range and work up.
   d. Misaligned meter jaws can be corrected by tapping them with a hammer.

4. Calculate the power in a circuit where the source of 240V produces 10A.

   a. 12W
   b. 240W
   c. 1,200W
   d. 2,400W

SUMMARY

Electricians often test and troubleshoot electrical circuits. This work can be done safely and more effectively if you know the theory of electricity and the relationships between voltage, current, resistance, and power. The basic tool for understanding these relationships is Ohm's law.

Testing and troubleshooting of electrical circuits involves the use of test instruments such as the multimeter, or VOM. The multimeter combines the voltmeter, ammeter, and ohmmeter into a single instrument. In analog meters, a pointer moves across a scale in proportion to the current flowing though the meter. In a digital meter, the measured value is displayed directly on the screen of the meter.

1. An electrical circuit contains, at minimum, a(n) ______.
   a. voltage source, load, and overcurrent device
   b. ammeter, load, and voltage source
   c. voltage source, load, and conductors
   d. conductor, switch, and load

2. A type of subatomic particle with a positive charge is a(n) ______.
   a. proton
   b. neutron
   c. electron
   d. nucleus

3. Which of the following substances is considered an insulator?
   a. Gold
   b. Copper
   c. Silver
   d. Porcelain

4. The voltage commonly supplied to a residence by the local utility is ______.
   a. 120V
   b. 240V
   c. 480V
   d. 208V

5. Current is measured in units called ______.
   a. joules
   b. coulombs
   c. amperes
   d. volt-amperes

6. Joules are ______.
   a. units of work
   b. the potential difference between two points
   c. the difference between EMF and current
   d. the rate of current flow

7. Another term used for voltage is ______.
   a. emf
   b. coulomb
   c. current
   d. joule

8. All conductors have ______.
   a. current flow
   b. some resistance
   c. EMF
   d. voltage potential

9. In order to calculate the current flowing in a circuit ______.
   a. multiply voltage by resistance ($E \times R$)
   b. divide resistance by power (R/P)
   c. multiply power by resistance ($P \times R$)
   d. divide voltage by resistance (E/R)

10. The color band that represents tolerance on a resistor is the ______.
    a. 4th band
    b. 3rd band
    c. 2nd band
    d. 1st band

11. In a parallel circuit, the total resistance is ______ the smallest resistance.
    a. greater than
    b. equal to
    c. less than
    d. proportional to

12. Ammeters measure ______.
    a. voltage
    b. resistance
    c. current
    d. power

13. Circuit continuity is checked using the ______ function of a multimeter.
    a. ammeter
    b. voltmeter
    c. ohmmeter
    d. wattmeter

14. Resistance is measured in _______.

    a. ohms
    b. amperes
    c. volt-amperes
    d. coulombs

15. The power in a circuit with 120 volts and 5 amps is _______.

    a. 5 watts
    b. 120 watts
    c. 240 watts
    d. 600 watts

Trade Terms Quiz

Fill in the blank with the correct term that you learned from your study of this module.

1. A(n) _________ is an instrument for measuring electrical current.

2. Measured in amperes, _________ is the flow of electrons in a circuit.

3. _________ are the force required to produce a current of one ampere through a resistance of one ohm.

4. Voltage is measured with a(n) _________.

5. _________ are the basic unit of measurement for electrical current.

6. One volt is the potential difference between two points for which one coulomb of electricity will do one _________ of work.

7. A(n) _________ is a quantity of electricity that is either positive or negative.

8. A(n) _________ is the common unit used for specifying the size of a given charge.

9. _________ is the driving force that makes current flow in a circuit.

10. _________ are the basic unit of measurement for electrical power.

11. _________ are the smallest particles of an element that will still retain the properties of that element.

12. The _________ is the center of an atom.

13. Both found in the nucleus of an atom, a _________ is an electrically positive particle and _________ are electrically neutral particles.

14. The outermost ring of electrons orbiting the nucleus of an atom is known as the _________.

15. _________ are negatively charged particles that orbit the nucleus of an atom.

16. The definition of _________ is any substance that has mass and occupies space.

17. The prefix used to indicate one thousand is _________.

18. The prefix used to indicate one million is _________.

19. Consisting of two or more cells, a _________ converts chemical energy into electrical energy.

20. A(n) _________ is a complete path for current flow.

21. _________ are materials through which it is relatively easy to maintain an electric current.

22. A(n) _________ is a material through which it is difficult to conduct an electric current.

23. _________ are the basic unit of measurement for resistance.

24. The instrument that is used to measure resistance is called a(n) _________.

25. _________ is a statement of the relationship between current, voltage, and resistance in an electrical circuit.

26. _________ is the rate of doing work or the rate at which energy is used or dissipated.

27. Measured in ohms, _________ is the electrical property that opposes the flow of current through a circuit.

28. _________ are components that normally oppose current flow in a DC circuit.

29. A(n) _________ is a drawing in which symbols are used to represent the components in a system.

30. A(n) _________ has only one route for current flow.

31. The change in voltage across a component is called __________.

32. __________ are electromechanical components used as switching devices.

33. Devices containing one or more coils of wire wrapped around a common core are called __________.

34. Electromagnetic devices used to control a mechanical device such as a valve are called __________.

Trade Terms

Ammeter	Electrons	Ohmmeter	Solenoids
Amperes (A)	Insulator	Ohm's law	Transformers
Atoms	Joule (J)	Power	Valence shell
Battery	Kilo	Proton	Volts (V)
Charge	Matter	Relays	Voltage
Circuit	Mega	Resistance	Voltage drop
Conductors	Neutrons	Resistors	Voltmeter
Coulomb	Nucleus	Schematic	Watts (W)
Current	Ohms (Ω)	Series circuit	

1. An atom that is missing only one or two electrons from its outer shell will __________.

2. To find voltage when both current and resistance are known, use the formula __________.

3. A resistor with a color code of yellow, orange, red, and silver has a tolerance of __________.

4. True or False? When using an in-line ammeter, it is important to connect the meter in series.

5. If a toaster draws 6.2A of current at 120V, how many kilowatt-hours of energy will be used in 3.5 hours?

   ____.

6. If a battery sends a current of 10A through a circuit for one hour, how many coulombs will flow through the circuit?

   ____.

7. The charged particles of an atom are called __________.

8. Conductors have __________ or fewer valence electrons.

9. The sum of the difference in potential of all the charges in an electrostatic field is called

   ____.

10. Electric charge is measured in __________.

11. Find the resistance when the voltage is 120V and the current is 6A.

   ____.

12. Give the resistance value and tolerance of a resistor where the color bands are red, yellow, brown, and silver.

   ____.

13. What is the power in a 120V circuit with a current of 12.5A?

   ____.

14. What is the voltage in a 30Ω circuit with a power rating of 480W?

   ____.

15. An ohmmeter is used to measure resistance and check for

   ____.

E.L. Jarrell
Associated Builders and Contractors

Eurlin Layne (E.L.) Jarrell is a prime example of a master electrician giving back to the electrical community by teaching and mentoring.

After serving in the US Army, E.L. went to work for Cities Services, now known as CITGO. He stayed at CITGO for 38 years, retiring in 1995. It was during his employment at CITGO that he first received apprenticeship training in the electrical field.

While at CITGO, E.L. worked as a process unit operator before moving to the electrical department. While there, he worked as a trainee electrician for three years until he became a first-class electrician. A few years later, he was promoted to temporary supervisor, planning and scheduling shut-down maintenance. In 1983, he passed the Block Master Electrician test for the City of Lake Charles, Louisiana. In 1997, E.L. became involved with Associated Builders and Contractors (ABC).

E.L. is currently the Electrical Department Head for the ABC Training Center, where he works in the lab, overseeing students doing hands-on electrical work. During his first semester teaching at the ABC Training Center, it became clear to E.L. that many students simply had no time to study because they worked 10-hour days, drove over 100 miles to work, and had family obligations. In response, E.L. began an in-class study guide. He encouraged students to form study groups, and he gave students time to study in class.

E.L. was an instrumental member of NCCER's Technical Review Committee, which completely rewrote all four levels of NCCER's Electrical curriculum. In addition, E.L. is currently a member of both NCCER's National Skills Assessment Written Test Committee and the Performance Verification Packet for Industrial Electricians Committee.

E.L. has decided to give back to the electrical community with his expertise and mentoring. Many of E.L.'s students have become his personal friends. He says, "At this point in my life, I just want to continue being the best electrical instructor that I can be and share some of my knowledge and experience with my students and hope that I can make a difference in their lives and careers."

Trade Terms Introduced in This Module

Ammeter: An instrument for measuring electrical current.

Amperes (A): The basic unit of measurement for electrical current, represented by the letter A

Atoms: The smallest particles to which an element may be divided and still retain the properties of the element.

Battery: A DC voltage source consisting of two or more cells that convert chemical energy into electrical energy.

Charge: A quantity of electricity that is either positive or negative.

Circuit: A complete path for current flow.

Conductors: Materials through which it is relatively easy to maintain an electric current.

Coulomb: A unit of electrical charge equal to 6.25×10^{18} electrons (or 6.25 quintillion electrons). A coulomb is the common unit of quantity used for specifying the size of a given charge.

Current: The movement, or flow, of electrons in a circuit. Current (I) is measured in amperes.

Electrons: Negatively charged particles that orbit the nucleus of an atom.

Insulator: A material through which it is difficult to conduct an electric current.

Joule (J): A unit of measurement for doing work, represented by the letter J. One joule is equal to one newton-meter (Nm).

Kilo: A prefix used to indicate one thousand (for example, one kilowatt is equal to one thousand watts).

Matter: Any substance that has mass and occupies space.

Mega: A prefix used to indicate one million; for example, one megawatt is equal to one million watts.

Neutrons: Electrically neutral particles (neither positive nor negative) that have the same mass as a proton and are found in the nucleus of an atom.

Nucleus: The center of an atom. It contains the protons and neutrons of the atom.

Ohms (Ω): The basic unit of measurement for resistance, represented by the symbol Ω.

Ohmmeter: An instrument used for measuring resistance.

Ohm's law: A statement of the relationships among current, voltage, and resistance in an electrical circuit: current (I) equals voltage (E) divided by resistance (R). Generally expressed as a mathematical formula: $I = E/R$.

Power: The rate of doing work, or the rate at which energy is used or dissipated. Electrical power is measured in watts.

Proton: The smallest positively charged particle of an atom. Protons are contained in the nucleus of an atom.

Relays: Electromechanical devices consisting of a coil and one or more sets of contacts. Used as a switching device.

Resistance: An electrical property that opposes the flow of current through a circuit. Resistance (R) is measured in ohms.

Resistors: Any devices in a circuit that resist the flow of electrons.

Schematic: A type of drawing in which symbols are used to represent the components in a system.

Series circuit: A circuit with only one path for current flow.

Solenoids: Electromagnetic coils used to control a mechanical device such as a valve.

Transformers: Devices consisting of one or more coils of wire wrapped around a common core. Transformers are commonly used to step voltage up or down.

Valence shell: The outermost ring of electrons that orbit about the nucleus of an atom.

Volts (V): The unit of measurement for voltage, represented by the letter V. One volt is equivalent to the force required to produce a current of one ampere through a resistance of one ohm.

Voltage: The driving force that makes current flow in a circuit. Voltage, often represented by the letter E, is also referred to as electromotive force (emf), difference of potential, or electrical pressure.

Voltage drop: The change in voltage across a component that is caused by the current flowing through it and the amount of resistance opposing it.

Voltmeter: An instrument for measuring voltage. The resistance of the voltmeter is fixed. When the voltmeter is connected to a circuit, the current passing through the meter will be directly proportional to the voltage at the connection points.

Watts (W): The basic unit of measurement for electrical power, represented by the letter W.

Additional Resources

This module presents thorough resources for task training. The following reference material is recommended for further study.

Electronics Fundamentals: Circuits, Devices, and Applications, Thomas L. Floyd. New York, NY: Pearson Education, Inc.

Principles of Electric Circuits, Thomas L. Floyd. New York, NY: Pearson Education, Inc.

Section Review Answer Key

Answer	Section Reference	Objective
Section One		
1. b	1.1.0	1a
2. a	1.2.0	1b
3. c	1.3.0	1c
4. b	1.4.0	1d
Section Two		
1. b	2.1.0	2a
2. a	2.2.0	2b
3. b	2.3.0	2c
4. c	2.4.0	2d
Section Three		
1. b	3.1.0	3a
2. c	3.2.1	3b
3. a	3.3.1	3c
4. d	3.4.1	3d

2.0.0 SECTION REVIEW

Question 4

Using Ohm's Law, solve for the voltage (E) using the current and resistance values:

$E = I \times R$
$E = 30A \times 4\Omega$
$E = 120V$

The applied voltage is **120V**.

3.0.0 SECTION REVIEW

Question 2

Find the sum of the resistances:

$R_T = 10\Omega + 10\Omega + 10\Omega = 30\Omega$

The total resistance is **30Ω**.

Question 4

Using the power equation, multiply the current by the voltage to find the power:

$P = IE$
$P = 10A \times 240V$
$P = 2400W$

The power is **2,400W**.

NCCER CURRICULA — USER UPDATE

NCCER makes every effort to keep its textbooks up-to-date and free of technical errors. We appreciate your help in this process. If you find an error, a typographical mistake, or an inaccuracy in NCCER's curricula, please fill out this form (or a photocopy), or complete the online form at **www.nccer.org/olf**. Be sure to include the exact module ID number, page number, a detailed description, and your recommended correction. Your input will be brought to the attention of the Authoring Team. Thank you for your assistance.

Instructors – If you have an idea for improving this textbook, or have found that additional materials were necessary to teach this module effectively, please let us know so that we may present your suggestions to the Authoring Team.

NCCER Product Development and Revision

13614 Progress Blvd., Alachua, FL 32615

Email: curriculum@nccer.org
Online: www.nccer.org/olf

❑ Trainee Guide   ❑ Lesson Plans   ❑ Exam   ❑ PowerPoints   Other _______________

Craft / Level: Copyright Date:

Module ID Number / Title:

Section Number(s):

Description:

Recommended Correction:

Your Name:

Address:

Email: Phone:

Basic Electrical Construction Drawings

OVERVIEW

In all large construction projects and in many of the smaller ones, an architect is commissioned to prepare complete working drawings and specifications for the project. These drawings include site plans, floor plans, detail drawings, lighting plans, power riser diagrams, equipment schedules, and specifications. This module describes how to interpret electrical drawings, and covers the use of architect's and engineer's scales.

Module 26110-17

Trainees with successful module completions may be eligible for credentialing through the NCCER Registry. To learn more, go to **www.nccer.org** or contact us at 1.888.622.3720. Our website has information on the latest product releases and training, as well as online versions of our *Cornerstone* magazine and Pearson's product catalog.

Your feedback is welcome. You may email your comments to **curriculum@nccer.org**, send general comments and inquiries to **info@nccer.org**, or fill in the User Update form at the back of this module.

This information is general in nature and intended for training purposes only. Actual performance of activities described in this manual requires compliance with all applicable operating, service, maintenance, and safety procedures under the direction of qualified personnel. References in this manual to patented or proprietary devices do not constitute a recommendation of their use.

Basic Electrical Construction Drawings

Objectives

When you have completed this module, you will be able to do the following:

1. Identify types of construction drawings.
   a. Identify the information found on site plans.
   b. Identify the information found on floor plans.
   c. Identify the information found on elevation drawings.
   d. Identify the information found on sectional views.
   e. Identify the information found on title blocks.
   f. Interpret drafting lines.
2. Work with scale drawings.
   a. Use an architect's scale.
   b. Use an engineer's scale.
   c. Use a metric scale.
3. Read electrical drawings.
   a. Interpret electrical symbols.
   b. Analyze a set of electrical drawings.
   c. Identify fixtures in a lighting floor plan.
   d. Read block and schematic diagrams.
   e. Interpret written specifications.

Performance Tasks

Under the supervision of the instructor, you should be able to do the following:

1. Using an architect's scale, state the actual dimensions of a given drawing component.
2. Make a material takeoff of the lighting fixtures specified in the provided drawing. The takeoff requires that all lighting fixtures be counted, and where applicable, the total number of lamps for each fixture type must be calculated. (Fill these in on the provided Lighting Fixture Takeoff worksheet.)

Trade Terms

Architect's scale	Dimensions	One-line diagram	Schematic diagram
Architectural drawings	Electrical drawing	Plan view	Sectional view
Block diagram	Elevation drawing	Power-riser diagram	Shop drawing
Blueprint	Engineer's scale	Scale	Site plan
Detail drawing	Floor plan	Schedule	Written specifications

Industry Recognized Credentials

If you are training through an NCCER-accredited sponsor, you may be eligible for credentials from NCCER's Registry. The ID number for this module is 26110-17. Note that this module may have been used in other NCCER curricula and may apply to other level completions. Contact NCCER's Registry at 888.622.3720 or go to **www.nccer.org** for more information.

Note

NFPA 70®, *National Electrical Code*® and *NEC*® are registered trademarks of the National Fire Protection Association, Quincy, MA.

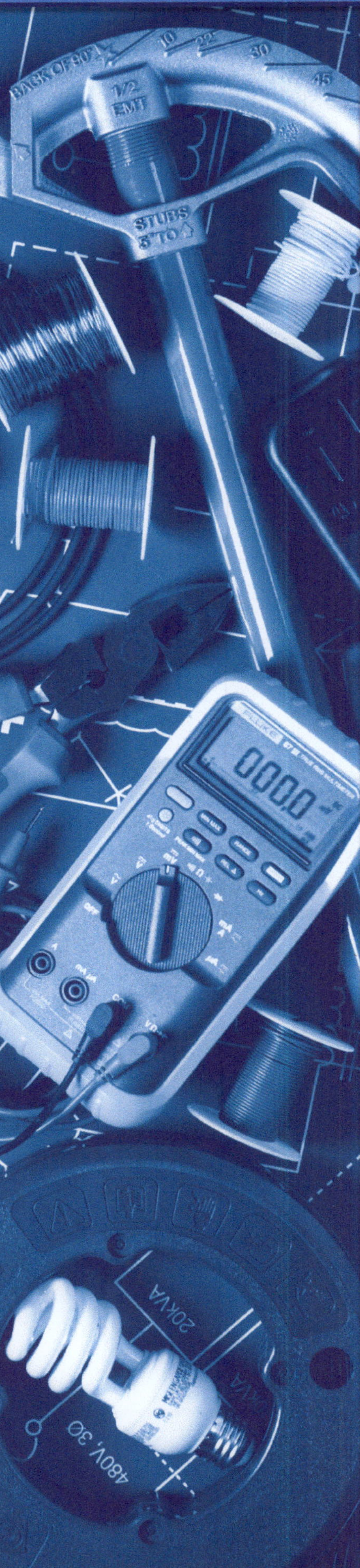

Contents

Figures

1.0.0 TYPES OF CONSTRUCTION DRAWINGS

Objective

Identify types of construction drawings.

a. Identify the information found on site plans.
b. Identify the information found on floor plans.
c. Identify the information found on elevation drawings.
d. Identify the information found on sectional views.
e. Identify the information found on title blocks.
f. Interpret drafting lines.

Trade Terms

Architectural drawings: Working drawings consisting of site plans, floor plans, elevations, sectional views, details, and other information necessary for the construction of a building.

Blueprint: An exact copy or reproduction of an original drawing.

Detail drawing: An enlarged, detailed view taken from an area of a drawing and shown in a separate view.

Dimensions: Sizes or measurements printed on a drawing.

Electrical drawing: A means of conveying a large amount of exact, detailed information in an abbreviated language. Consists of lines, symbols, dimensions, and notations to accurately convey an engineer's designs to electricians who install the electrical system on a job.

Elevation drawing: An architectural drawing showing height, but not depth; usually the front, rear, and sides of a building or object.

Floor plan: A drawing of a building as if a horizontal cut were made through a building at about window level, and the top portion removed. The floor plan is what would appear if the remaining structure were viewed from above.

Plan view: A drawing made as though the viewer were looking straight down (from above) on an object.

Scale: On a drawing, the size relationship between an object's actual size and the size it is drawn. Scale also refers to the measuring tool used to determine this relationship.

Sectional view: A cutaway drawing that shows the inside of an object or building.

Site plan: A drawing showing the location of a building or buildings on the building site. Such drawings frequently show topographical lines, electrical and communication lines, water and sewer lines, sidewalks, driveways, and similar information.

In all large construction projects and in many of the smaller ones, an architect is commissioned to prepare complete working drawings and specifications for the project. The set of drawings usually includes a site plan, floor plans, elevation drawings, and sectional views. Depending on the complexity of the installation, a **detail drawing** may also be required.

For larger projects, the architect usually hires consulting engineers to prepare structural, electrical, and various types of mechanical drawings, including pipefitting, instrumentation, plumbing, and heating, ventilating, and air conditioning (HVAC) drawings.

1.1.0 Site Plans

A **site plan** indicates the location of the building on the property. This type of plan of the building site looks as if the site is viewed from an airplane and shows the property boundaries, the existing contour lines, the new contour lines (after grading), the location of the building on the property, new and existing roadways, all utility lines, and other pertinent details. The drawing **scale** is also shown. Descriptive notes may also be found on the site (plot) plan listing names of adjacent property owners, the land surveyor, and the date of the survey. A legend or symbol list is also included so that anyone who must work with the site plan can readily read the information. See *Figure 1*.

It is usually the owner's responsibility to furnish the architect/engineer with property and topographic surveys, which are made by a certified land surveyor or civil engineer. These surveys show:

- All property lines
- Existing public utilities and their location on or near the property (e.g., electrical lines, sanitary sewer lines, gas lines, water-supply lines, storm sewers, manholes, telephone lines, etc.)

A land surveyor does the property survey from information obtained from a deed description of the property. A property survey shows only the property lines and their lengths, as if the property were perfectly flat.

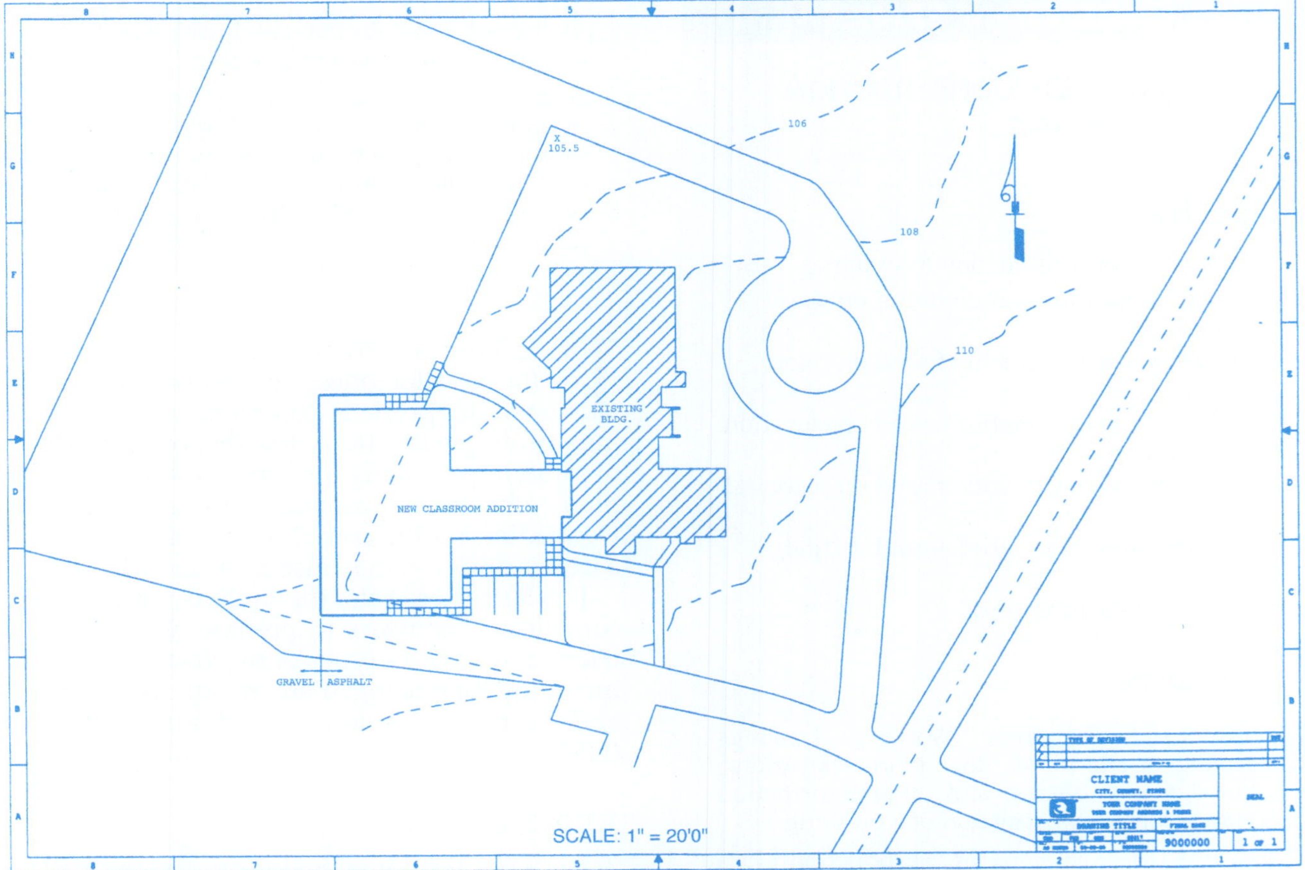

Figure 1 Typical site plan.

The topographic survey shows both the property lines and the physical characteristics of the land by using contour lines, notes, and symbols. The physical characteristics may include:

- The direction of the land slope
- Whether the land is flat, hilly, wooded, swampy, high, or low, and other features of its physical nature

All of this information is necessary so that the architect can properly design a building to fit the property. The electrical engineer also needs this information to locate existing electrical utilities, route the new service to the building, and provide outdoor lighting and circuits.

1.2.0 Floor Plans

A **floor plan** shows the walls and partitions for each floor or level. The **plan view** of any object is a drawing showing the outline and all details as seen when looking directly down on the object. It shows only two **dimensions**, length and width. The floor plan of a building is drawn as if a horizontal cut were made through the building—at about window height—and then the top portion removed to reveal the bottom part. A cut through the second floor windows would be called the second floor plan or upper level. Refer to *Figure 2* and *Figure 3*.

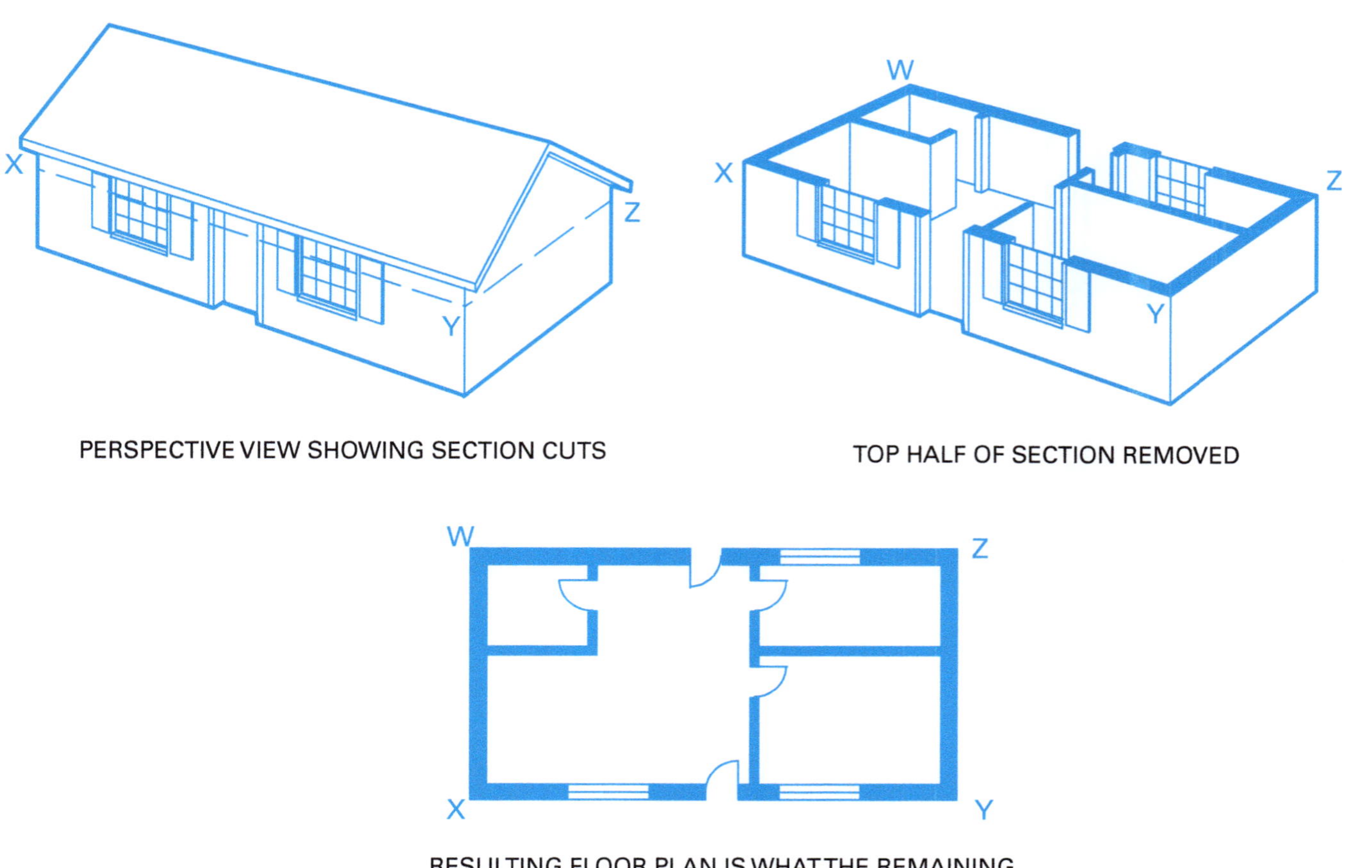

Figure 2 Principles of floor plan layout.

Using a Drawing Set

Always treat a drawing set with care. It is best to keep two sets, one for the office and one for field use. Be sure to use the most current revision. After you use a sheet from a set of drawings, refold the sheet with the title block facing up.

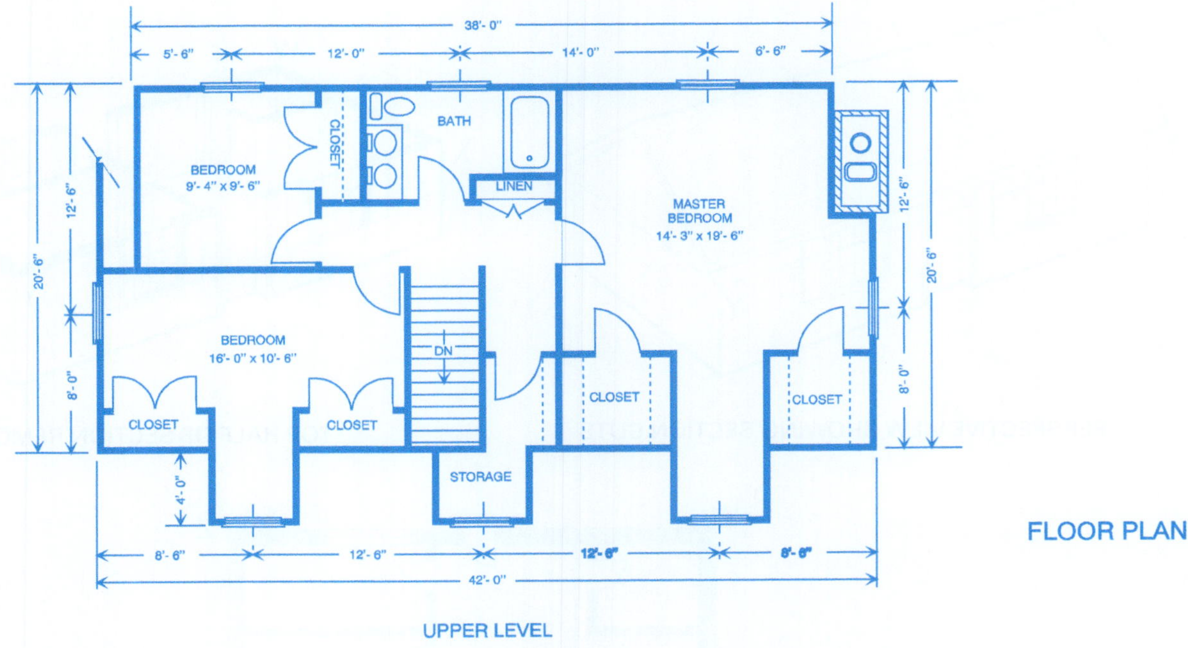

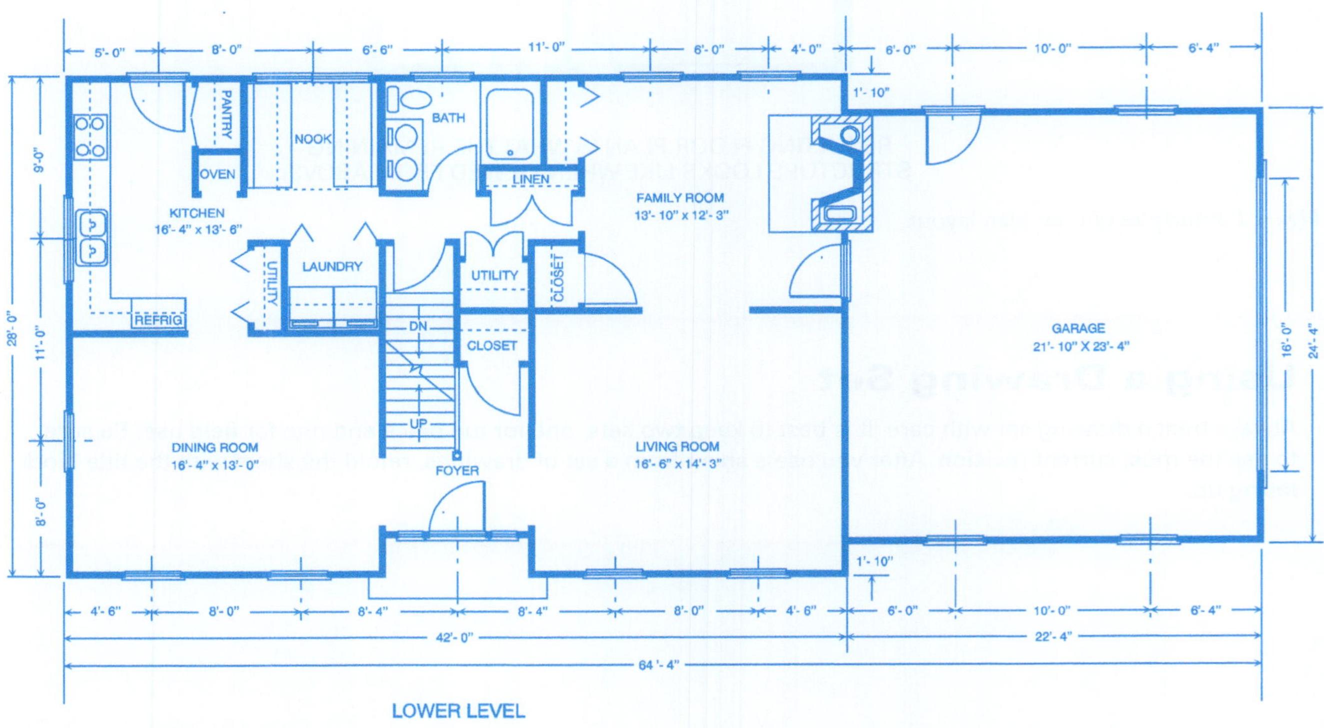

Figure 3 Floor plans of a building.

1.3.0 Elevation Drawings

An **elevation drawing** shows all exterior faces of the building. The elevation is an outline of an object that shows heights and may show the length or width of a particular side, but not depth. Refer to *Figure 4* and *Figure 5*.

1.4.0 Sectional Views

A section or **sectional view** (*Figure 6*) is a cutaway view that shows the inside of a structure. Sectional views are used to illustrate floor levels and details of footings, foundations, walls, floors, ceilings, and roof construction. A longitudinal section is taken lengthwise, while a cross section is usually taken straight across the width of an object. Sometimes, a section is taken along a zigzag line to show important parts of an object.

The point on the plan or elevation showing where the imaginary cut has been made is indicated by a section line, which is usually a dashed line. Arrow points are placed at the ends of the section lines to indicate which part is represented in the sectional drawing. If a drawing shows two or more sections, they are distinguished using different letters at the ends of the lines. The section is named according to these letters (e.g., Section A-A, Section B-B, and so forth).

Wall sections are nearly always made vertically so that the cut edge is exposed from top to bottom. Wall sections show how each wall is constructed and usually indicate the material to be used. This information is important when determining wiring methods.

FRONT ELEVATION

REAR ELEVATION

Figure 4 Front and rear elevations.

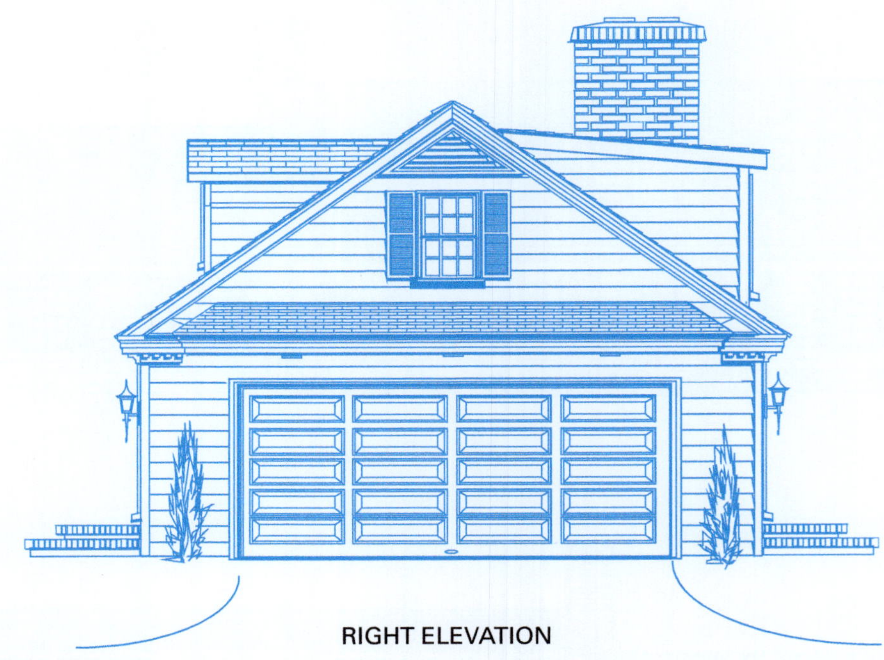

Figure 5 Left and right elevations.

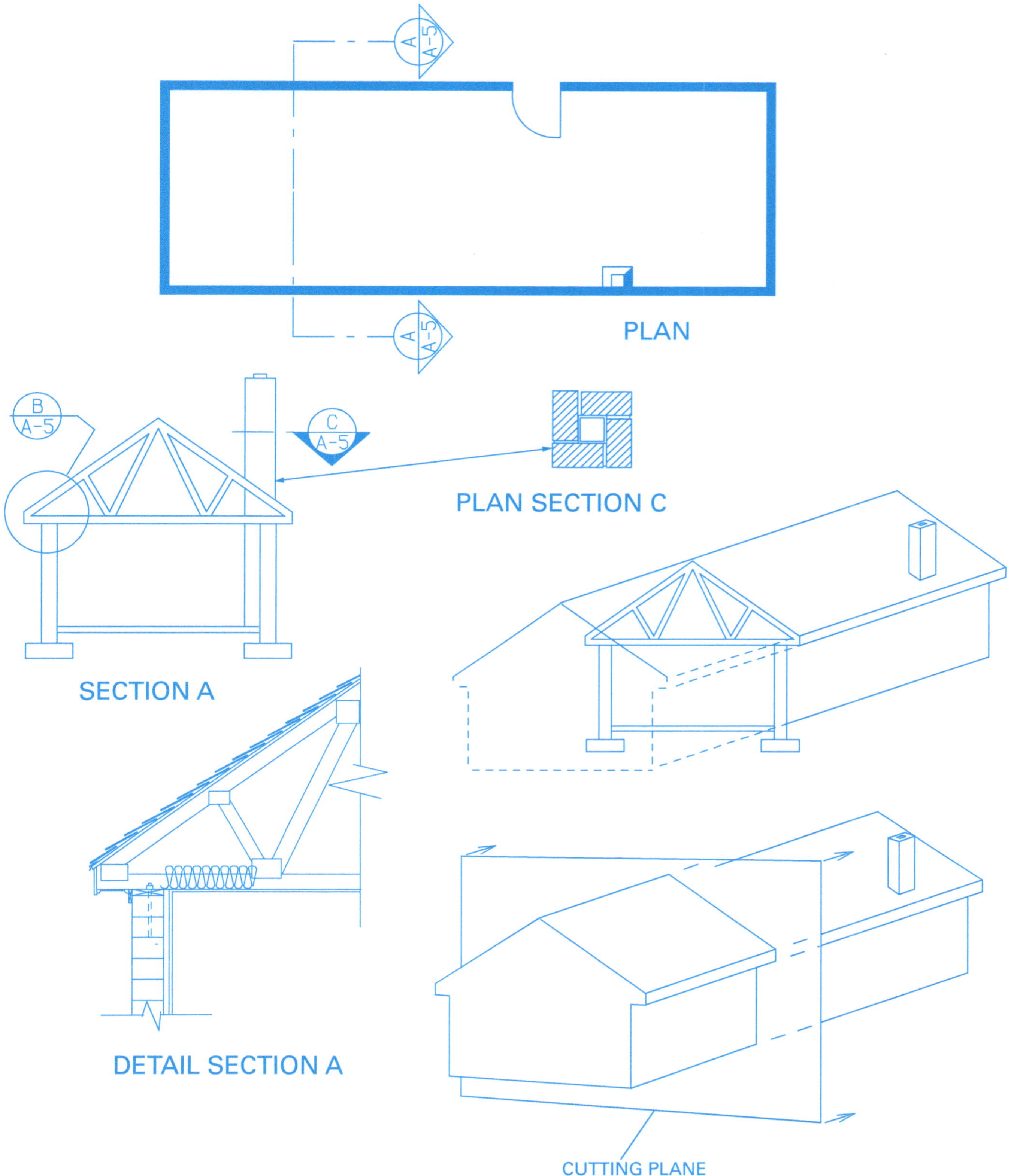

Figure 6 Sectional drawings.

Basic Electrical Construction Drawings 7

1.5.0 Title Blocks

Although a strong effort has been made to standardize drawing practices in the building construction industry, a drawing or **blueprint** prepared by an architectural or engineering firm will rarely be identical to a drawing prepared by another firm. Similarities, however, will exist between most sets of drawings, and with a little experience, you should have no trouble interpreting any set of drawings that might be encountered.

Most drawings used for building construction projects will be drawn on sheets in various sizes. Each drawing sheet has border lines framing the overall drawing and one or more title blocks, as shown in *Figure 7*. The type and size of title blocks varies with each firm preparing the drawings. In addition, some drawing sheets will also contain a revision block near the title block, and perhaps an approval block. This information is normally found on each drawing sheet, regardless of the type of project or the information contained on the sheet.

The architect's title block for a drawing is usually boxed in the lower right-hand corner of the drawing sheet; the size of the block varies with the size of the drawing and with the information required (*Figure 8*).

In general, the title block of an **electrical drawing** should contain the following information:

- Name of the project
- Address of the project
- Name of the owner or client
- Name of the architectural firm
- Date of completion
- Scale(s)
- Initials of the drafter, checker, and designer, with dates under each
- Job number
- Sheet number
- General description of the drawing

Often, the consulting engineering firm will also be listed, which means that an additional title block will be applied to the drawing, usually next to the architect's title block. *Figure 9* shows completed architectural and engineering title blocks as they appear on an actual drawing.

Check the Title Block

Always refer to the title block for scale and be aware that it often changes from drawing to drawing within the same set. Also check for the current revision, and when you replace a drawing with a new revision, be sure to remove and file the older version.

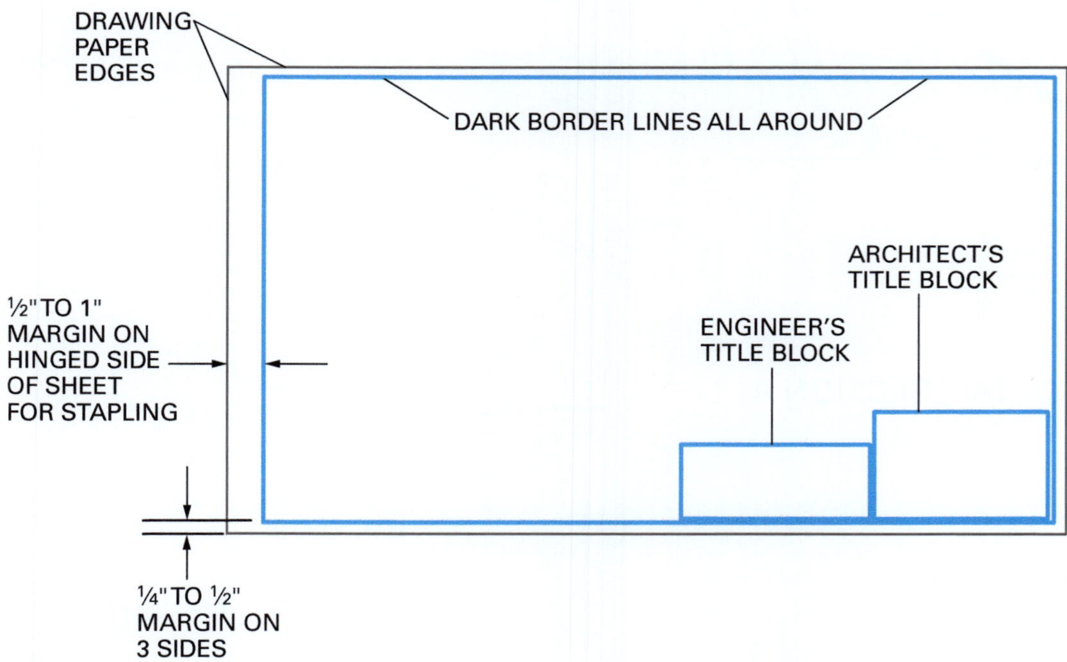

Figure 7 Typical drawing layout.

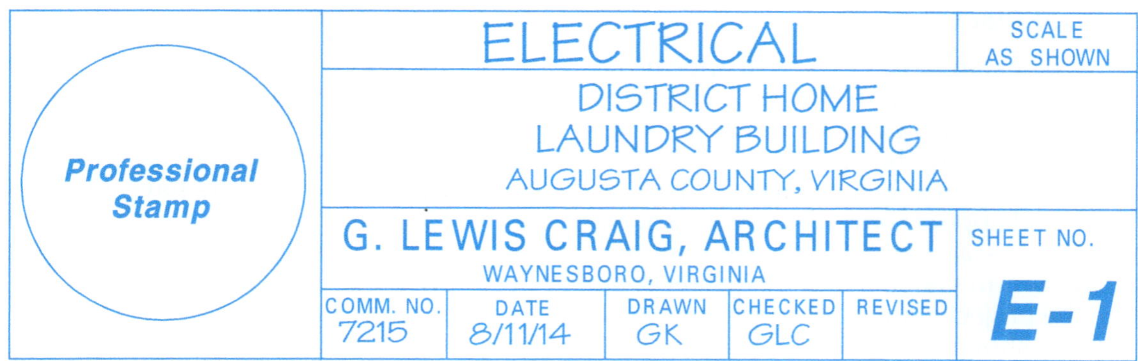

Figure 8 Typical architect's title block.

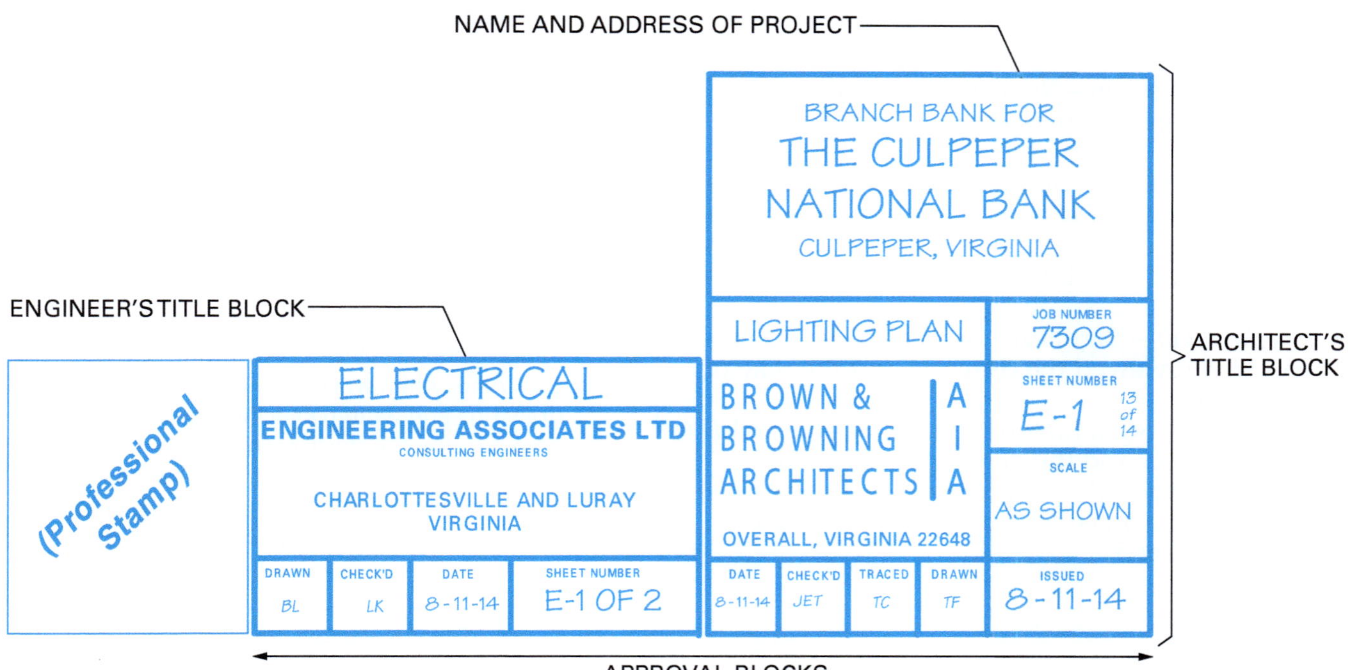

Figure 9 Title blocks.

Then and Now

Years ago, blueprints were created by placing a hand drawing against light-sensitive paper and then exposing it to ultraviolet light. The light would turn the paper blue except where lines were drawn on the original. The light-sensitive paper was then developed, and the resulting print had white lines against a blue background. Modern blueprints usually have blue or black lines against a white background and are generated using computer-aided design programs. Newer programs offer three-dimensional modeling and other enhanced features.

1.5.1 Approval Block

The approval block, in most cases, will appear on the drawing sheet as shown in *Figure 10*. The various types of approval blocks (drawn, checked, etc.) will be initialed by the appropriate personnel. This type of approval block is usually part of the title block and appears on each drawing sheet.

On some projects, authorized signatures are required before certain systems may be installed, or even before the project begins. An approval block such as the one shown in *Figure 11* indicates that all required personnel have checked the drawings for accuracy, and that the set meets with everyone's approval. Such an approval block usually appears on the front sheet of the blueprint set and may include:

- *Professional stamp* – Registered seal of approval by the licensed architect or consulting engineer.
- *Design supervisor* – Signature of the person who is overseeing the design.
- *Drawn (by)* – Signature or initials of the person who drafted the drawing and the date it was completed.
- *Checked (by)* – Signature or initials of the person who reviewed the drawing and the date of approval.
- *Approved* – Signature or initials of the architect/engineer and the date of the approval.
- *Owner's approval* – Signature of the project owner or the owner's representative along with the date signed.

1.5.2 Revision Block

Sometimes electrical drawings will have to be partially redrawn or modified during the construction of a project. It is extremely important that such modifications are noted and dated on the drawings to ensure that the workers have an up-to-date set of drawings to work from. In some situations, sufficient space is left near the title block for dates and descriptions of revisions, as shown in *Figure 12*. In other cases, a revision block is provided (again, near the title block), as shown in *Figure 13*. The area on the drawing where the revision has been made will often be circled with a cloud shape.

COMM. NO.	DATE	DRAWN	CHECKED	REVISED
7215	8/11/14	GK	GLC	

Figure 10 Typical approval block.

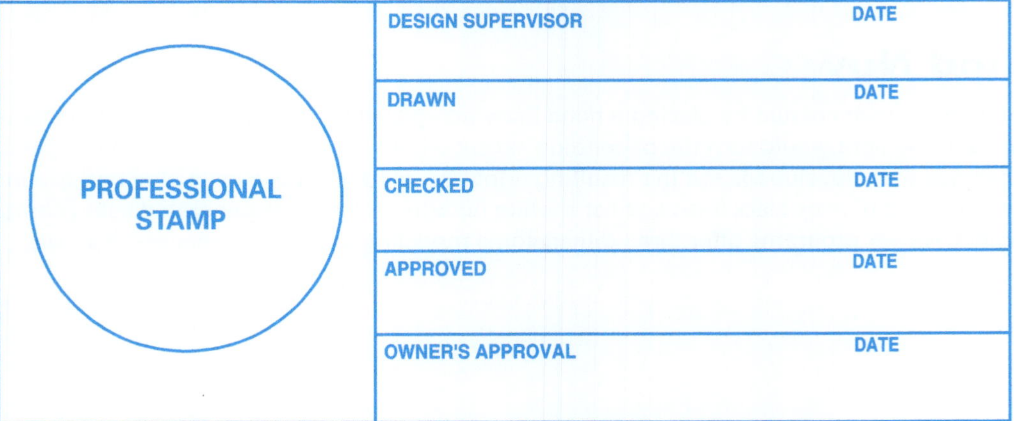

Figure 11 Alternate approval block.

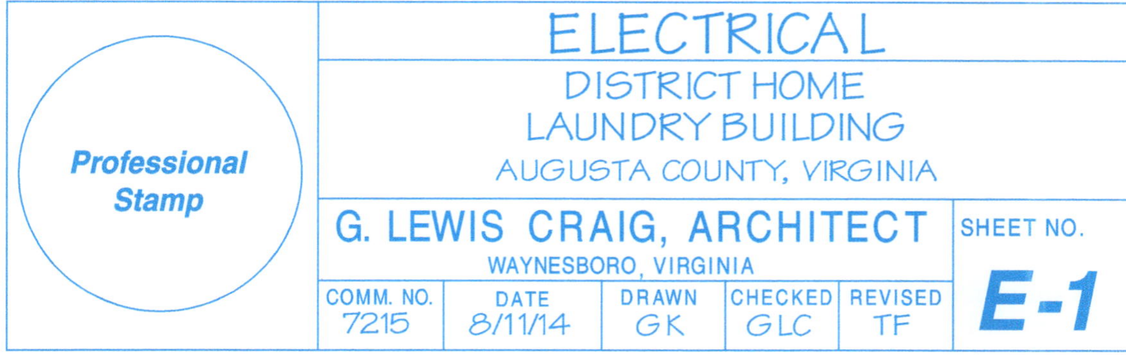

Figure 12 One method of showing revisions on working drawings.

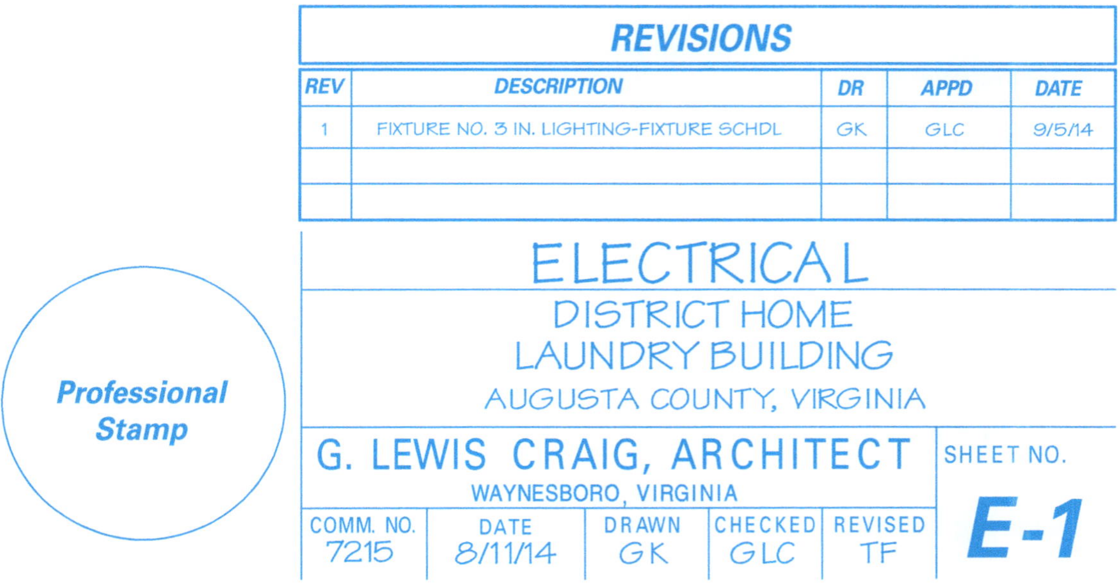

Figure 13 Alternative method of showing revisions on working drawings.

Applying Your Skills

Once you learn how to interpret construction drawings, you can apply that knowledge to any type of construction, from simple residential applications to large industrial complexes.

1.6.0 Drafting Lines

You will encounter many types of drafting lines. To specify the meaning of each type of line, contrasting lines can be made by varying the width of the lines or breaking the lines in a uniform way. *Figure 14* shows common lines used on **architectural drawings**. However, these lines can vary. Architects and engineers have strived for a common standard for the past century, but unfortunately, their goal has yet to be reached. Therefore, you will find variations in lines and symbols from drawing to drawing, so always consult the legend or symbol list when referring to any drawing. Also, carefully inspect each drawing to ensure that line types are used consistently.

The drafting lines shown in *Figure 14* are used as follows:

- *Light full line* – This line is used for section lines, building background (outlines), and similar uses where the object to be drawn is secondary to the system being shown (e.g., HVAC or electrical).
- *Medium full line* – This type of line is frequently used for lettering on drawings. It is further used for some drawing symbols, circuit lines, etc.
- *Heavy full line* – This line is used for borders around title blocks, schedules, and for lettering drawing titles. Some types of symbols are frequently drawn with a heavy full line.
- *Extra heavy full line* – This line is used for border lines on architectural/engineering drawings.
- *Centerline* – A centerline is a broken line made up of alternately spaced long and short dashes. It indicates the centers of objects such as holes, pillars, or fixtures. Sometimes, the centerline indicates the dimensions of a finished floor.

- *Hidden line* – A hidden line consists of a series of short dashes that are closely and evenly spaced. It shows the edges of objects that are not visible in a particular view. The object outlined by hidden lines in one drawing is often fully pictured in another drawing.
- *Dimension line* – These are thin lines used to show the extent and direction of dimensions. The actual dimension is usually placed in a break inside the dimension lines. Normal practice is to place the dimension lines outside the object's outline. However, it may sometimes be necessary to draw the dimensions inside the outline.
- *Short break line* – This line is used for short breaks.
- *Long break line* – This line is used for long breaks.
- *Match line* – This line is used to show the position of the cutting plane. Therefore, it is also called the cutting plane line. A match or cutting plane line is a heavy line with long dashes alternating with two short dashes. It is used on drawings of large structures to show where one sheet of the drawing stops and the next begins.
- *Secondary line* – This line is frequently used to outline pieces of equipment or to indicate reference points of a drawing that are secondary to the drawing's purpose.
- *Property line* – This is a light line made up of one long and two short dashes that are alternately spaced. It indicates land boundaries on the site plan.

Other uses of the lines just mentioned include the following:

- *Extension lines* – Extension lines are lightweight lines that start about $\frac{1}{16}$" (1.6 mm) away from the edge of an object and extend out. A common use of extension lines is to create a boundary for dimension lines. Dimension lines meet extension lines with arrowheads, slashes, or dots. Extension lines that point from a note or other reference to a particular feature on a drawing are called leaders. They usually end in either an arrowhead or a dot and may include an explanatory note at the end.
- *Section lines* – These are often referred to as cross-hatch lines. Drawn at a 45° angle, these lines show where an object has been cut away to reveal the inside.

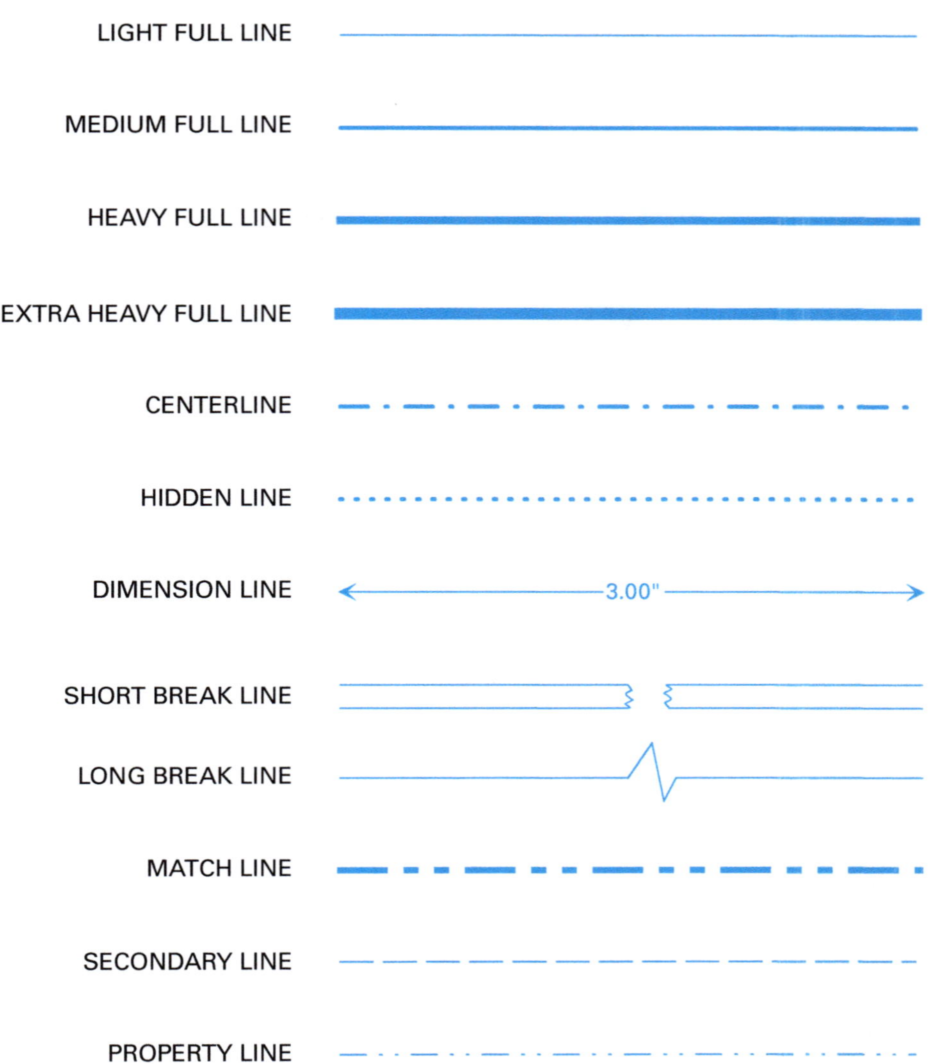

Figure 14　Typical drafting lines.

- *Phantom lines* – Phantom lines are solid, light lines that show where an object will be installed. A future door opening or a future piece of equipment can be shown with phantom lines.
- *Electrical drafting lines* – Electrical engineers, designers, and drafters use additional lines to represent circuits and their related components. These lines may vary from drawing to drawing, so check the symbol list or legend for the exact meaning of lines on the drawing with which you are working. *Figure 15* shows lines used on some electrical drawings.

Interpreting Electrical Drawings

A good example of when an electrician must interpret the drawings is when wiring a log cabin. The drawings will show the receptacle and switch locations in branch circuits as usual, but the electrician must figure out how to route wires and install boxes where there is no hollow wall and sometimes no ceiling space.

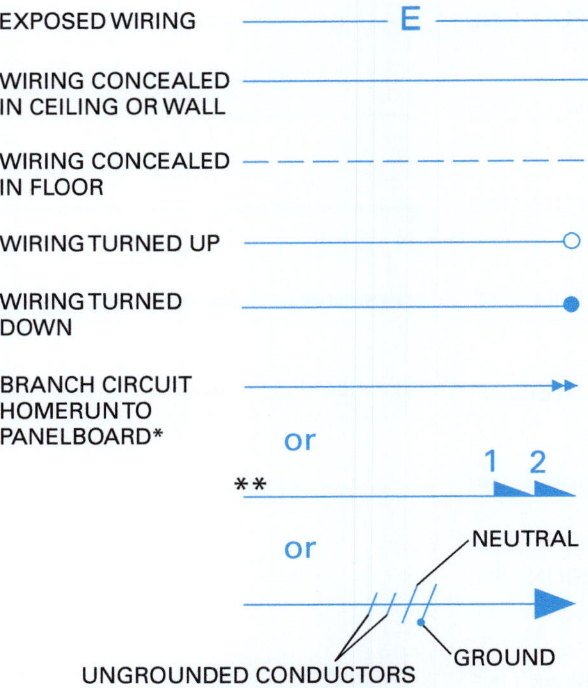

Figure 15   Electrical drafting lines.

Check the Legend

Be sure to check the legend on every drawing set. Symbols and abbreviations often vary widely from drawing set to drawing set.

Orient Yourself

When reading a drawing, find the north arrow to orient yourself to the structure. Knowing where north is enables you to accurately describe the locations of walls and other parts of the building.

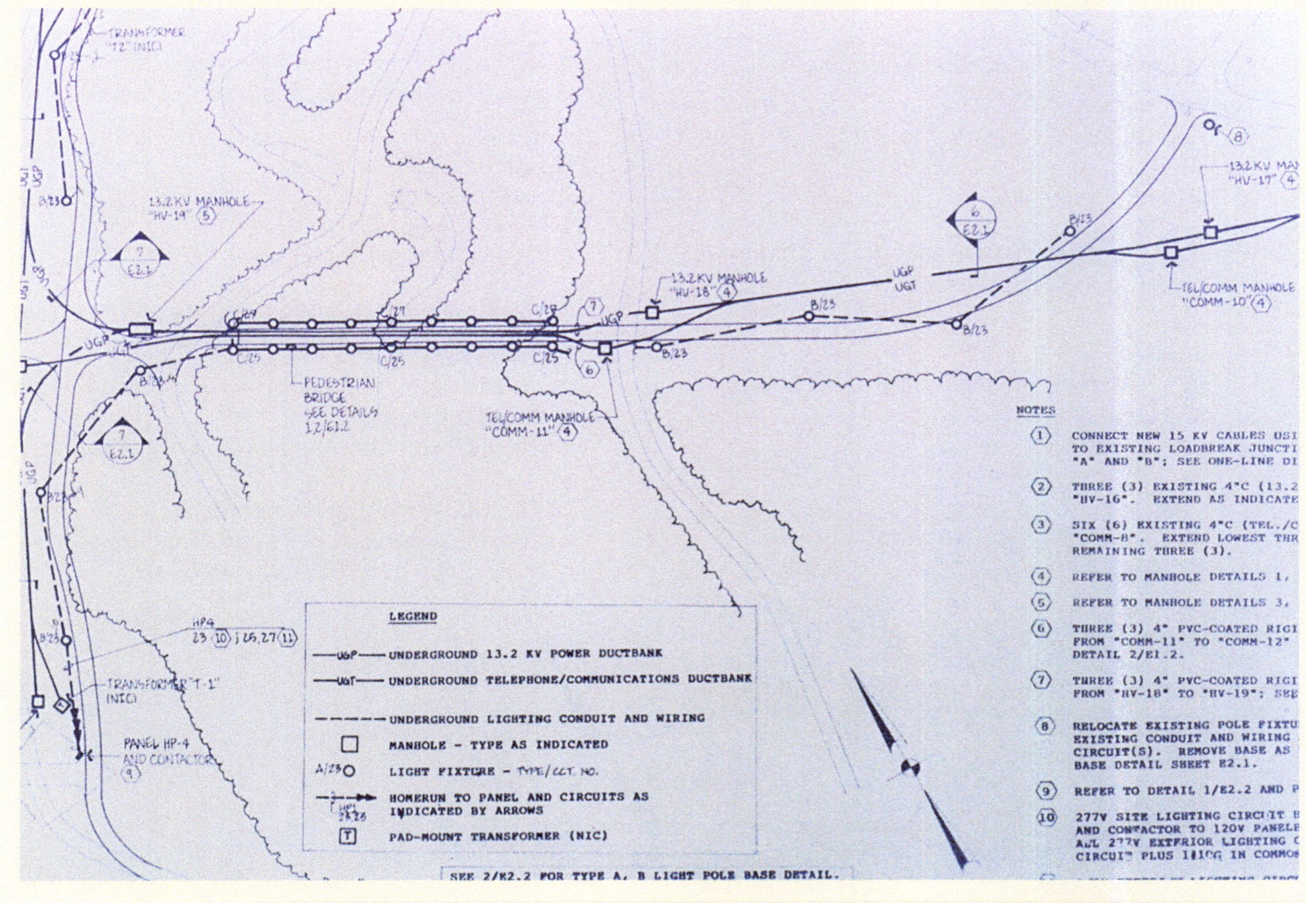

National Electrical Code® Handbook, Latest Edition. Quincy, MA: National Fire Protection Association.

1.0.0 Section Review

1. Contour lines are typically found on a(n) _____.

   a. schematic
   b. floor plan
   c. site plan
   d. section view

2. Floor plans are drawn as though an imaginary cut has been made at the level of the _____.

   a. floor
   b. windows
   c. roof
   d. foundation

3. Which of the following shows the face of a building?

   a. section view
   b. floor plan
   c. site plan
   d. elevation

4. A complicated architectural detail might be illustrated with a(n) _____.

   a. section view
   b. schematic
   c. site plan
   d. floor plan

5. A drawing contains an area circled in a cloud shape. This information is often detailed in the _____.

   a. architect's title block
   b. approval block
   c. revision block
   d. engineer's title block

6. A drawing contains a match line. What does this mean?

   a. It indicates that the drawing continues on another sheet.
   b. It indicates that the structure is symmetrical.
   c. It indicates a duplex receptacle.
   d. It indicates reference points that are secondary to the drawing's purpose.

2.0.0 SCALE DRAWINGS

Objective

Work with scale drawings.
  a. Use an architect's scale.
  b. Use an engineer's scale.
  c. Use a metric scale.

Performance Task

  1. Using an architect's scale, state the actual dimensions of a given drawing component.

Trade Terms

Architect's scale: A special ruler with various measurement scales that can be used when drafting or making measurements on architectural drawings. Architect's scales measure distances in inches.

Engineer's scale: A special ruler with various measurement scales that can be used when drafting or making measurements on architectural drawings. Engineer's scales measure distances in decimal units.

In most electrical drawings, the components are so large that it would be impossible to draw them actual size. Consequently, drawings are made to some reduced scale; that is, all the distances are drawn smaller than the actual dimensions of the object itself, with all dimensions being reduced in the same proportion. For example, if a floor plan of a building is drawn to a scale of $\frac{1}{4}$" = 1'–0", each $\frac{1}{4}$" on the drawing would equal 1 foot on the building itself; if the scale is $\frac{1}{8}$" = 1'–0", each $\frac{1}{8}$" on the drawing equals 1 foot on the building, and so forth.

When architectural and engineering drawings are produced, the selected scale is very important. Where dimensions must be held to extreme accuracy, the scale drawings should be made as large as practical with dimension lines added. Where dimensions require only reasonable accuracy, the object may be drawn to a smaller scale (with dimension lines possibly omitted).

In dimensioning drawings, the dimensions written on the drawing are the actual dimensions of the building, not the distances that are measured on the drawing. To further illustrate this point, look at the floor plan in *Figure 16*; it is drawn to a scale of $\frac{1}{2}$" = 1'–0". One of the walls is drawn to an actual length of $3\frac{1}{2}$" on the drawing paper, but since the scale is $\frac{1}{2}$" = 1'–0" and since $3\frac{1}{2}$" contains 7 halves of an inch (7 × $\frac{1}{2}$ = $3\frac{1}{2}$"), the dimension shown on the drawing will therefore be 7'–0" on the actual building.

As shown in the previous example, the most common method of reducing all the dimensions (in feet and inches) in the same proportion is to choose a certain distance and let that distance represent one foot. This distance can then be divided into 12 parts, each of which represents an inch. If half inches are required, these twelfths are further subdivided into halves, etc. Now the scale represents the common foot rule with its subdivisions into inches and fractions, except that the scaled foot is smaller than the distance known as a foot and, likewise, its subdivisions are proportionately smaller.

When a measurement is made on the drawing, it is made with the reduced foot rule or scale; when a measurement is made on the building, it is made with the standard foot rule. The most common reduced foot rules or scales used in electrical drawings are the **architect's scale** and the **engineer's scale**. Drawings may sometimes be encountered that use a metric scale, but using this scale is similar to using the architect's or engineer's scales.

2.1.0 Architect's Scale

Figure 17 shows two configurations of architect's scales. The one on the top is designed so that 1" = 1'–0", and the one on the bottom has graduations spaced to represent $\frac{1}{8}$" = 1'–0".

Note that on the one-inch scale in *Figure 18*, the longer marks to the right of the zero (with a numeral beneath) represent feet. Therefore, the distance between the zero and the numeral 1 equals one foot. The shorter mark between the zero and 1 represents $\frac{1}{2}$ of a foot, or six inches.

Referring again to *Figure 18*, look at the marks to the left of the zero. The numbered marks are spaced three scaled inches apart and have the numerals 0, 3, 6, and 9 for use as reference points.

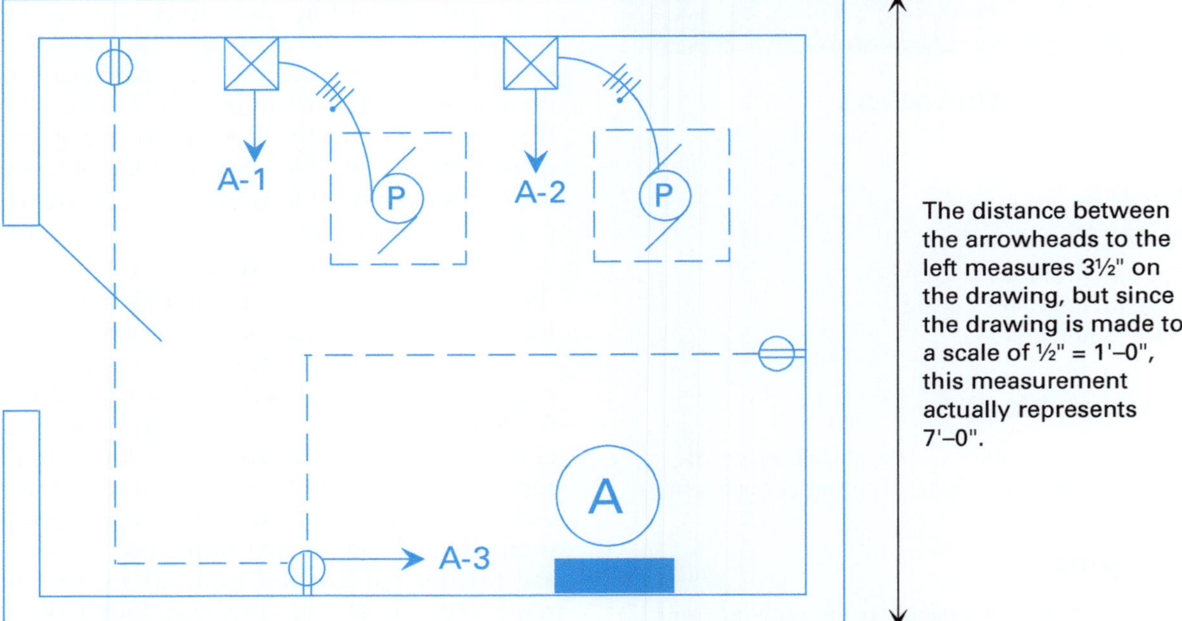

Figure 16 Typical floor plan showing drawing scale.

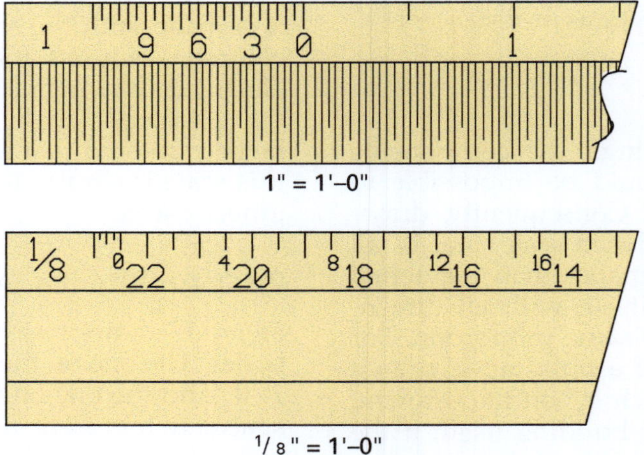

Figure 17   Two different configurations of architect's scales.

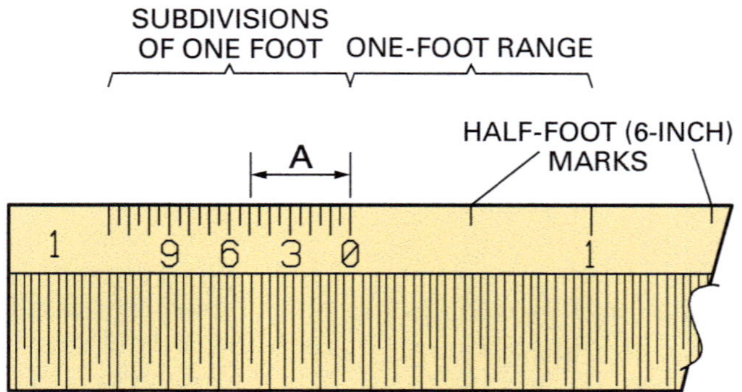

Figure 18 One-inch architect's scale.

The other lines of the same length also represent scaled inches, but are not marked with numerals. In use, you can count the number of long marks to the left of the zero to find the number of inches, but after some practice, you will be able to tell the exact measurement at a glance. For example, the measurement A represents five inches because it is the fifth inch mark to the left of the zero; it is also one inch mark short of the six-inch line on the scale.

The lines that are shorter than the inch line are the half-inch lines. On smaller scales, the basic unit is not divided into as many divisions. For example, the smallest subdivision on some scales represents two inches.

Architect's scales are available in several types, but the most common include the triangular scale (*Figure 19*) and the flat scale. The quality of architect's scales also varies from cheap plastic scales (costing a dollar or two) to high-quality wooden-laminated tools that are calibrated to precise standards.

The triangular scale is frequently found in drafting and estimating departments or engineering and electrical contracting firms, while the flat scales are more convenient to carry on the job site.

Triangular architect's scales have 12 different scales—two on each edge—as follows:

- Common foot rule (12 inches)
- $\frac{1}{16}$" = 1'–0"
- $\frac{3}{32}$" = 1'–0"
- $\frac{3}{16}$" = 1'–0"
- $\frac{1}{8}$" = 1'–0"
- $\frac{1}{4}$" = 1'–0"
- $\frac{3}{8}$" = 1'–0"
- $\frac{3}{4}$" = 1'–0"
- 1" = 1'–0"
- $\frac{1}{2}$" = 1'–0"
- 1 $\frac{1}{2}$" = 1'–0"
- 3" = 1'–0"

Two separate scales on one face may seem confusing at first, but after some experience, reading these scales becomes second nature.

In all but one of the scales on the triangular architect's scale, each face has one of the scales placed opposite to the other. For example, on the one-inch face, the one-inch scale is read from left to right, starting from the zero mark. The half-inch scale is read from right to left, again starting from the zero mark.

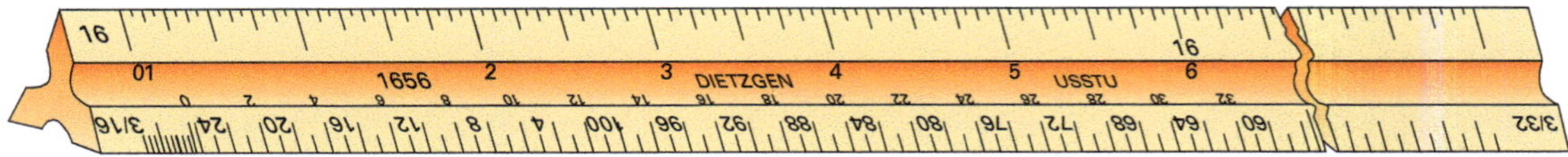

Figure 19 Typical triangular architect's scale.

On the remaining foot-rule scale ($\frac{1}{16}$" = 1'–0") each $\frac{1}{16}$" mark on the scale represents one foot.

Figure 20 shows all the scales found on the triangular architect's scale. The flat architect's scale shown in *Figure 21* is ideal for workers on most projects. It is easily and conveniently carried in the shirt pocket, and the four scales ($\frac{1}{8}$", $\frac{1}{4}$", $\frac{1}{2}$", and 1") are adequate for most projects that will be encountered.

The partial floor plan shown in *Figure 21* is drawn to a scale of $\frac{1}{8}$" = 1'–0". The dimension in question is found by placing the $\frac{1}{8}$" architect's scale on the drawing and reading the figures. It can be seen that the dimension reads 25'–6".

Every drawing should have the scale to which it is drawn plainly marked on it as part of the drawing title. However, it is not uncommon to have several different drawings on one blueprint sheet—all with different scales. Therefore, always check the scale of each different view found on a drawing sheet.

2.2.0 Engineer's Scale

The civil engineer's scale is used in basically the same manner as the architect's scale, with the principal difference being that the graduations on the engineer's scale are decimal units rather than feet as on the architect's scale.

The engineer's scale is used by placing it on the drawing with the working edge away from the user. The scale is then aligned in the direction of the required measurement. Then, by looking down at the scale, the dimension is read.

Civil engineer's scales commonly show the following graduations:

- 1" = 10 units
- 1" = 20 units
- 1" = 30 units
- 1" = 40 units
- 1" = 60 units
- 1" = 80 units
- 1" = 100 units

The purpose of this scale is to transfer the relative dimensions of an object to the drawing or vice versa. It is used mainly on site plans to determine distances between property lines, manholes, duct runs, direct-burial cable runs, and the like.

Site plans are drawn to scale using the engineer's scale rather than the architect's scale. On small lots, a scale of 1 inch = 10 feet or 1 inch = 20 feet is used. For a 1:10 scale, this means that one inch (the actual measurement on the drawing) is equal to 10 feet on the land itself.

On larger drawings, where a large area must be covered, the scale could be 1 inch = 100 feet or 1 inch = 1,000 feet, or any other integral power of 10. On drawings with the scale in multiples of 10, the engineer's scale marked 10 is used. If the scale is 1 inch = 200 feet, the engineer's scale marked 20 is used, and so on.

Although site plans appear reduced in scale, depending on the size of the object and the size of the drawing sheet to be used, the actual dimensions must be shown on the drawings at all times. When you are reading the drawing plans to scale, think of each dimension in its full size and not in the reduced scale it happens to be on the drawing (*Figure 22*).

2.3.0 Metric Scale

Metric scales are calibrated in units of 10 (*Figure 23*). The two common length measurements used in the metric scale on architectural drawings are the meter and the millimeter, the millimeter being $\frac{1}{1,000}$ of a meter. On drawings drawn to scales between 1:1 and 1:100, the millimeter is typically used. On drawings drawn to scales between 1:200 and 1:2,000, the meter is generally used. Many contracting firms that deal in international trade have adopted a dual-dimensioning system expressed in both metric and English symbols. Drawings prepared for government projects may also require metric dimensions. A metric conversion chart is provided in the *Appendix*.

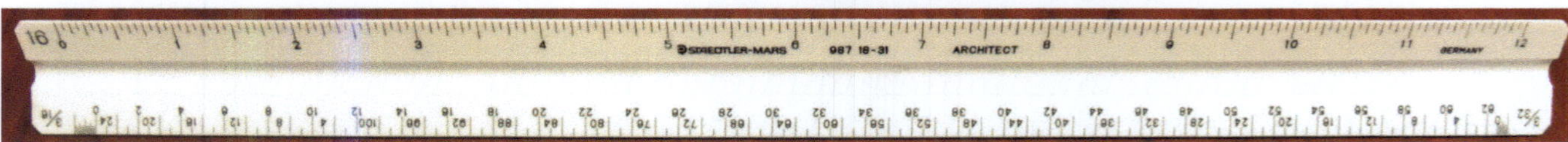

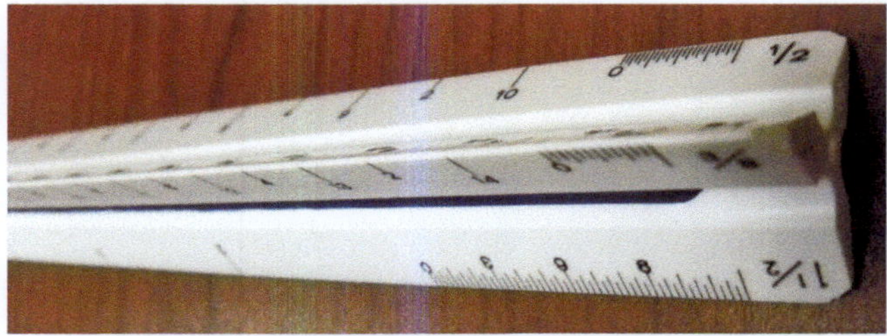

Figure 20   Various scales on a triangular architect's scale.

NCCER – *Maritime Electrical*

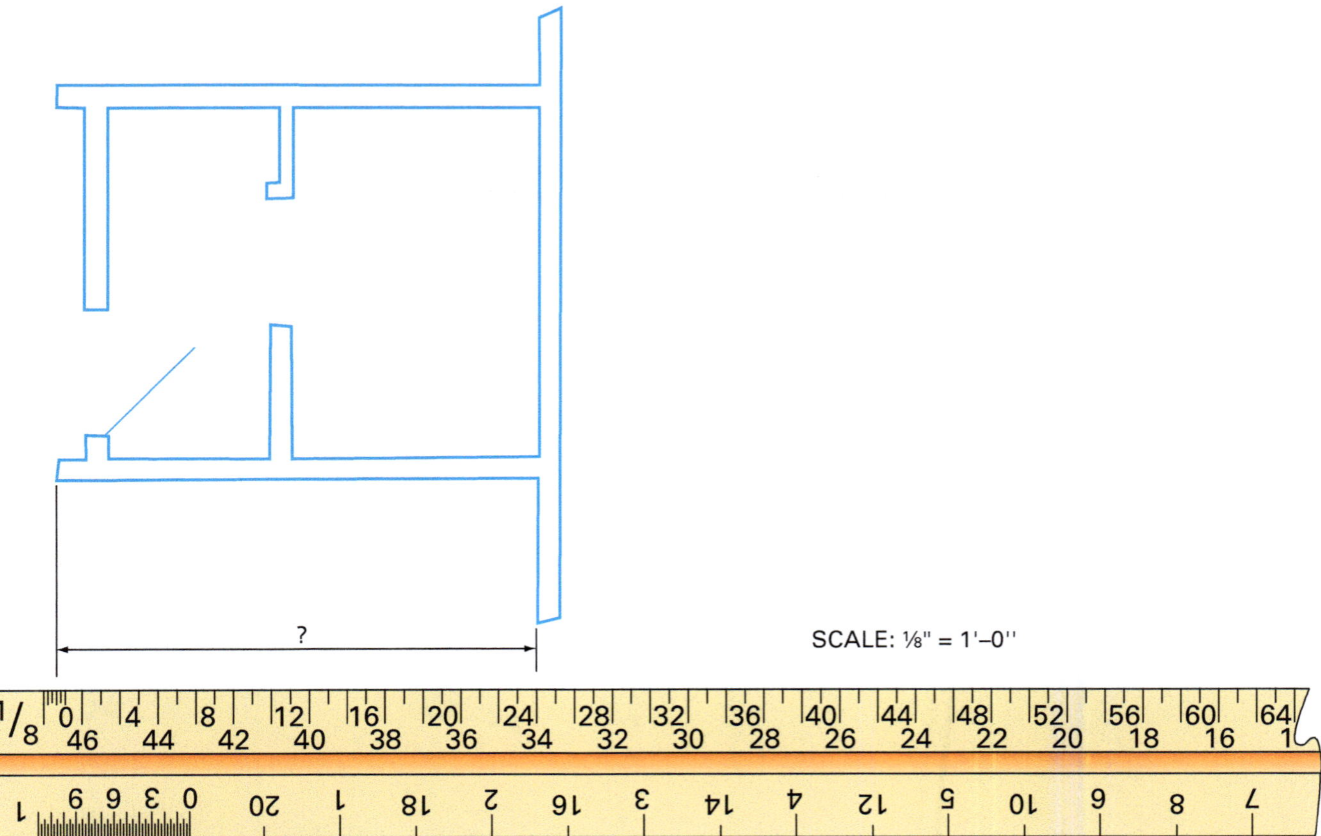

Figure 21   Using the ⅛" architect's scale to determine the dimensions on a drawing.

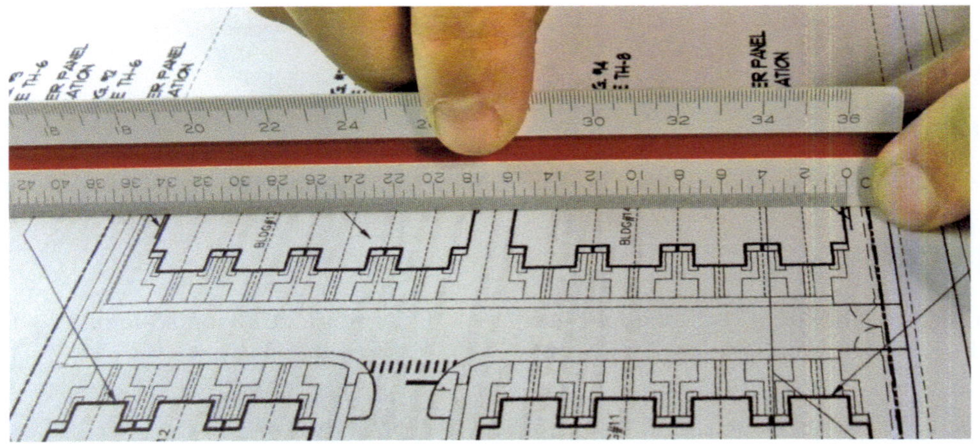

Figure 22   Practical use of the engineer's scale.

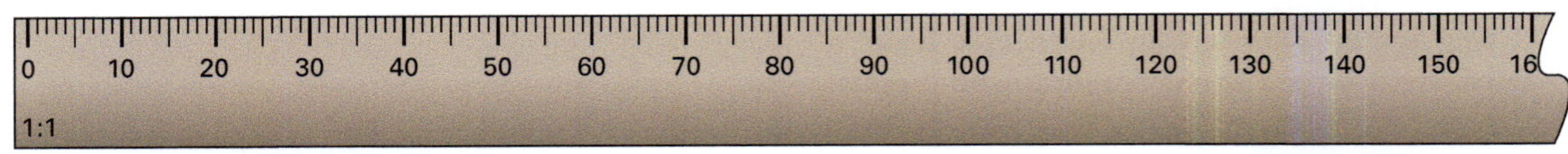

Figure 23   Typical metric scale.

Architect's Scale

Measurements are usually made on architectural drawings using an architect's scale rather than a standard ruler. Architect's scales, like the ones on the left, are divided into feet and inches and usually consist of several scales on one rule. Architect's scales also come in other forms such as tapes or with wheels, like the one shown on the right.

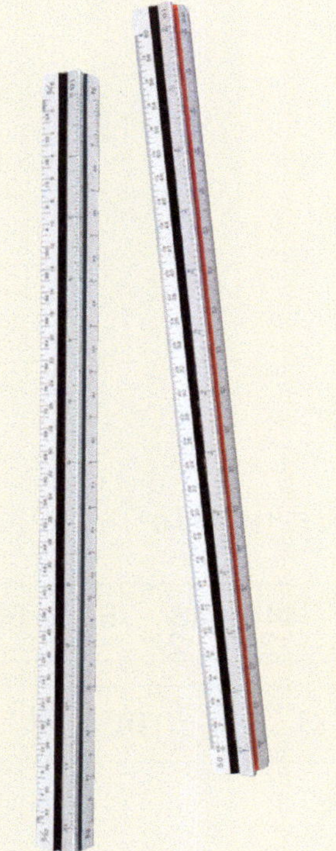

Figure Credit: STAEDTLER Mars Limited

Figure Credit: Courtesy of Calculated Industries

2.0.0 Section Review

1. If a floor plan of a building is drawn to a scale of ⅛" = 1'–0", each 1 ⅜" on the drawing would equal _____.

   a. 1'–3" on the building itself
   b. 8'–0" on the building itself
   c. 11'–0" on the building itself
   d. 13'–8" on the building itself

2. Using a 1:20 scale on an engineer's scale, a measurement of 2" on a drawing would equal _____.

   a. 5 feet
   b. 10 feet
   c. 20 feet
   d. 40 feet

3. A metric drawing scale of 1:1,000 typically depicts units in _____.

   a. millimeters
   b. centimeters
   c. decimeters
   d. meters

3.0.0 ELECTRICAL DRAWINGS

Objective

Read electrical drawings.
  a. Interpret electrical symbols.
  b. Analyze a set of electrical drawings.
  c. Identify fixtures in a lighting floor plan.
  d. Read block and schematic diagrams.
  e. Interpret written specifications.

Performance Task

2. Make a material takeoff of the lighting fixtures specified in the provided drawing. The takeoff requires that all lighting fixtures be counted, and where applicable, the total number of lamps for each fixture type must be calculated. (Fill these in on the provided Lighting Fixture Takeoff worksheet.).

Trade Terms

Block diagram: A single-line diagram used to show electrical equipment and related connections. See power-riser diagram.

One-line diagram: A drawing that shows, by means of lines and symbols, the path of an electrical circuit or system of circuits along with the various circuit components. Also called a single-line diagram.

Power-riser diagram: A single-line block diagram used to indicate the electric service equipment, service conductors and feeders, and sub-panels. Notes are used on power-riser diagrams to identify the equipment; indicate the size of conduit; show the number, size, and type of conductors; and list related materials. A panelboard schedule is usually included with power-riser diagrams to indicate the exact components (panel type and size), along with fuses, circuit breakers, etc., contained in each panelboard.

Schedule: A systematic method of presenting equipment lists on a drawing in tabular form.

Schematic diagram: A detailed diagram showing complicated circuits, such as control circuits.

Shop drawing: A drawing that is usually developed by manufacturers, fabricators, or contractors to show specific dimensions and other pertinent information concerning a particular piece of equipment and its installation methods.

Written specifications: A written description of what is required by the owner, architect, and engineer in the way of materials and workmanship. Together with working drawings, the specifications form the basis of the contract requirements for construction.

An electrical drawing shows in a clear, concise manner exactly what is required of the electricians. The amount of data shown on such drawings should be sufficient, but not overdone. This means that a complete set of electrical drawings could consist of only one 8½" × 11" sheet, or it could consist of several dozen 24" × 36" (or larger) sheets, depending on the size and complexity of a given project. A shop drawing, for example, may contain details of only one piece of equipment, while a set of working drawings for an industrial installation may contain dozens of drawing sheets detailing the electrical system for lighting and power, along with equipment, motor controls, and wiring diagrams. A schematic diagram and equipment schedule may contain a host of other pertinent data.

In general, the electrical working drawings for a given project serve three distinct functions:

- They provide an exact description of the project so that electrical contractors can estimate materials and labor to calculate a total cost of the project for bidding purposes.
- They provide workers on the project with instructions as to how the electrical system is to be installed.
- They provide a map of the electrical system once the job is completed to aid in maintenance and troubleshooting for years to come.

Electrical drawings from consulting engineering firms will vary in quality from sketchy, incomplete drawings to neat, precise drawings that are easy to understand. Few, however, will cover every detail of the electrical system. Therefore, a good knowledge of installation practices must go hand-in-hand with interpreting electrical working drawings.

Sometimes electrical contractors will have electrical drafters prepare special supplemental drawings for use by the contractors' employees. On certain projects, these supplemental drawings can save supervision time in the field once the project has begun.

3.1.0 Electrical Symbols

The electrician must be able to correctly read and understand electrical working drawings. This skill requires a thorough knowledge of electrical symbols and their applications.

An electrical symbol is a figure or mark that stands for a component used in the electrical system. *Figure 24* shows a list of electrical symbols that are currently recommended by the American National Standards Institute (ANSI). It is evident from this list of symbols that many have the same basic form, but, because of some slight difference, their meaning changes. For example, the receptacle symbols in *Figure 25* each have the same basic form (a circle), but the addition of a line or an abbreviation gives each an individual meaning. A good procedure to follow in learning symbols is to first learn the basic form and then apply the variations for obtaining different meanings.

It would be much simpler if all architects, engineers, electrical designers, and drafters used the same symbols; however, this is not the case. Although standardization is getting closer to a reality, existing symbols are still modified, and new symbols are created for almost every new project.

The electrical symbols described in the following paragraphs represent those found on actual electrical working drawings throughout the United States and Canada. Many are similar to those recommended by ANSI and the Consulting Engineers Council/US; others are not. Understanding how these symbols were devised will help you to interpret unknown electrical symbols in the future.

Some of the symbols used on electrical drawings are abbreviations, such as WP for weatherproof and AFF for above finished floor. Others are simplified pictographs, such as those shown in *Figure 26*.

In some cases, the symbols are combinations of abbreviations and pictographs, such as in *Figure 26* for a fusible safety switch, a nonfusible safety switch, and a double-throw safety switch. In each example, a pictograph of a switch enclosure has been combined with an abbreviation: F (fusible), NF (nonfusible), and DT (double-throw), respectively.

Lighting outlet symbols have been devised that represent incandescent, fluorescent, and high-intensity discharge lighting; a circle usually represents an incandescent fixture, and a rectangle is used to represent a fluorescent fixture. These symbols are designed to indicate the physical shape of a particular fixture, and while the circles representing incandescent lamps are frequently enlarged somewhat, symbols for fluorescent fixtures are usually drawn as close to scale as possible. The type of mounting used for all lighting fixtures is usually indicated in a lighting fixture schedule, which is shown on the drawings or in the written specifications.

The type of lighting fixture is identified by a numeral placed inside a triangle or other symbol, and placed near the fixture to be identified. A complete description of the fixtures identified by the symbols must be given in the lighting fixture schedule and should include the manufacturer, catalog number, number and type of lamps, voltage, finish, mounting, and any other information needed for proper installation of the fixture.

Switches used to control lighting fixtures are also indicated by symbols (usually the letter S followed by numerals or letters to define the exact type of switch). For example, S3 indicates a three-way switch; S4 identifies a four-way switch; and SP indicates a single-pole switch with a pilot light. A subscript letter is often used to identify the fixtures that are controlled by that switch.

Main distribution centers, panelboards, transformers, safety switches, and other similar electrical components are indicated by electrical symbols on floor plans and by a combination of symbols and semipictorial drawings in riser diagrams.

A detailed description of the service equipment is usually given in the panelboard schedule or in the written specifications. However, on small projects, the service equipment is sometimes indicated only by notes on the drawings.

Circuit and feeder wiring symbols are getting closer to being standardized. Most circuits concealed in the ceiling or wall are indicated by a solid line; a broken line is used for circuits concealed in the floor or ceiling below; and exposed raceways are indicated by short dashes or else the letter E placed in the same plane with the circuit line at various intervals. The number of conductors in a conduit or raceway system may be indicated in the panelboard schedule under the appropriate column, or the information may be shown on the floor plan.

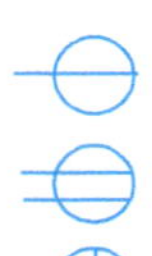
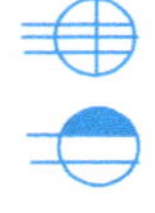

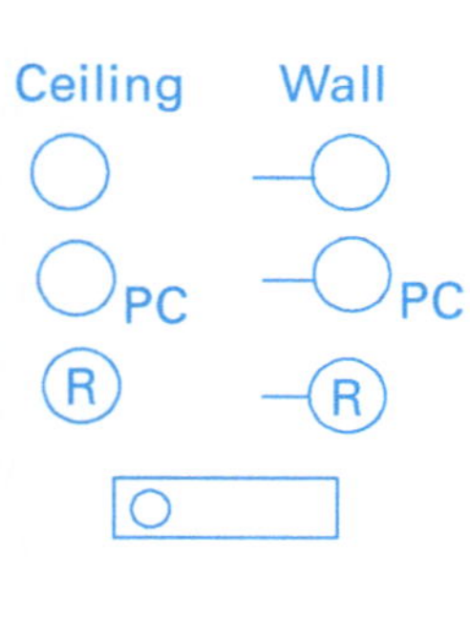

SWITCH OUTLETS

Single-Pole Switch

Double-Pole Switch

Three-Way Switch

Four-Way Switch

Key-Operated Switch

Switch w/Pilot

Low-Voltage Switch

Switch & Single Receptacle

Switch & Duplex Receptacle

Door Switch

Momentary Contact Switch

RECEPTACLE OUTLETS

Single Receptacle

Duplex Receptacle

Triplex Receptacle

Split-Wired Duplex Recep.

Single Special-Purpose Recep.

Duplex Special-Purpose Recep.

Range Receptacle

Special Purpose Connection
or Provision for Connection.
Subscript letters indicate
Function (DW - Dishwasher;
CD - Clothes Dryer, etc.)

Clock Receptacle w/Hanger

Fan Receptacle w/Hanger

Single Floor Receptacle

Note: A numeral or letter within
the symbol or as a sub-
script keyed to the list of
symbols indicates type of
receptacle or usage.

LIGHTING OUTLETS

Ceiling   **Wall**

Surface Fixture

Surface Fixt. w/Pull Chain

Recessed Fixture

Surface or Pendant
Fluorescent Fixture

Recessed Fluor. Fixture

Surface or Pendant
Continuous Row Fluor.
Fixtures

Recessed Continuous
Row Fluorescent Fixtures

Surface Exit Light

Recessed Exit Light

Blanked Outlet

Junction Box

CIRCUITING

Wiring Concealed in
Ceiling or Wall

Wiring Concealed in Floor

Wiring Exposed

Branch Circuit Homerun to
Panelboard. Number of
arrows indicates number of
circuits in run. Note: Any
circuit without further
identification is 2-wire.
A greater number of wires
is indicated by cross lines
as shown below. Wire size
is sometimes shown with
numerals placed above or
below cross lines.

3-Wire

4-Wire

Figure 24 ANSI electrical symbols.

Symbols for communication and signal systems, as well as symbols for light and power, are drawn to an appropriate scale and accurately located with respect to the building. This reduces the number of references made to the architectural drawings. Where extreme accuracy is required in locating outlets and equipment, exact dimensions are given on larger-scale drawings and shown on the plans.

Each different category in an electrical system is usually represented by a basic distinguishing symbol. To further identify items of equipment or outlets in the category, a numeral or other identifying mark is placed within the open basic symbol. In addition, all such individual symbols used on the drawings should be included in the symbol list or legend. The electrical symbols shown in *Figure 27* were modified by a consulting engineering firm for use on a small industrial electrical installation. The symbols shown in *Figure 28A* through *Figure 28G* are those recommended by the Consulting Engineers Council/US. You should become familiar with these symbols.

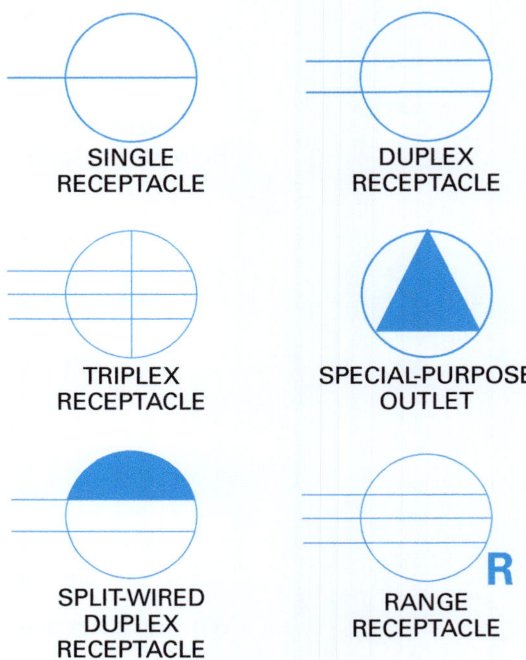

Figure 25   Various receptacle symbols used on electrical drawings.

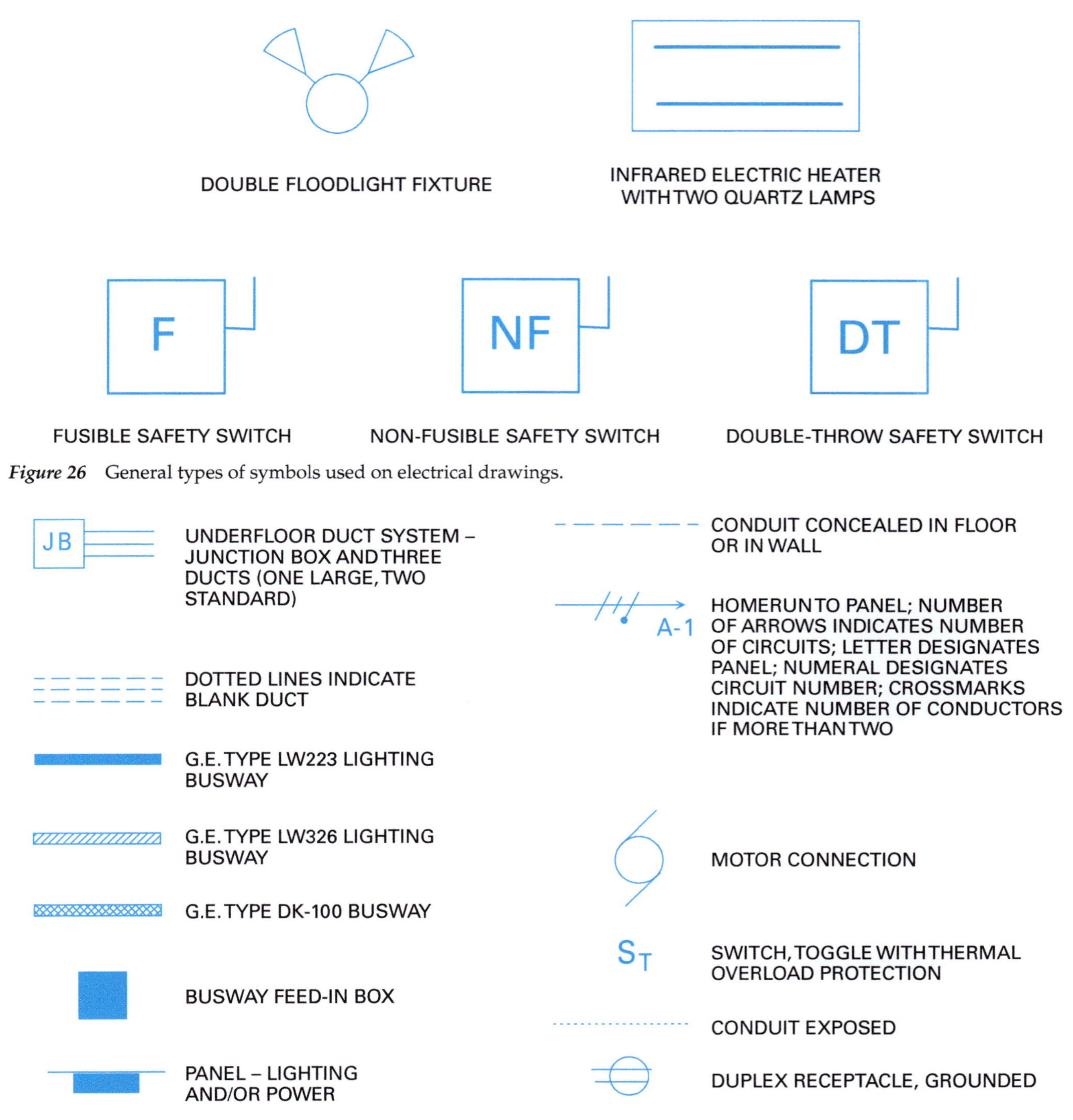

Figure 26 General types of symbols used on electrical drawings.

Figure 27 Electrical symbols used by one consulting engineering firm.

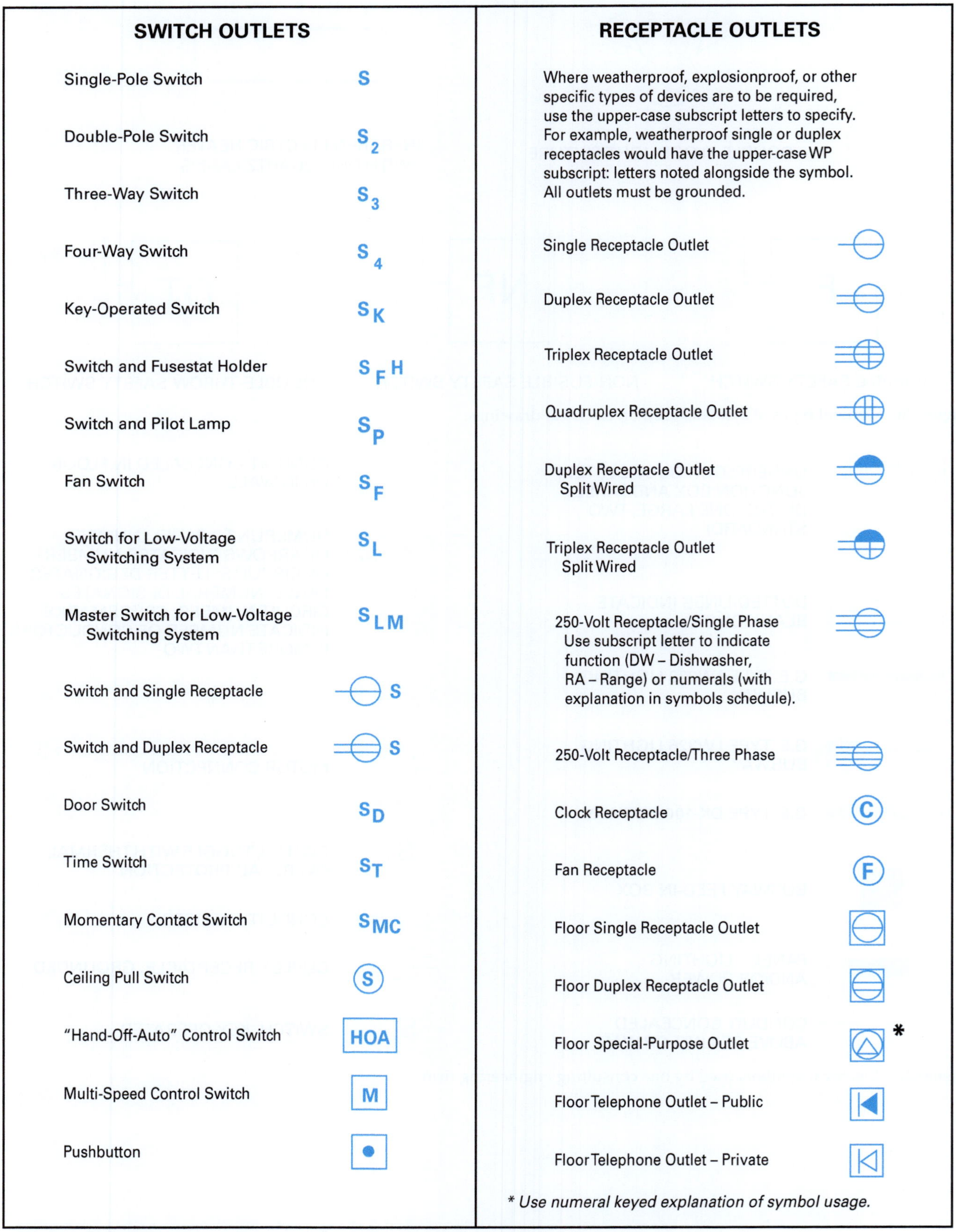

<table>
<tr><th>SWITCH OUTLETS</th><th>RECEPTACLE OUTLETS</th></tr>
</table>

SWITCH OUTLETS

Single-Pole Switch	S
Double-Pole Switch	S_2
Three-Way Switch	S_3
Four-Way Switch	S_4
Key-Operated Switch	S_K
Switch and Fusestat Holder	$S_F H$
Switch and Pilot Lamp	S_P
Fan Switch	S_F
Switch for Low-Voltage Switching System	S_L
Master Switch for Low-Voltage Switching System	S_{LM}
Switch and Single Receptacle	S
Switch and Duplex Receptacle	S
Door Switch	S_D
Time Switch	S_T
Momentary Contact Switch	S_{MC}
Ceiling Pull Switch	S
"Hand-Off-Auto" Control Switch	HOA
Multi-Speed Control Switch	M
Pushbutton	

RECEPTACLE OUTLETS

Where weatherproof, explosionproof, or other specific types of devices are to be required, use the upper-case subscript letters to specify. For example, weatherproof single or duplex receptacles would have the upper-case WP subscript: letters noted alongside the symbol. All outlets must be grounded.

Single Receptacle Outlet	
Duplex Receptacle Outlet	
Triplex Receptacle Outlet	
Quadruplex Receptacle Outlet	
Duplex Receptacle Outlet Split Wired	
Triplex Receptacle Outlet Split Wired	
250-Volt Receptacle/Single Phase Use subscript letter to indicate function (DW – Dishwasher, RA – Range) or numerals (with explanation in symbols schedule).	
250-Volt Receptacle/Three Phase	
Clock Receptacle	C
Fan Receptacle	F
Floor Single Receptacle Outlet	
Floor Duplex Receptacle Outlet	
Floor Special-Purpose Outlet	*
Floor Telephone Outlet – Public	
Floor Telephone Outlet – Private	

** Use numeral keyed explanation of symbol usage.*

Figure 28A   Recommended electrical symbols (Sheet 1 of 7).

Example of the use of several floor outlet symbols to identify a 2, 3, or more gang outlet:

Underfloor duct and junction box for triple, double, or single duct system as indicated by the number of parallel lines

Example of the use of various symbols to identify the location of different types of outlets or connections for underfloor duct or cellular floor systems:

Cellular Floor Heater Duct

CIRCUITING

Wiring Exposed (not in conduit)

Wiring Concealed in Ceiling or Wall

Wiring Concealed in Floor

Wiring Existing*

Wiring Turned Up

Wiring Turned Down

Branch Circuit Homerun to Panelboard

Number of arrows indicates number of circuits. (A number at each arrow may be used to identify the circuit number.)**

BUSDUCTS AND WIREWAYS

Trolley Duct***   T   T

Busway (Service, Feeder or Plug-in)***   B   B

Cable Trough Ladder or Channel***   C   C

Wireway***   W   W

PANELBOARDS, SWITCHBOARDS AND RELATED EQUIPMENT

Flush-Mounted Panelboard and Cabinet***

Surface-Mounted Panelboard and Cabinet***

Switchboard, Power Control Center, Unit Substation (Should be drawn to scale)***

Flush-Mounted Terminal Cabinet (In small scale drawings the TC may be indicated alongside the symbol)***

Surface-Mounted Terminal Cabinet (In small scale drawings the TC may be indicated alongside the symbol)***

Pull Box (Identify in relation to wiring system section and size)

Motor or Other Power Controller May be a starter or contactor***

Externally Operated Disconnection Switch***

Combination Controller and Disconnection Means***

*Note: Use heavy-weight line to identify service and feeders. Indicate empty conduit by notation CO.

**Note: Any circuit without further identification indicates two-wire circuit. For a greater number of wires, indicate with cross lines, e.g.:

3 wires   4 wires, etc.

Neutral and ground wires may be shown longer. Unless indicated otherwise, the wire size of the circuit is the minimum size required by the specification. Identify different functions of wiring system (e.g., signaling system) by notation or other means.

***Identify by notation or schedule.

Figure 28B Recommended electrical symbols (Sheet 2 of 7).

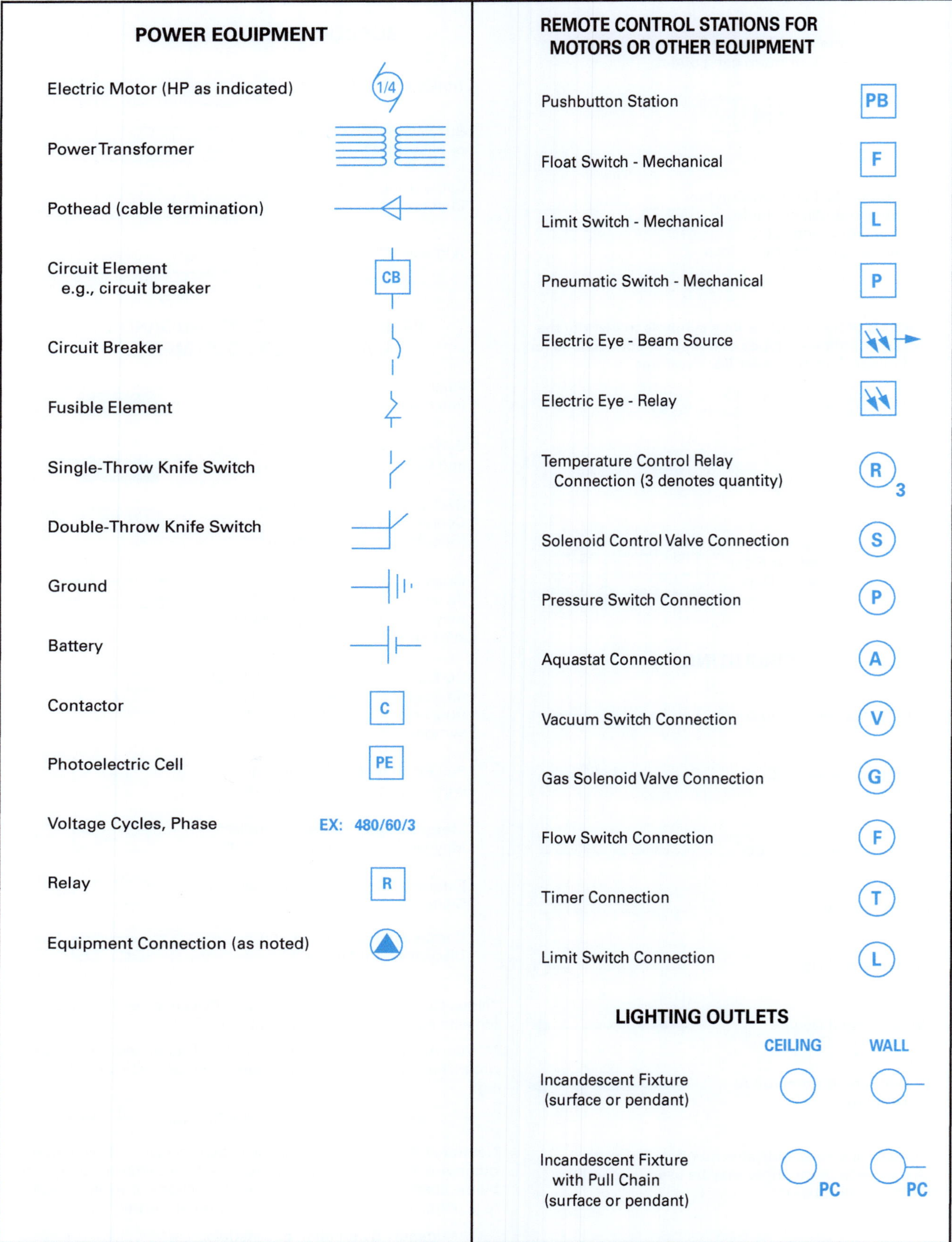

Figure 28C Recommended electrical symbols (Sheet 3 of 7).

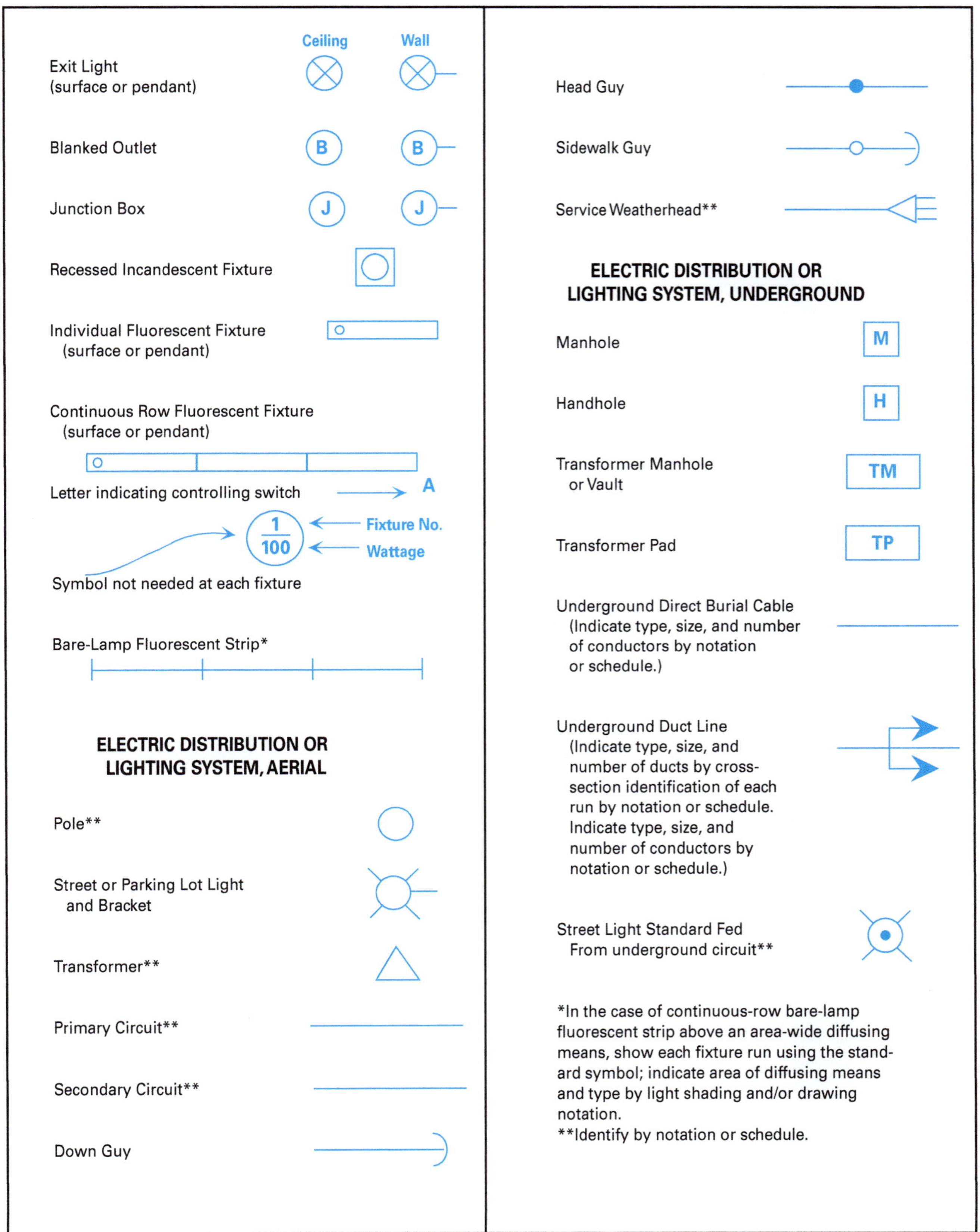

Figure 28D Recommended electrical symbols (Sheet 4 of 7).

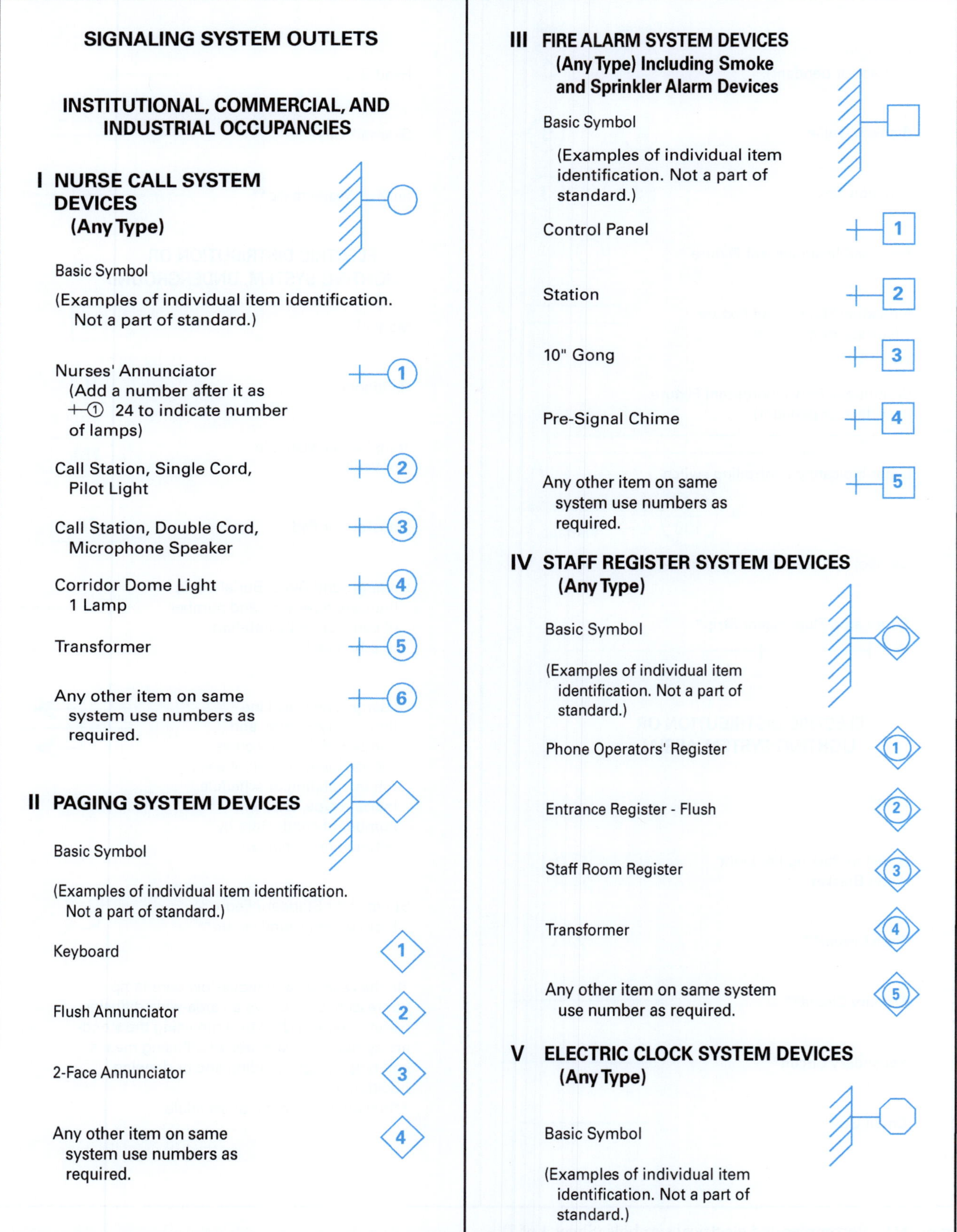

Figure 28E   Recommended electrical symbols (Sheet 5 of 7).

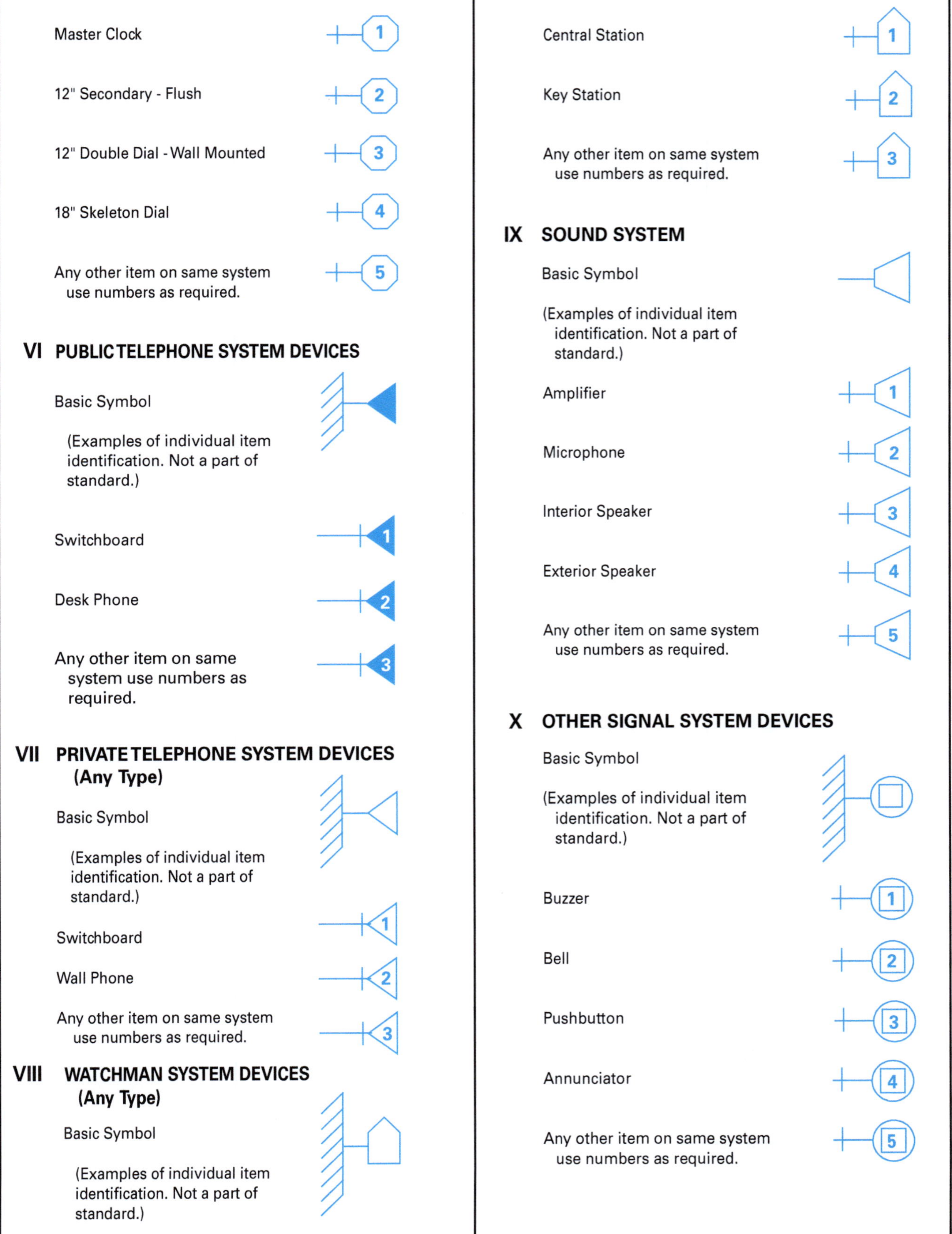

Figure 28F Recommended electrical symbols (Sheet 6 of 7).

RESIDENTIAL OCCUPANCIES

Signaling system symbols identify standardized residential-type signal system items on residential drawings where a descriptive symbol list is not included on the drawing. When other signal system items are to be identified, use the basic symbols below for such items together with a descriptive symbol list.

Pushbutton	
Buzzer	
Bell	
Combination Bell - Buzzer	
Chime	CH
Annunciator	
Electric Door Opener	D
Maid's Signal Plug	M
Interconnection Box	
Bell-Ringing Transformer	BT
Outside Telephone	
Interconnecting Telephone	
Television Outlet	TV

Figure 28G   Recommended electrical symbols (Sheet 7 of 7).

3.2.0 Analyzing Electrical Drawings

The most practical way to learn how to read electrical construction documents is to analyze an existing set of drawings prepared by consulting or industrial engineers.

Engineers or electrical designers are responsible for the complete layout of electrical systems for most projects. Electrical drafters then transform the engineer's designs into working drawings, using either manual drafting instruments or computer-aided design (CAD) systems. The following is a brief outline of what usually takes place in the preparation of electrical design and working drawings:

- The engineer meets with the architect and owner to discuss the electrical needs of the building or project and to discuss various recommendations made by all parties.
- After that, an outline of the architect's floor plan is laid out.
- The engineer then calculates the required power and lighting outlets for the project; these are later transferred to the working drawings.
- All communications and alarm systems are located on the floor plan, along with lighting and power panelboards.
- Circuit calculations are made to determine wire size and overcurrent protection.
- The main electric service and related components are determined and shown on the drawings.
- Schedules are then placed on the drawings to identify various pieces of equipment.
- Wiring diagrams are made to show the workers how various electrical components are to be connected.
- A legend or electrical symbol list is drafted and shown on the drawings to identify all symbols used to indicate electrical outlets or equipment.
- Various large-scale electrical details are included, if necessary, to show exactly what is required of the electricians.
- Written specifications are then made to give a description of the materials and installation methods.

3.2.1 Electrical Site Plans

Electrical site work is sometimes shown on the architect's plot plan. However, when site work involves many trades and several utilities (e.g., gas, telephone, electric, television, water, and sewage), it can become confusing if all details are shown on one drawing sheet. In cases like these, it is best to have a separate drawing devoted entirely to the electrical work, as shown in *Figure 29*. This project is an office/warehouse building for Virginia Electric, Inc. The electrical drawings consist of four large drawing sheets, along with a set of written specifications, which will be discussed later in this module.

The electrical site or plot plan shown in *Figure 29* has the conventional architect's and engineer's title blocks in the lower right-hand corner of the drawing. These blocks identify the project and project owners, the architect, and the engineer. They also show how this drawing sheet relates to the entire set of drawings. Note the engineer's professional stamp of approval to the left of the engineer's title block. Similar blocks appear on all four of the electrical drawing sheets.

When examining a set of electrical drawings for the first time, always look at the area around the title block. This is where most revision blocks or revision notes are placed. If revisions have been made to the drawings, make certain that you have a clear understanding of what has taken place before proceeding with the work.

Refer again to the drawing in *Figure 29* and note the north arrow in the upper left corner. A north arrow shows the direction of true north to help you orient the drawing to the site. Look directly down from the north arrow to the bottom of the page and notice the drawing title, *Plot Utilities*. Directly beneath the drawing title you can see that the drawing scale of 1" = 30' is shown. This means that each inch on the drawing represents 30 feet on the actual job site. This scale holds true for all drawings on the page unless otherwise noted.

An outline of the proposed building is indicated on the drawing along with a callout, *Proposed Bldg. Fin. Flr. Elev. 590.0*. This means that the finished floor level of the building is to be 590 feet above sea level, which in this part of the country will be about two feet above finished grade around the building. This information helps the electrician locate conduit sleeves and stub-ups to the correct height before the finished concrete floor is poured.

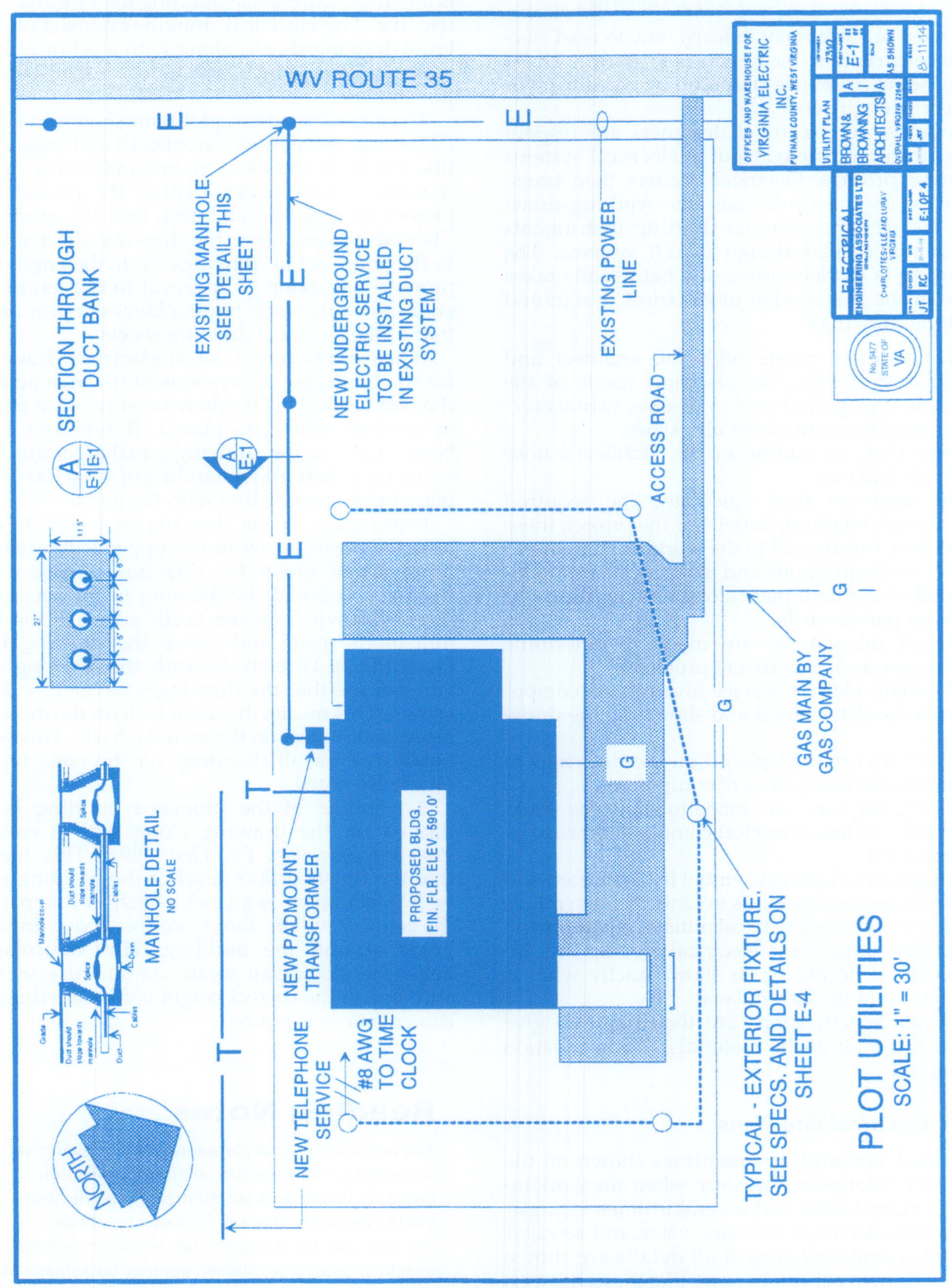

Figure 29 Typical electrical site plan.

NCCER – *Maritime Electrical*

The shaded area represents asphalt paving for the access road, drives, and parking lot. Note that the access road leads into a highway, which is designated Route 35. This information further helps workers to orient the drawing to the building site.

Existing manholes are indicated by a solid circle, while an open circle is used to show the position of the five new pole-mounted lighting fixtures that are to be installed around the new building. Existing power lines are shown with a light solid line with the letter E placed at intervals along the line. The new underground electric service is shown in the same way, except the lines are somewhat wider and darker on the drawing. Note that this new high-voltage cable terminates into a pad-mounted transformer near the proposed building. New telephone lines are similar except the letter T is used to identify the telephone lines.

The direct-burial underground cable supplying the exterior lighting fixtures is indicated with dashed lines on the drawing, which is shown connecting the open circles. A homerun for this circuit is also shown to a time clock.

The manhole detail shown to the right of the north arrow may seem to serve very little purpose on this drawing since the manholes have already been installed. However, the dimensions and details of their construction will help the electrical contractor or supervisor to better plan the pulling of the high-voltage cable. The same is true of the cross section shown of the duct bank. The electrical contractor knows that three empty ducts are available if it is discovered that one of them is damaged when the work begins.

Although the electrical work will not involve working with gas, the main gas line is shown on the electrical drawing to let the electrical workers know its approximate location while they are installing the direct-burial conductors for the exterior lighting fixtures.

3.2.2 Power Plans

The electrical power plan (*Figure 30*) shows the complete floor plan of the office/warehouse building with all interior partitions drawn to scale. Sometimes, the physical locations of all wiring and outlets are shown on one drawing; that is, outlets for lighting, power, signal and communications, special electrical systems, and related equipment are shown on the same plan. However, on complex installations, the drawing would become cluttered if both lighting and power were shown on the same floor plan. Therefore, most projects will have a separate drawing for power and another for lighting. Riser diagrams and details may be shown on yet another drawing sheet or, if room permits, they may be shown on the lighting or power floor plan sheets.

A closer look at this drawing reveals the title blocks in the lower right corner of the drawing sheet. These blocks list both the architectural and engineering firms, along with information to identify the project and drawing sheet. Also note that the floor plan is titled *Floor Plan "B"—Power* and is drawn to a scale of $\frac{1}{8}$" = 1'–0". There are no revisions shown on this drawing sheet. Two important items to note on the power plan are the key plan and the electrical symbols list.

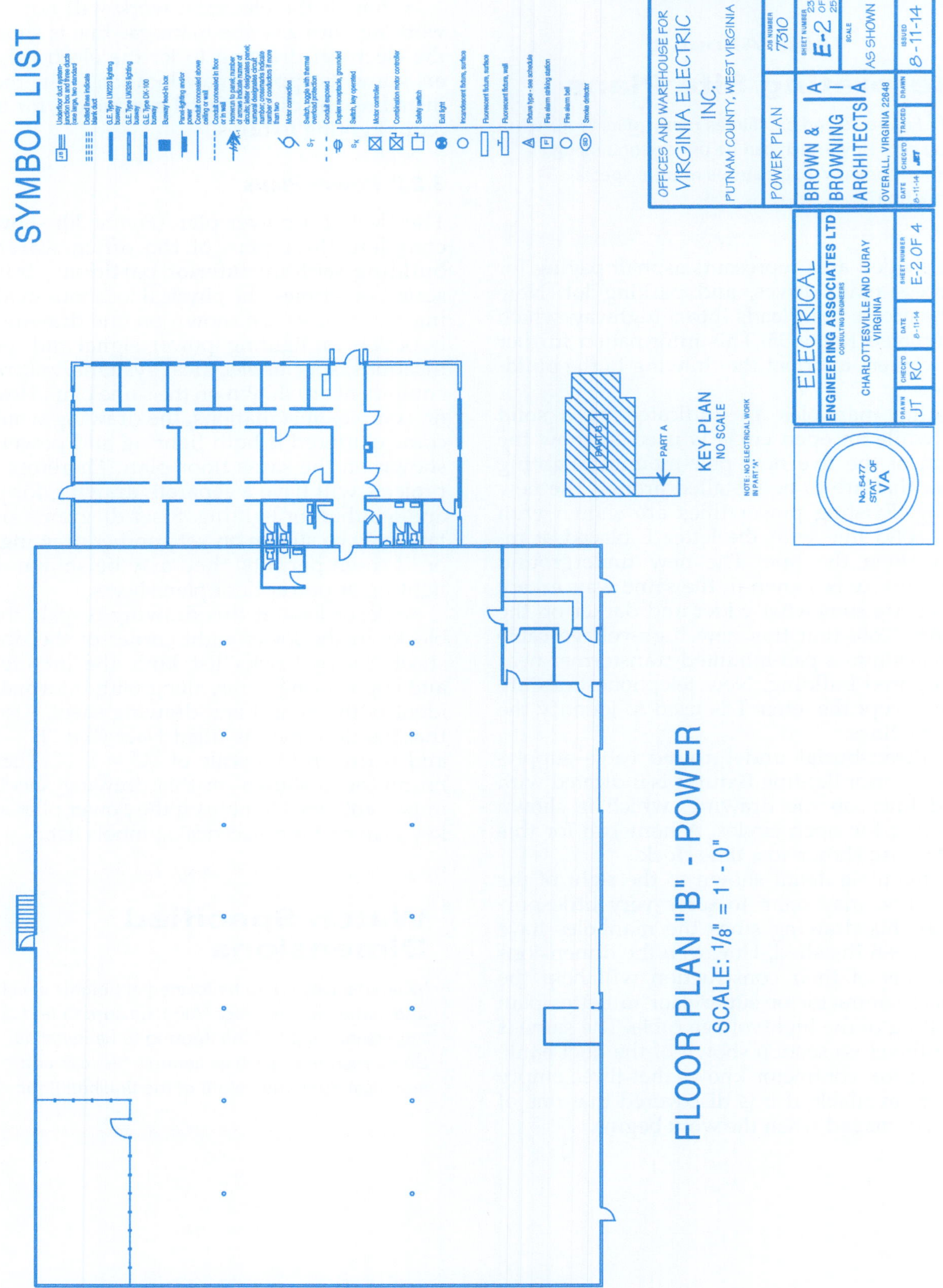

Figure 30 Electrical power plan.

A key plan appears on the drawing sheet immediately above the engineer's title block (*Figure 31*). The purpose of this key plan is to identify that part of the project to which this sheet applies. In this case, the project involves two buildings: Building A and Building B. Since the outline of Building B is cross-hatched in the key plan, this is the building to which this drawing applies. Note that this key plan is not drawn to scale; only its approximate shape is shown.

Although Building A is also shown on this key plan, a note below the key plan title states that there is no electrical work required in Building A.

On some larger installations, the overall project may involve several buildings requiring appropriate key plans on each drawing to help the workers orient the drawings to the appropriate building. In some cases, separate drawing sheets may be used for each room or area in an industrial project—again requiring key plans on each drawing sheet to identify applicable drawings for each room.

A symbol list also appears on the electrical power plan (immediately above the architect's title block) to identify the various symbols used for both power and lighting on this project. In most cases, the only symbols listed are those that apply to the particular project. In other cases, however, a standard list of symbols is used for all projects with the following note:

"These are standard symbols and may not all appear on the project drawings; however, wherever the symbol on the project drawings occurs, the item shall be provided and installed."

Only electrical symbols that are actually used for the office/warehouse drawings are shown in the list on the example electrical power plan. A close-up look at these symbols appears in *Figure 32*.

3.2.3 Floor Plan

A somewhat enlarged view of the electrical floor plan drawing is shown in *Figure 33*. Due to the size and complexity of the drawing, it is still difficult to see very much detail. This illustration is meant to show the overall layout of the floor plan and how the symbols and notes are arranged.

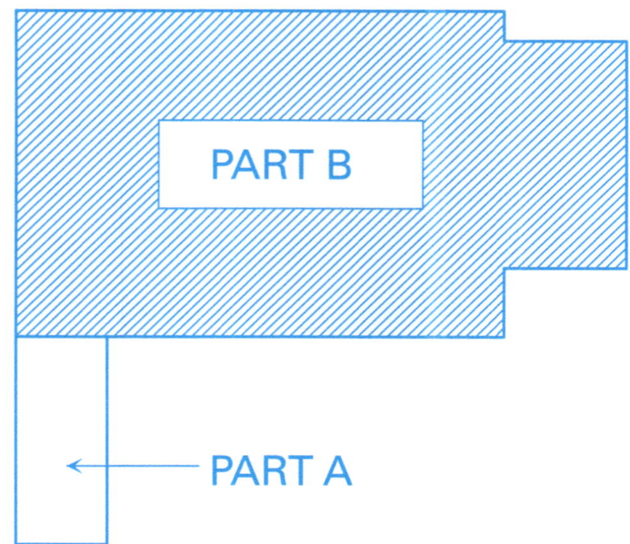

Figure 31   Key plan appearing on electrical power plan.

Symbols

Never assume that you know the meaning of any electrical symbol. Although great efforts have been made in recent years to standardize drawing symbols, architects, consulting engineers, and electrical drafters still modify existing symbols or devise new ones to meet their own needs. Always consult the symbol list or legend on electrical working drawings for an exact interpretation of the symbols used.

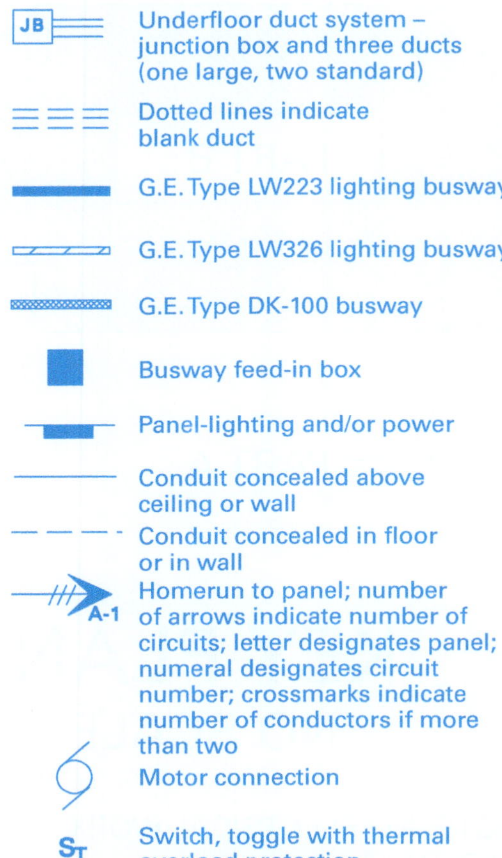

Figure 32   Electrical symbols list for the example building.

In general, this plan shows the service equipment (in plan view), receptacles, underfloor duct system, motor connections, motor controllers, electric heat, busways, and similar details. The electric panels and other service equipment are drawn close to scale. The locations of other electrical outlets and similar components are only approximated on the drawings because they have to be exaggerated to show up on the prints. To illustrate, a common duplex receptacle is only about 3" (76 mm) wide. If such a receptacle were to be located on the floor plan of this building (drawn to a scale of $\frac{1}{8}$" = 1'–0"), even a small dot on the drawing would be too large to draw the receptacle exactly to scale. Therefore, the receptacle symbol is exaggerated. When such receptacles are scaled on the drawings to determine the proper location, a measurement is usually taken to the center of the symbol to determine the distance between outlets. Junction boxes, switches, and other electrical connections shown on the floor plan will be exaggerated in a similar manner.

The office/warehouse project utilizes three types of busways: two types of lighting busways and one power busway. Only the power busway is shown on the floor plan; the lighting busways will appear on the lighting plan.

Figure 33 shows two runs of busways: one running the length of the building on the south end (top wall on drawing), and one running the length of the north wall. The symbol list in *Figure 32* shows this busway as General Electric Type DK-100. These busways are fed from the main distribution panel (circuits MDP-1 and MDP-2) through GE No. DHIBBC41 tap boxes.

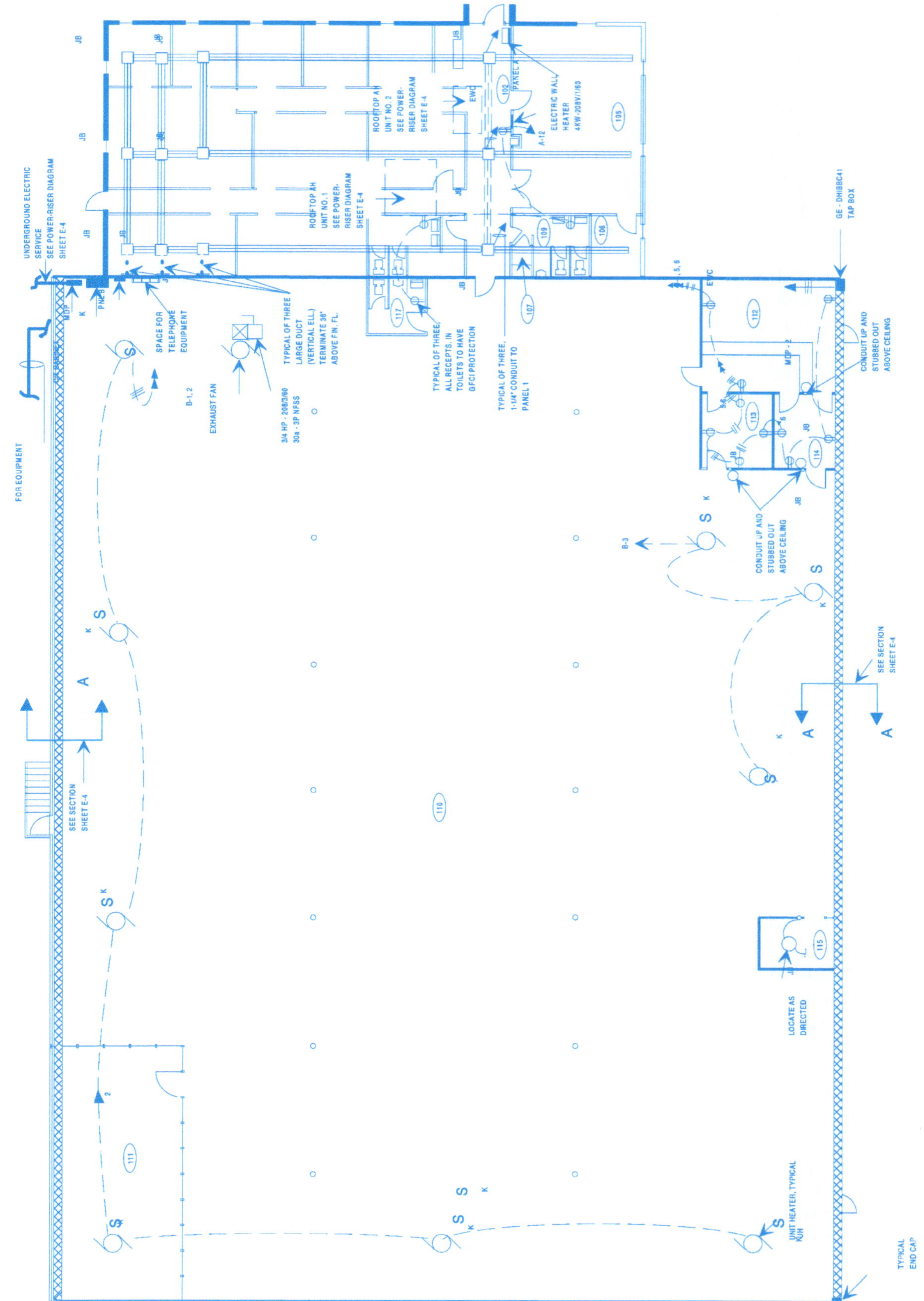

Figure 33 Floor plan for an office/warehouse building.

NEC Article 368 defines a busway as a metal enclosure containing factory-mounted, bare or insulated conductors, which are usually copper or aluminum bars, rods, or tubes.

The relationship of the busway and hangers to the building construction should be checked prior to commencing the installation so that any problems due to space conflicts, inadequate or inappropriate supporting structure, openings through walls, etc., are worked out in advance so as not to incur lost time.

For example, the drawings and specifications may call for the busway to be suspended from brackets clamped or welded to steel columns. However, the spacing of the columns may be such that additional supplementary hanger rods suspended from the ceiling or roof structure may be necessary for the adequate support of the busway. To offer more assistance to workers on the office/warehouse project, the engineer may also provide an additional drawing that shows how the busway is to be mounted.

There are also several notes appearing at various places on the floor plan. These notes offer additional information to clarify certain aspects of the drawing. For example, only one electric heater is to be installed by the electrical contractor; this heater is located in the building's vestibule. Rather than have a symbol in the symbol list for this one heater, a note is used to identify it on the drawing. Other notes on this drawing describe how certain parts of the system are to be installed. In the office area of the drawing (rooms 112, 113, and 114), you will see the following note: *"CONDUIT UP AND STUBBED OUT ABOVE CEILING"*. This empty conduit is for telephone/communications cables that will be installed later by the telephone company.

Other details include the general arrangement of the underfloor duct system, junction boxes and feeder conduit for the underfloor duct system, and plan views of the service and telephone equipment, along with duplex receptacle outlets. A note on the drawing requires all receptacles in the toilets to be provided with ground fault circuit interrupter (GFCI) protection.

A section of the drawing in *Figure 33* is shown in an enlarged view with additional details in *Figure 34*. The letters EWC next to the receptacle in the vestibule designate this receptacle for use with an electric water cooler. Notice the numbers placed inside an oval symbol in each room.

These numbered ovals represent the room name or type and correspond to a room schedule in the architectural drawings. For example, on the room schedule for the floor plan in *Figure 33* (this room schedule is not provided), room number 112 is designated as the lobby, room number 113 is designated as office No. 1, and so on. On some drawings, these room symbols are omitted and the room names are written out on the drawings.

When viewing an electrical drawing, outlets are indicated by symbols (usually a small circle with appropriate markings to indicate the type of outlet). The receptacles in the sample building are duplex receptacles, as shown in *Figure 35*.

3.2.4 Branch Circuit Drawings

The point at which electrical equipment is connected to the wiring system is commonly called an outlet. There are many classifications of outlets: lighting, receptacle, motor, appliance, and so forth. This section, however, deals with the power outlets normally found in residential electrical wiring systems.

In the past, with the exception of very large residences and tract-development houses, the size of the average residential electrical system was not large enough to justify the expense of preparing complete electrical working drawings and specifications. Such electrical systems were either laid out by the architect in the form of a sketchy outlet arrangement, or laid out by the electrician on the job as the work progressed. However, many technical developments in residential electrical use—such as electric heat with sophisticated control wiring, increased use of electrical appliances, various electronic alarm systems, new lighting techniques, and the need for energy conservation techniques—have greatly expanded the demand and extended the complexity of today's residential electrical systems.

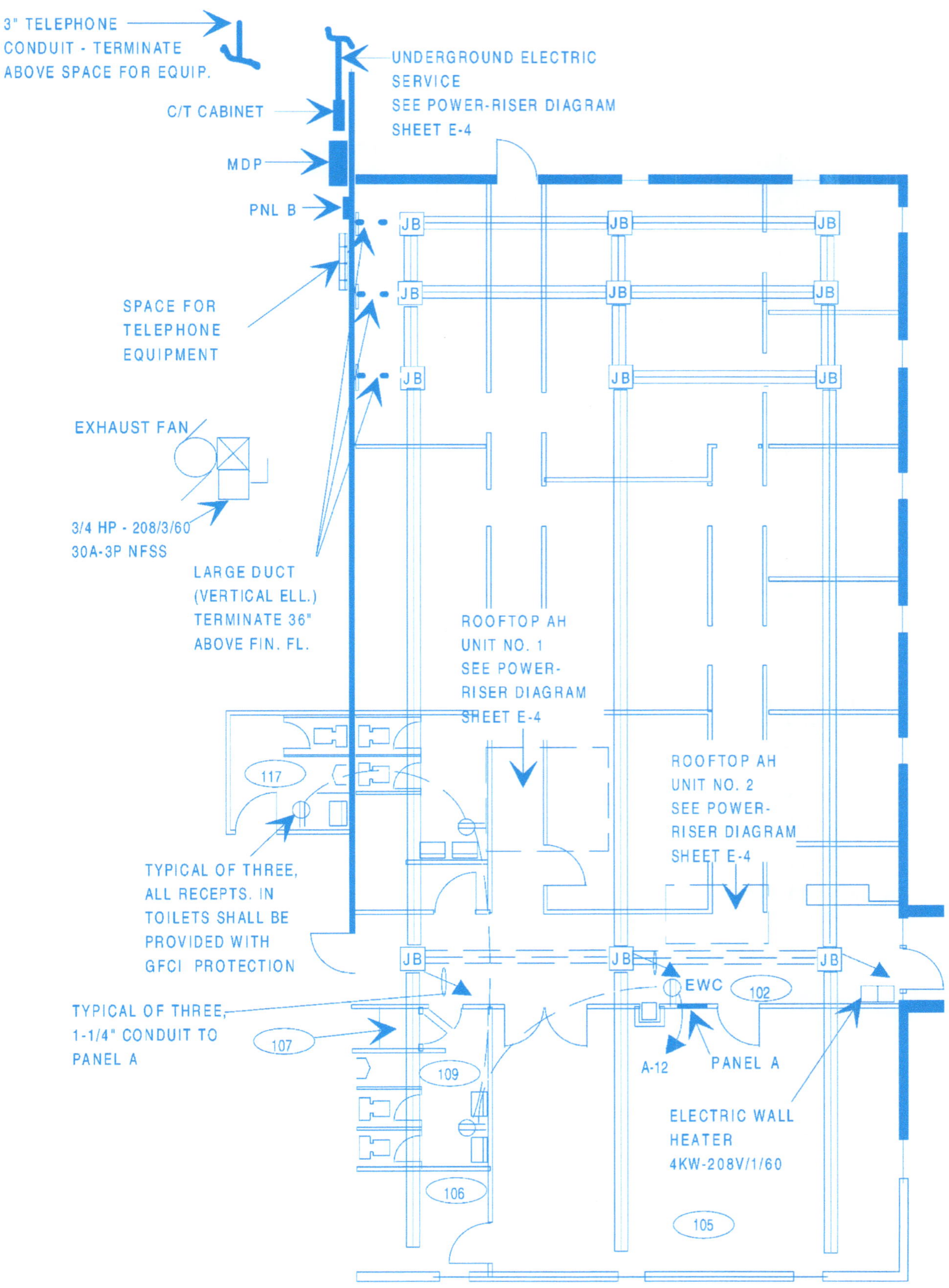

Figure 34 Partial floor plan for office/warehouse building.

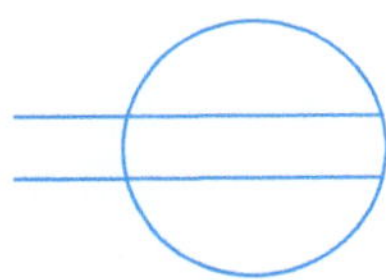

Figure 35   Duplex receptacle symbol.

Each year, the number of homes with electrical systems designed by consulting engineering firms increases. Such homes are provided with complete electrical working drawings and specifications, similar to those frequently provided for commercial and industrial projects. Still, these are more the exception than the rule. Most residential projects will not have a complete set of drawings.

Circuit layout is provided on the drawings to follow for several reasons:

- They provide a visual layout of house wiring circuitry.
- They provide a sample of electrical residential drawings that are prepared by consulting engineering firms, although the number may still be limited.
- They introduce the method of showing electrical systems on working drawings to provide a foundation for tackling advanced electrical systems.

Branch circuits are shown on electrical drawings by means of a single line drawn from the panelboard (or by homerun arrowheads indicating that the circuit goes to the panelboard) to the outlet, or from outlet to outlet where there is more than one outlet on the circuit.

The lines indicating branch circuits can be solid to show that the conductors are to be run concealed in the ceiling or wall; dashed to show that the conductors are to be run in the floor or ceiling below; or dotted to show that the wiring is to be run exposed. *Figure 36* shows examples of these three types of branch circuit lines.

In *Figure 36*, No. 12 indicates the wire size. The slash marks shown through the circuits in *Figure 36* indicate the number of current-carrying conductors in the circuit. Although two slash marks are shown for the current-carrying conductors (along with one slash mark for the ground), in actual practice, a branch circuit containing only two conductors usually contains no slash marks; that is, any circuit with no slash marks is assumed to have two conductors. However, three or more conductors are always indicated on electrical working drawings—either by slash marks for each conductor, or else by a note.

NEC Section 210.52(A) states the minimum requirements for the location of receptacles in dwelling units. It specifies that in each kitchen, family room, and dining room, receptacle outlets shall be installed so that no point along the floor line in any wall space is more than 6' (1.8 m), measured horizontally, from an outlet in that space, including any wall space 2' (600 mm) or more in width and the wall space occupied by fixed panels in walls, but excluding sliding panels. This means that the outlets will be no more than 12' (3.7 m) apart. When spaced in this manner, a 6' (1.8 m) extension cord will reach a receptacle at any point along the wall line. Receptacle outlets shall, insofar as practicable, be spaced equal distances apart. Receptacle outlets in floors shall not be counted as part of the required number of receptacle outlets unless located within 18" (450 mm) of the wall.

NEC Section 210.52(A)(2) defines wall space as a wall that is unbroken along the floor line by doorways, fireplaces, or similar openings and fixed cabinets that do not have countertops or similar work surfaces. Each wall space that is 2' (600 mm) or more in width must be treated individually and separately from other wall spaces within the room.

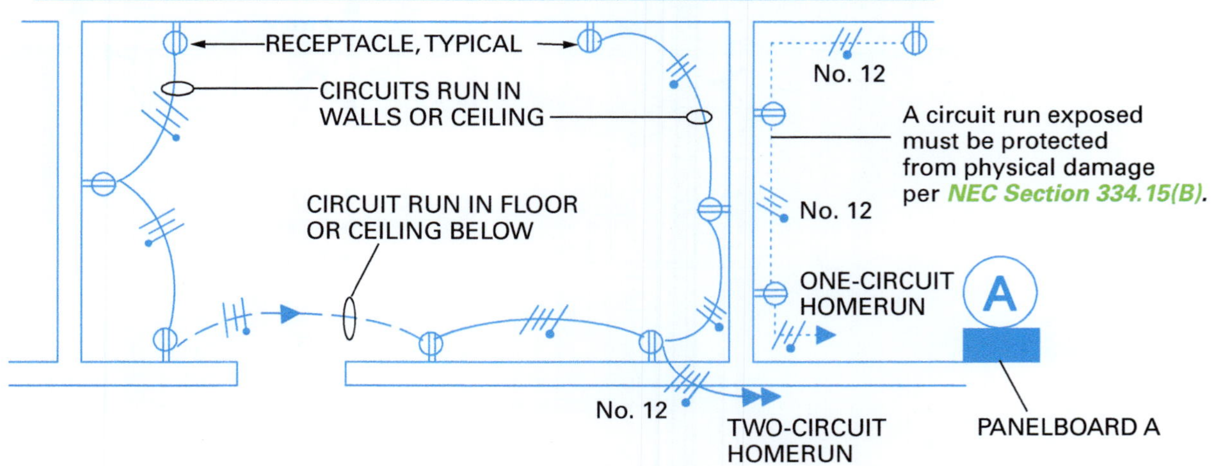

Figure 36   Types of branch circuit lines shown on electrical working drawings.

NCCER – *Maritime Electrical*

The purpose of *NEC Section 210.52(A)* is to minimize the use of cords across doorways, fireplaces, and similar openings.

Figure 37 shows the outlets for a sample residence. In laying out these receptacle outlets, the floor line of the wall is measured (also around corners), but not across doorways, fireplaces, passageways, or other spaces where a flexible cord extended across the space would be unsuitable.

3.3.0 Identifying Fixtures in a Lighting Floor Plan

A skeleton view of a lighting floor plan is shown in *Figure 38*. Again, the architect's/engineer's title blocks appear in the lower right corner of the drawing. A key plan appears above the engineer's title block. This plan is drawn to a scale of $\frac{1}{8}$" = 1'-0". Symbols for the various lighting fixtures (luminaires) are shown in the upper right corner of the drawing.

The lighting outlet symbols found on the drawing for the office/warehouse building represent both incandescent and fluorescent types; a circle on most electrical drawings usually represents an incandescent fixture, and a rectangle represents a fluorescent one. All of these symbols are designed to indicate the physical shape of a particular fixture and are usually drawn to scale.

The type of mounting used for all lighting fixtures is usually indicated in a lighting fixture schedule, which in this case is shown on the drawings. On some projects, the schedule may be found only in the written specifications.

The type of lighting fixture is identified by a numeral placed inside a triangle near each lighting fixture. If one type of fixture is used exclusively in one room or area, the triangular indicator need only appear once with the word ALL lettered at the bottom of the triangle.

3.3.1 Drawing Schedules

A schedule is a systematic method of presenting notes or lists of equipment on a drawing in tabular form. When properly organized and thoroughly understood, schedules are powerful timesaving devices for both those preparing the drawings and workers on the job.

For example, the lighting fixture schedule shown in *Figure 39* lists the fixture and identifies each fixture type on the drawing by number. The manufacturer and catalog number of each type are given along with the number, size, and type of lamp for each.

At times, all of the same information found in schedules will be duplicated in the written specifications, but combing through page after page of written specifications can be time consuming. Workers do not always have access to the specifications while on the job, whereas they usually do have access to the working drawings. Therefore, the schedule is an excellent means of providing essential information in a clear and accurate manner, allowing the workers to carry out their assignments in the least amount of time.

Other schedules that are frequently found on electrical working drawings include:

- Connected load schedule
- Panelboard schedule
- Electric heat schedule
- Kitchen equipment schedule
- Schedule of receptacle types

There are also other schedules found on electrical drawings, depending upon the type of project. However, most will deal with lists of equipment such as motors, motor controllers, and similar items.

Don't Just Check the Electrical Plan

Always review all of the drawings in a drawing set, not just the electrical plan. Several drawings in the set will have information of relevance to the electrician. For example, you should review:

- Site plans for utility lines and elevation information
- Mechanical drawings for routing, clearances, and HVAC equipment and controls
- Architectural drawings for the type of construction (block, wood, metal stud, etc.), fire ratings, and special details.
- Finish drawings (e.g., reflected ceiling plans) for locations of fixtures, fans, and other devices
- Room finish schedules for ceiling heights and floor and wall finishing details
- Plumbing drawings for pumps, water service, and sprinklers

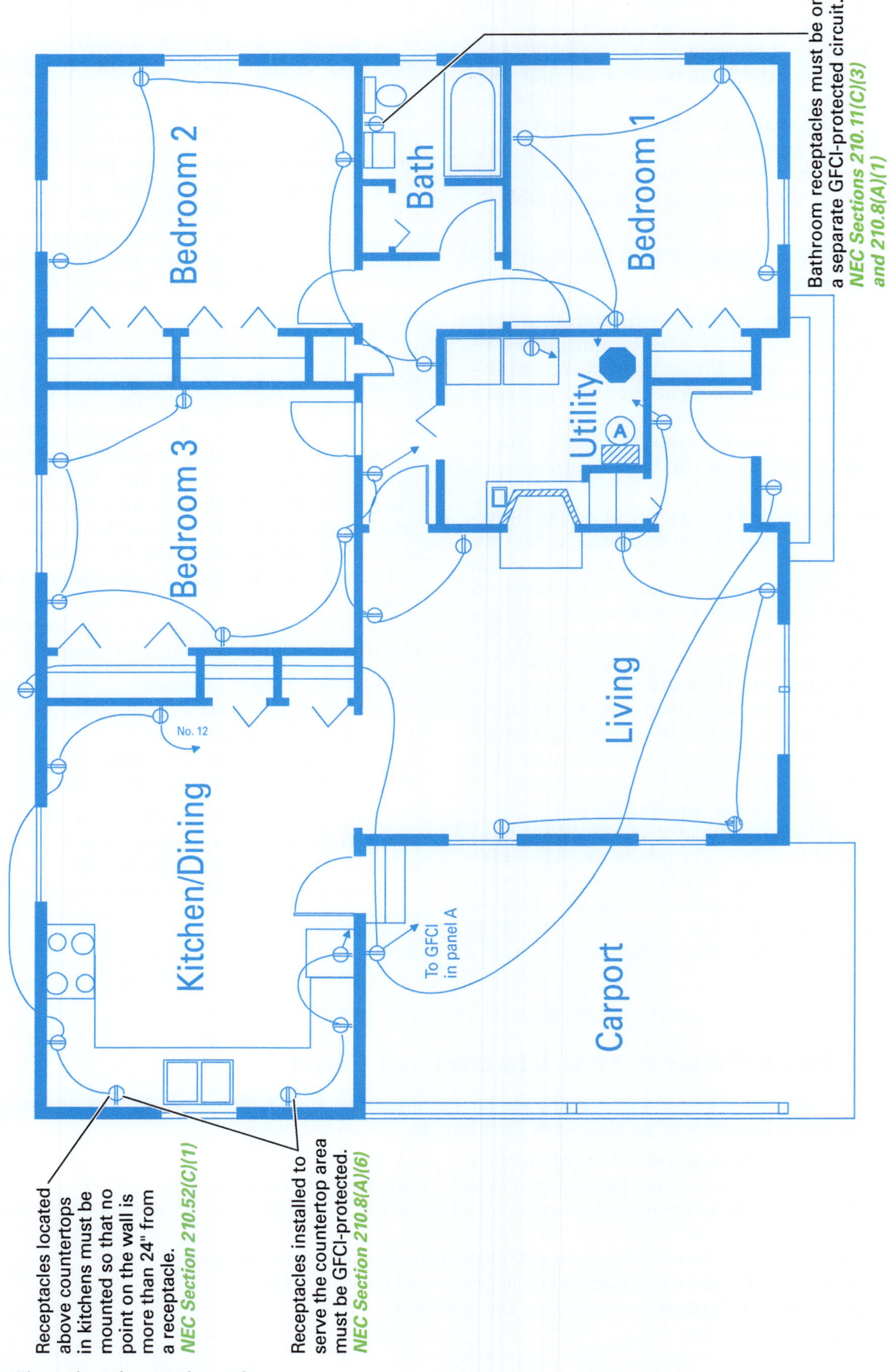

Figure 37   Floor plan of a sample residence.

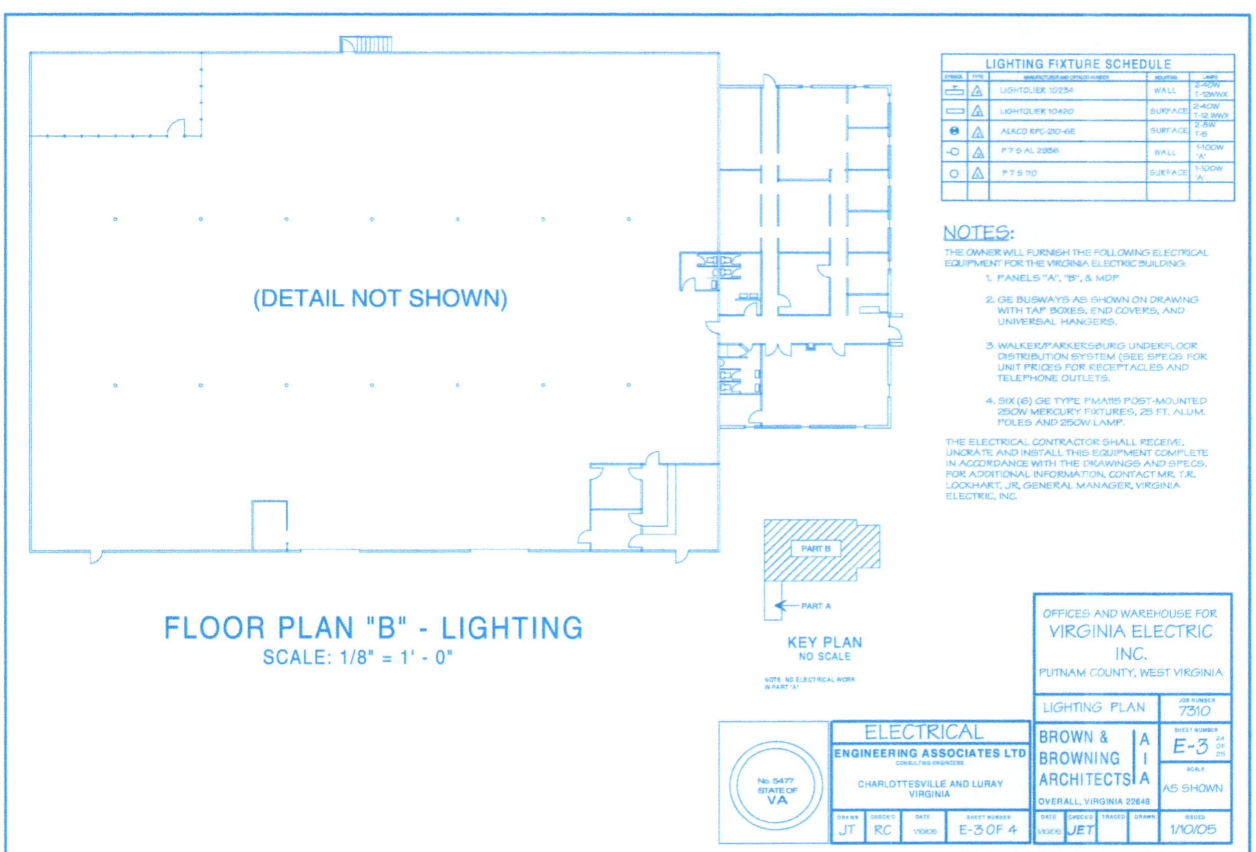

Figure 38 Sample lighting plan.

LIGHTING FIXTURE SCHEDULE

SYMBOL	TYPE	MANUFACTURER AND CATALOG NUMBER	MOUNTING	LAMPS
	A	LIGHTOLIER 10234	WALL	2-40W T-12 WWX
	B	LIGHTOLIER 10420	SURFACE	2-40W T-12 WWX
	C	ALKCO RPC-210-6E	SURFACE	2-8W T-5
	D	P 7 S AL 2936	WALL	1-100W 'A'
	E	P 7 S 110	SURFACE	1-100W 'A'

Figure 39 Lighting fixture (luminaire) schedule.

3.3.2 Branch Circuit Layout for Lighting

A simple lighting branch circuit requires two conductors to provide a continuous path for current flow. The usual lighting branch circuit operates at either 120V or 277V; the white (grounded) circuit conductor is therefore connected to the neutral bus in the panelboard, while the black (ungrounded) circuit conductor is connected to an overcurrent protection device.

Lighting branch circuits and outlets are shown on electrical drawings by means of lines and symbols; that is, a single line is drawn from outlet to outlet and then terminated with an arrowhead to indicate a homerun to the panelboard. Several methods are used to indicate the number and size of conductors, but the most common is to indicate the number of conductors in the circuit by using slash marks through the circuit lines and then indicate the wire size by a notation adjacent to these slash marks.

The circuits used to feed residential lighting must conform to standards established by the *NEC*® as well as by local and state ordinances. Most of the lighting circuits should be calculated to include the total load. However, this is sometimes not possible because the electrician cannot be certain of the exact wattage that might be used by the homeowner. For example, an electrician may install four porcelain lampholders for the unfinished basement area, each to contain one 100-watt (100W) incandescent lamp. The homeowners may eventually replace the original lamps with others rated at 150W, or even 200W. For this reason, if the electrician initially loads the lighting circuit to full capacity, the circuit will probably become overloaded in the future.

It is recommended that no residential branch circuit be loaded to more than 80% of its rated capacity. Since most circuits used for lighting are rated at 15A, the total ampacity (in volt-amperes) for the circuit is as follows:

$$15A \times 120V = 1,800VA$$

Therefore, if the circuit is to be loaded to only 80% of its rated capacity, the maximum initial connected load should be no more than 1,440VA, as shown by the calculation below:

$$80\% \text{ of } 1,800VA =$$
$$0.8 \times 1,800VA =$$
$$1,440VA$$

Figure 40 shows one possible lighting arrangement for the sample residence discussed earlier. All lighting fixtures are shown in their approximate physical location as they should be installed.

Electrical symbols are used to show the fixture types. Switches and lighting branch circuits are also shown by appropriate lines and symbols. The meanings of the symbols used on this drawing are explained in the symbol list in *Figure 41*.

In actual practice, the location of lighting fixtures and their related switches will probably be the extent of the information shown on working drawings. The circuits shown in *Figure 40* are meant to illustrate how lighting circuits are routed, not to imply that such drawings are typical for residential construction. If incandescent fixtures are used in a closet, they must meet the requirements of **NEC Section 410.16** and be completely enclosed.

3.4.0 Block and Schematic Diagrams

Electrical diagrams are drawings that are intended to show electrical components and their related connections. They show the electrical association of the different components, but are seldom, if ever, drawn to scale.

3.4.1 Block Diagrams

A single-line **block diagram**, also called a **one-line diagram**, is typically used to show the arrangement of electric service equipment. A **power-riser diagram** (*Figure 42*) is a common example of such drawings. Power-riser diagrams show all pieces of electrical equipment as well as the connecting lines used to indicate service-entrance conductors and feeders. Notes are used to identify the equipment, indicate the size of conduit necessary for each feeder, and show the number, size, and type of conductors in each conduit.

A panelboard schedule (*Figure 43*) is included with the power-riser diagram to indicate the exact components contained in each panelboard. This panelboard schedule is for the main distribution panel. On the actual drawings, schedules would also be shown for the other two panels (PNL A and PNL B).

In general, panelboard schedules usually indicate the panel number, type of cabinet (either flush- or surface-mounted), panel mains (ampere and voltage rating), phase (single- or three-phase),

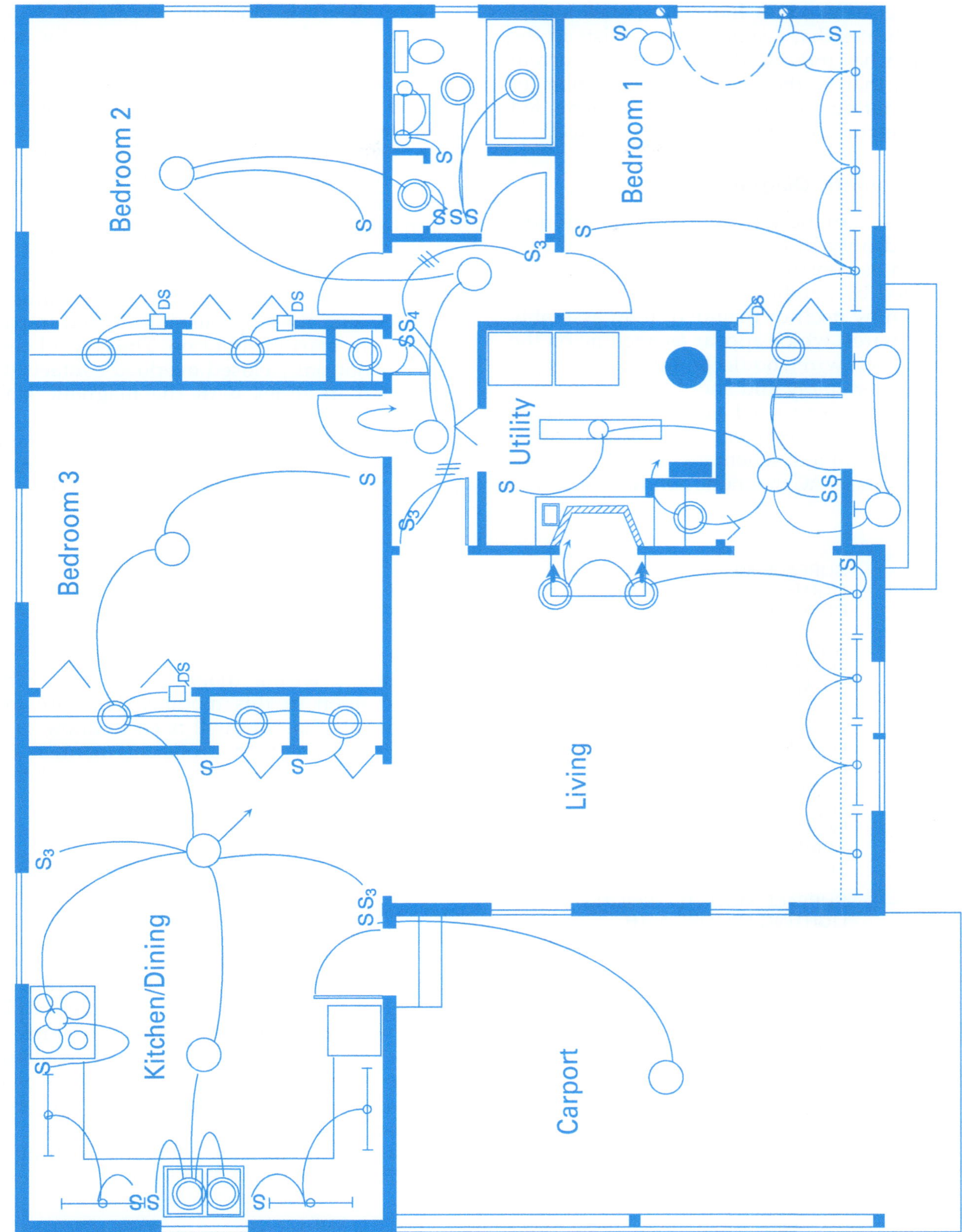

Figure 40 Lighting layout of the sample residence.

and number of wires. A four-wire panel, for example, indicates that a solid neutral exists in the panel. Branches indicate the type of overcurrent protection; that is, they indicate the number of poles, trip rating, and frame size. The items fed by each overcurrent device are also indicated.

3.4.2 Schematic Diagrams

Complete schematic wiring diagrams are normally used only in complicated electrical systems, such as control circuits. Components are represented by symbols, and every wire is either shown by itself or included in an assembly of several wires, which appear as one line on the drawing. Each wire should be numbered when it enters an assembly and should keep the same number when it comes out again to be connected to some electrical component in the system. *Figure 44* shows a complete schematic wiring diagram for a three-phase, AC magnetic non-reversing motor starter.

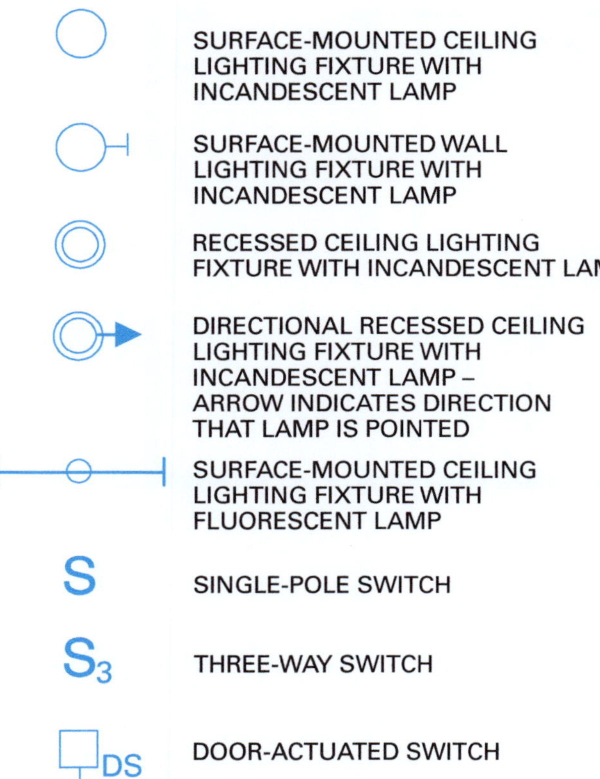

Figure 41 Electrical symbols list for the sample residence.

Note that this diagram shows the various devices in symbol form and indicates the actual connections of all wires between the devices. The three-wire supply lines are indicated by L_1, L_2, and L_3; the motor terminals of motor M are indicated by T_1, T_2, and T_3. Lines L_1, L_2, and L_3 each have a thermal overload protection device (OL) connected in series with normally open line contacts C_1 and C_3, respectively, which are both controlled by the magnetic starter coil, C. The control station, consisting of start pushbutton 1 and stop pushbutton 2, is connected across lines L_1 and L_2. Auxiliary contacts (C_4) are connected in series with the stop pushbutton and in parallel with the start pushbutton. The control circuit also has normally closed overload contacts (OC) connected in series with the magnetic starter coil (C).

Any number of additional pushbutton stations may be added to this control circuit similarly to the way in which three-way and four-way switches are added to control a lighting circuit. When adding pushbutton stations, the stop buttons are always connected in series and the start buttons are always connected in parallel. *Figure 45* shows the same motor starter circuit in *Figure 44*, but this time it is controlled by two sets of start/stop buttons.

Schematic wiring diagrams have only been touched upon in this module; there are many other details that you will need to know to perform your work in a proficient manner. Later modules cover wiring diagrams in more detail.

3.4.3 Drawing Details

A detail drawing is a drawing of a separate item or portion of an electrical system, giving a complete and exact description of its use and all the details needed to show the electrician exactly what is required for its installation. For example, the power plan for the office/warehouse has a sectional cut through the bus duct. This is a good example of where an extra, detailed drawing is desirable.

A set of electrical drawings will sometimes require large-scale drawings of certain areas that are not indicated with sufficient clarity on the small-scale drawings. For example, the site plan may show exterior pole-mounted lighting fixtures that are to be installed by the contractor.

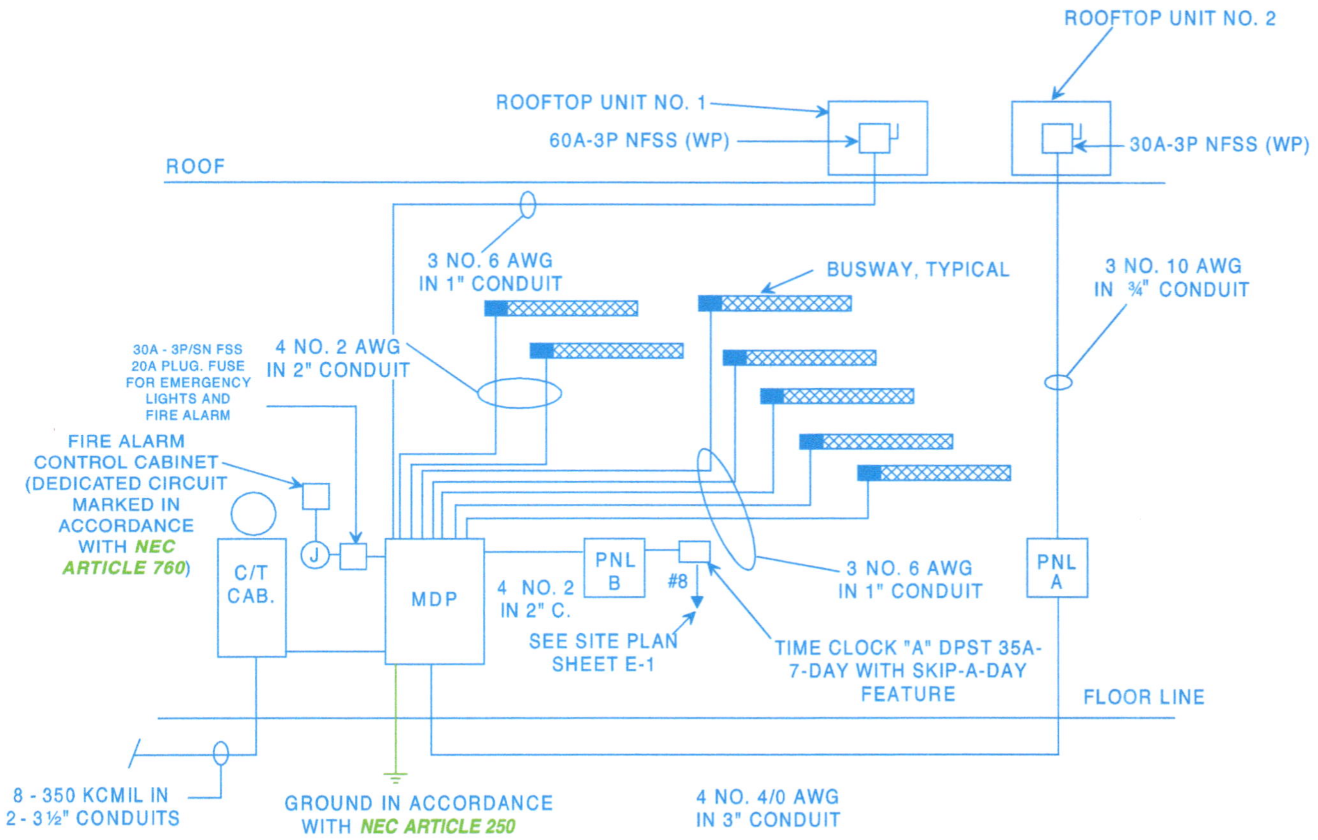

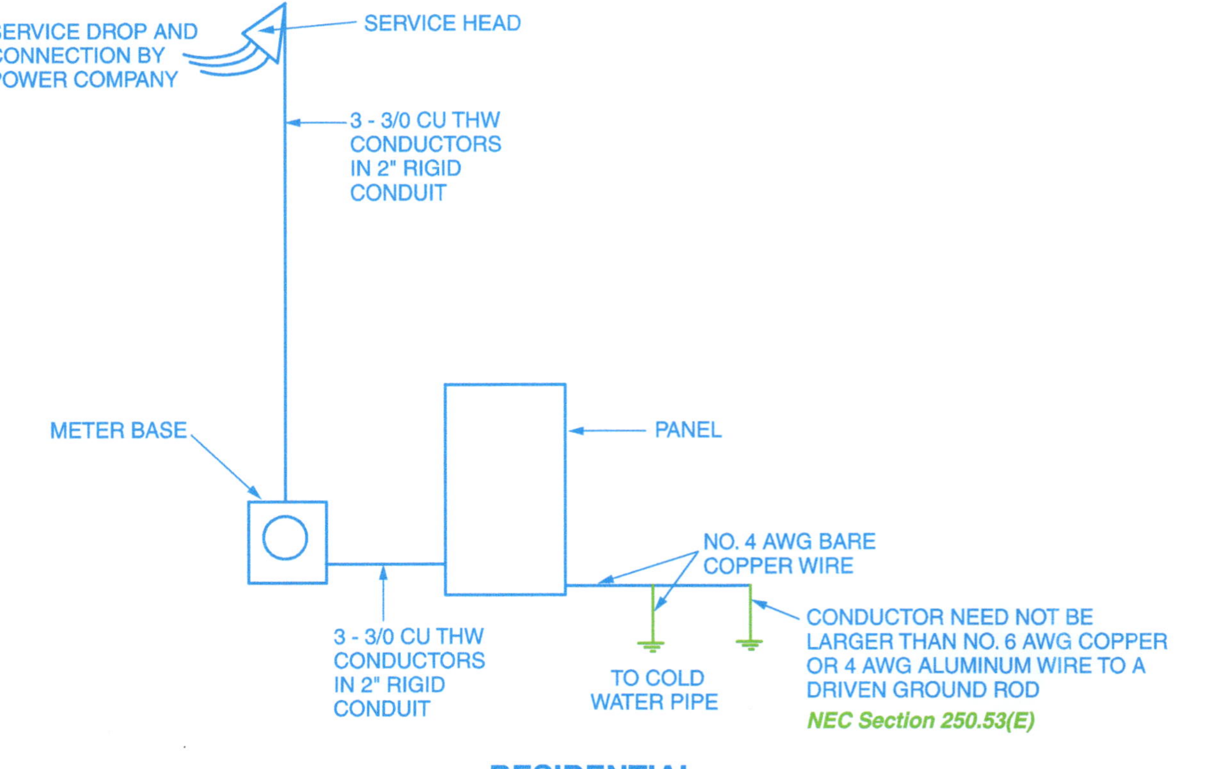

Figure 42 Typical power-riser diagrams.

PANELBOARD SCHEDULE

PANEL No.	CABINET TYPE	PANEL MAINS			BRANCHES					ITEMS FED OR REMARKS
		AMPS	VOLTS	PHASE	1P	2P	3P	PROT.	FRAME	
MDP	SURFACE	600A	120/208	3φ,4-W	–	–	1	225A	25,000	PANEL "A"
					–	–	1	100A	18,000	PANEL "B"
					–	–	1	100A		POWER BUSWAY
					–	–	1	60A		LIGHTING BUSWAY
					–	–	1	70A		ROOFTOP UNIT #1
					–	–	1	70A		SPARE
					–	–	1	600A	42,000	MAIN CIRCUIT BRKR

Figure 43 Typical panelboard schedule.

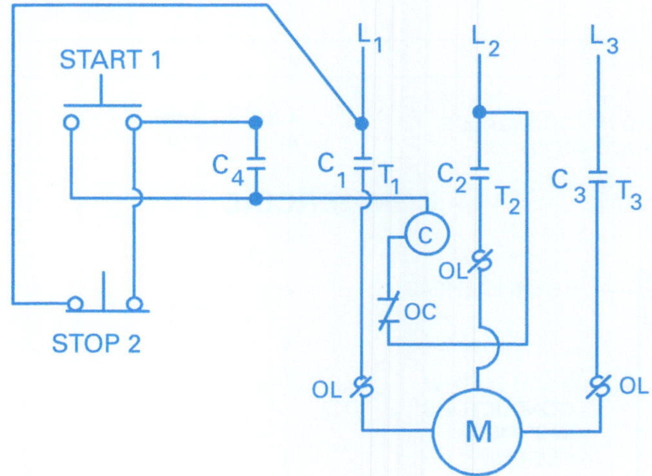

Figure 44 Wiring diagram.

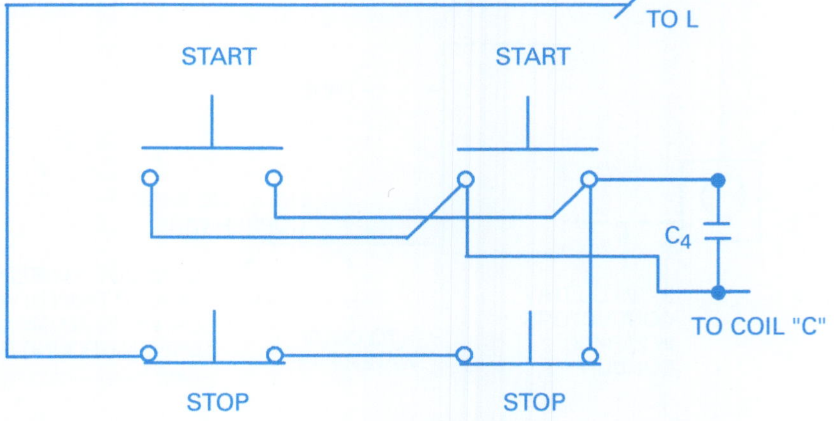

Figure 45 Circuit being controlled by two sets of start/stop buttons.

NCCER – *Maritime Electrical*

3.5.0 Written Specifications

The written specifications for a building or project are the written descriptions of work and duties required of the owner, architect, and consulting engineer. Together with the working drawings, these specifications form the basis of the contract requirements for the construction of the building or project. Those who use the construction drawings and specifications must always be alert to discrepancies between the working drawings and the written specifications. These are some situations where discrepancies may occur:

- Architects or engineers use standard or prototype specifications and attempt to apply them without any modification to specific working drawings.
- Previously prepared standard drawings are changed or amended by reference in the specifications only and the drawings themselves are not changed.
- Items are duplicated in both the drawings and specifications, but an item is subsequently amended in one and overlooked in the other contract document.

In such instances, the person in charge of the project has the responsibility to ascertain whether the drawings or the specifications take precedence. Such questions must be resolved, preferably before the work begins, to avoid added costs to the owner, architect/engineer, or contractor.

3.5.1 How Specifications are Written

Writing accurate and complete specifications for building construction is a serious responsibility for those who design the buildings because the specifications, combined with the working drawings, govern practically all important decisions that are made during the construction span of every project. Compiling and writing these specifications is not a simple task, even for those who have had considerable experience in preparing such documents. A set of written specifications for a single project will usually contain thousands of products, parts, and components, and the methods of installing them, all of which must be covered in the drawings or in the specifications, or both. No one can memorize all of the necessary items required to describe accurately the various areas of construction. One must rely upon reference materials such as manufacturer's data, catalogs, checklists, and, most of all, a high-quality master specification.

3.5.2 Format of Specifications

For convenience in writing, speed in estimating, and ease of reference, the most suitable organization of the specifications is a series of sections dealing with the construction requirements, products, and activities that is easily understandable by the different trades. Those people who use the specifications must be able to find all the information they need without spending too much time looking for it.

The most commonly used specification-writing format used in North America is the MasterFormat®. This standard was developed jointly by the Construction Specifications Institute (CSI) and Construction Specifications Canada (CSC). For many years prior to 2004, the organization of construction specifications was based on a standard with 16 sections, otherwise known as divisions. In 2004, the MasterFormat® standard was expanded to four major groupings and 49 divisions, with some divisions reserved for future expansion. MasterFormat® was again updated in 2010 (*Figure 46*). In addition, the numbering system was changed to 6 digits to allow for more subsections in each division, which allowed for finer definition.

In the new numbering system, the first two digits represent the division number. The next two digits represent subsections of the division and the two remaining digits represent the third level sub-subsection numbers. The fourth level, if required, is a decimal and number added to the end of the last two digits. For example, the number 132013.04 represents division 13, subsection 20, sub-subsection 13, and sub-sub-subsection 04. Under the new standard, the Facility Services Subgroup contains the divisions that are most important to the electrician. These include the following divisions:

- *Division 25* – Integrated Automation
- *Division 26* – Electrical
- *Division 27* – Communications
- *Division 28* – Electronic Safety and Security

Figure 47A through *Figure 47M* contain a detailed breakdown of the electrical division.

Specifications

Written specifications supplement the related working drawings in that they contain details not shown on the drawings. Specifications define and clarify the scope of the job. They describe the specific types and characteristics of the components that are to be used on the job and the methods for installing some of them. Many components are identified specifically by the manufacturer's model and part numbers. This type of information is used to purchase the various items of hardware needed to accomplish the installation in accordance with the contractual requirements.

MasterFormat GROUPS, SUBGROUPS, AND DIVISIONS

PROCUREMENT AND CONTRACTING REQUIREMENTS GROUP

Division 00 – Procurement and Contracting
  Requirements
    Introductory Information
    Procurement Requirements
    Contracting Requirements

SPECIFICATIONS GROUP

GENERAL REQUIREMENTS SUBGROUP

Division 01 – General Requirements

FACILITY CONSTRUCTION SUBGROUP

Division 02 – Existing Conditions
Division 03 – Concrete
Division 04 – Masonry
Division 05 – Metals
Division 06 – Wood, Plastics, and Composites
Division 07 – Thermal and Moisture Protection
Division 08 – Openings
Division 09 – Finishes
Division 10 – Specialties
Division 11 – Equipment
Division 12 – Furnishings
Division 13 – Special Construction
Division 14 – Conveying Equipment
Division 15 – Reserved for Future Expansion
Division 16 – Reserved for Future Expansion
Division 17 – Reserved for Future Expansion
Division 18 – Reserved for Future Expansion
Division 19 – Reserved for Future Expansion

FACILITY SERVICES SUBGROUP

Division 20 – Reserved for Future Expansion
Division 21 – Fire Suppression

Division 22 – Plumbing
Division 23 – Heating, Ventilating, and Air-Conditioning
  (HVAC)
Division 24 – Reserved for Future Expansion
Division 25 – Integrated Automation
Division 26 – Electrical
Division 27 – Communications
Division 28 – Electronic Safety and Security
Division 29 – Reserved for Future Expansion

SITE AND INFRASTRUCTURE SUBGROUP

Division 30 – Reserved for Future Expansion
Division 31 – Earthwork
Division 32 – Exterior Improvements
Division 33 – Utilities
Division 34 – Transportation
Division 35 – Waterway and Marine Construction
Division 36 – Reserved for Future Expansion
Division 37 – Reserved for Future Expansion
Division 38 – Reserved for Future Expansion
Division 39 – Reserved for Future Expansion

PROCESS EQUIPMENT SUBGROUP

Division 40 – Process Interconnections
Division 41 – Material Processing and Handling
  Equipment
Division 42 – Process Heating, Cooling, and Drying
  Equipment
Division 43 – Process Gas and Liquid Handling,
  Purification, and Storage Equipment
Division 44 – Pollution and Waste Control Equipment
Division 45 – Industry-Specific Manufacturing
  Equipment
Division 46 – Water and Wastewater Equipment
Division 47 – Reserved for Future Expansion
Division 48 – Electrical Power Generation
Division 49 – Reserved for Future Expansion

19

Introduction and Applications Guide

Figure 46   2016 MasterFormat ®.

DIVISION **26** – ELECTRICAL

26 00 00 Electrical

may be used as division level section title.

See: 02 41 19 for selective demolition of existing electrical systems.
03 30 00 for cast-in-place concrete equipment bases.
07 84 00 for firestopping.
07 92 00 for joint sealants.
08 31 00 for access doors and panels.
09 91 00 for field painting.
31 23 33 for trenching and backfilling.

---

26 01 00 Operation and Maintenance of Electrical Systems

Includes: maintenance, repair, rehabilitation, replacement, restoration, preservation, etc. of electrical systems. medium voltage: 2400 V to 69 kV. low voltage: 600 V and less.

Notes: Definitions medium voltage: 2400 V to 69 kV. low voltage: 600 V and less.

Level 4 Numbering Recommendation: following numbering is recommended for the creation of Level 4 titles:
.51-.59 for maintenance.
.61-.69 for repair.
.71-.79 for rehabilitation.
.81-.89 for replacement.
.91-.99 for restoration.

26 01 10	Operation and Maintenance of Medium-Voltage Electrical Distribution
26 01 20	Operation and Maintenance of Low-Voltage Electrical Distribution
26 01 26	Maintenance Testing of Electrical Systems
26 01 30	Operation and Maintenance of Facility Electrical Power Generating and Storing Equipment
26 01 40	Operation and Maintenance of Electrical and Cathodic Protection Systems
26 01 50	Operation and Maintenance of Lighting
26 01 50.51	Luminaire Relamping
26 01 50.81	Luminaire Replacement

26 05 00 Common Work Results for Electrical

Includes: subjects common to multiple titles in Division 26. raceway and boxes includes conduit, tubing, surface raceways, and electrical boxes. medium voltage: 2400 V to 69 kV. low voltage: 600 V and less. control voltage: 50 V

311

Figure 47A Detailed breakdown of the electrical division (Sheet 1 of 13).

NUMBER	TITLE	EXPLANATION

and less.

Alternate Terms/Abbreviations: EMT: electrical metallic tubing.

Notes: Definitions medium voltage: 2400 V to 69 kV. low voltage: 600 V and less. control voltage: 50 V and less.

See 01 80 00 for performance requirements of subjects common to multiple titles.
05 35 00 for raceway decking assemblies.
05 45 16 for electrical metal supports.
13 48 00 for sound, vibration, and seismic control.
25 05 13 for conductors and cables for integrated automation.
25 05 26 for grounding and bonding for integrated automation.
25 05 28 for pathways for integrated automation.
25 05 48 for vibration and seismic control for integrated automation.
25 05 53 for identification for integrated automation.
27 05 28 for pathways for communications systems.
27 05 46 for utility poles for communications systems.
27 05 48 for vibration and seismic controls for communications.
27 05 53 for identification for communications.
28 05 13 for conductors and cables for electronic safety and security.
28 05 26 for grounding and bonding for electronic safety and security.
28 05 28 for pathways for electronic safety and security.
28 05 48 for vibration and seismic controls for electronic safety and security.
28 05 53 for identification for electronic safety and security.
33 71 16 for electrical utility poles.
33 71 19 for electrical utility underground ducts and manholes.

NUMBER	TITLE
26 05 13	Medium-Voltage Cables
	26 05 13.13 Medium-Voltage Open Conductors
	26 05 13.16 Medium-Voltage, Single- and Multi-Conductor Cables
26 05 19	Low-Voltage Electrical Power Conductors and Cables
	26 05 19.13 Undercarpet Electrical Power Cables
	26 05 19.23 Manufactured Wiring Assemblies
26 05 23	Control-Voltage Electrical Power Cables
26 05 26	Grounding and Bonding for Electrical Systems
26 05 29	Hangers and Supports for Electrical Systems
26 05 33	Raceway and Boxes for Electrical Systems
	26 05 33.13 Conduit for Electrical Systems
	26 05 33.16 Boxes for Electrical Systems
	26 05 33.23 Surface raceways for Electrical Systems

312

Figure 47B   Detailed breakdown of the electrical division (Sheet 2 of 13).

NUMBER	TITLE	EXPLANATION

26 05 36   Cable Trays for Electrical Systems
26 05 39   Underfloor Raceways for Electrical Systems
26 05 43   Underground Ducts and Raceways for Electrical Systems
26 05 46   Utility Poles for Electrical Systems
26 05 48   Vibration and Seismic Controls for Electrical Systems
26 05 53   Identification for Electrical Systems
26 05 73   Overcurrent Protective Device Coordination Study
26 05 83   Wiring Connections

26 06 00   Schedules for Electrical

Notes: a schedule may be included on drawings, in the project manual, or a project book.

Definitions: medium voltage: 2400 V to 69 kV. low voltage: 600 V and less.

Includes: schedules of items common to multiple titles in Division 26.

26 06 10   Schedules for Medium-Voltage Electrical Distribution
26 06 20   Schedules for Low-Voltage Electrical Distribution
       26 06 20.13   Electrical Switchboard Schedule
       26 06 20.16   Electrical Panelboard Schedule
       26 06 20.19   Electrical Motor-Control Center Schedule
       26 06 20.23   Electrical Circuit Schedule
       26 06 20.26   Wiring Device Schedule
26 06 30   Schedules for Facility Electrical Power Generating and Storing Equipment
26 06 40   Schedules for Electrical and Cathodic Protection Systems
26 06 50   Schedules for Lighting
       26 06 50.13   Lighting Panelboard Schedule
       26 06 50.16   Lighting Fixture Schedule

26 08 00   Commissioning of Electrical Systems

Includes: commissioning of items common to multiple titles in Division 26.

See: 01 91 00 for commissioning of subjects common to multiple divisions.

26 09 00   Instrumentation and Control for Electrical Systems

Includes: instrumentation and control associated with electrical systems.

See: 13 50 00 for special instrumentation.
25 36 00 for integrated automation instrumentation and terminal devices for electrical systems.
25 56 00 for integrated automation control of electrical systems.

313

Figure 47C   Detailed breakdown of the electrical division (Sheet 3 of 13).

25 96 00 for integrated automation control sequences for electrical systems.
33 09 70 for instrumentation and control for electrical utilities.

26 09 13	Electrical Power Monitoring
26 09 15	Peak Load Controllers
26 09 16	Electrical Controls and Relays
26 09 17	Programmable Controllers
26 09 19	Enclosed Contactors
26 09 23	Lighting Control Devices

Includes: clock and calendar, photoelectric switches, occupancy sensors, and light-leveling control devices. control- and low-voltage lighting control devices connected through computers. addressable lighting control devices and lighting components (ballasts) connected through computers.

See: 11 61 00 for theater and stage equipment.
26 50 00 for lighting.
26 55 61 for theatrical lighting.

See Also: 11 61 00 for theatrical lighting controls.

26 09 26	Lighting Control Panelboards
26 09 33	Central Dimming Controls
	26 09 33.13	Multichannel Remote-Controlled Dimmers
	26 09 33.16	Remote-Controlled Dimming Stations
26 09 36	Modular Dimming Controls
	26 09 36.13	Manual Modular Dimming Controls
	26 09 36.16	Integrated Multipreset Modular Dimming Controls
26 09 43	Network Lighting Controls
	26 09 43.13	Digital-Network Lighting Controls
	26 09 43.16	Addressable Fixture Lighting Control
26 09 61	Theatrical Lighting Controls

26 10 00 Medium-Voltage Electrical Distribution

Includes: substations, transformers, switchgear, and circuit protection devices to distribute medium-voltage electrical power from the facility service point to the point of delivery.

Notes: Definitions medium voltage: 2400 V to 69 kV.

See 26 05 13 for medium-voltage cables.
26 20 00 for low-voltage electrical distribution.
26 30 00 for facility electrical power generating and storing equipment.
33 71 00 for electrical utility distribution.

314

Figure 47D   Detailed breakdown of the electrical division (Sheet 4 of 13).

NUMBER	TITLE	EXPLANATION

26 11 00   Substations

Includes: assembly of switches, circuit breakers, buses, and transformers to switch circuits and convert power from one voltage to another.

See: 33 72 00 for utility substations.
34 21 16 for traction power substations.

26 11 13   Primary Unit Substations
26 11 16   Secondary Unit Substations

26 12 00   Medium-Voltage Transformers

Includes: transformers for medium-voltage applications.

See: 26 22 00 for low-voltage transformers.
33 73 00 for utility transformers.
34 21 23 for traction power transformer-rectifier units.

26 12 13   Liquid-Filled, Medium-Voltage Transformers
26 12 16   Dry-Type, Medium-Voltage Transformers
26 12 19   Pad-Mounted, Liquid-Filled, Medium-Voltage Transformers

26 13 00   Medium-Voltage Switchgear

Includes: switchgear for medium-voltage applications.

See: 26 23 00 for low-voltage switchgear.
33 77 00 for medium-voltage utility switchgear.
34 21 19 for traction power switchgear.

26 13 13   Medium-Voltage Circuit Breaker Switchgear
26 13 16   Medium-Voltage Fusible Interrupter Switchgear
26 13 19   Medium-Voltage Vacuum Interrupter Switchgear
26 13 23   Medium-Voltage Metal-Enclosed Switchgear
26 13 26   Medium-Voltage Metal-Clad Switchgear
26 13 29   Medium-Voltage Compartmentalized Switchgear

26 16 00   Medium-Voltage Metering

26 18 00   Medium-Voltage Circuit Protection Devices

Includes: circuit protection devices for medium-voltage applications.

See: 26 28 00 for low-voltage circuit protective devices.
26 41 23 for lightning protection surge arresters and suppressors.
33 77 00 for medium-voltage utility circuit protection devices.

26 18 13   Medium-Voltage Cutouts

315

Figure 47E   Detailed breakdown of the electrical division (Sheet 5 of 13).

NUMBER	TITLE	EXPLANATION
26 18 16	Medium-Voltage Fuses	
26 18 19	Medium-Voltage Lightning Arresters	
26 18 23	Medium-Voltage Surge Arresters	
26 18 26	Medium-Voltage Reclosers	
26 18 29	Medium-Voltage Enclosed Bus	
26 18 33	Medium-Voltage Enclosed Fuse Cutouts	
26 18 36	Medium-Voltage Enclosed Fuses	
26 18 39	Medium-Voltage Motor Controllers	

26 20 00 Low-Voltage Electrical Transmission

Includes: overhead power systems, transformers, switchgear, switchboards, panelboards, enclosed bus assemblies, power distribution units, controllers, wiring devices, and circuit protection devices to distribute low-voltage electrical power from the point of voltage transformation to the point of use. typical voltages: 120, 208, 230, 240, 277, 460, and 480.

Notes: Definitions low voltage: 600 V and less.

See 26 05 19 for low-voltage electrical power conductors and cables.
26 10 00 for medium-voltage electrical distribution.
26 30 00 for facility electrical power generating and storing equipment.

26 21 00 Low-Voltage Electrical Service Entrance

See: 26 05 19 for low-voltage electrical power conductors and cables.
26 05 46 for utility poles for electrical systems.
33 71 13 for site electrical transmission towers.
33 71 16 for electrical utility poles.

| 26 21 13 | Low-Voltage Overhead Electrical Service Entrance |
| 26 21 16 | Low-Voltage Underground Electrical Service Entrance |

26 22 00 Low-Voltage Transformers

Includes: transformers for low-voltage applications.

See: 26 12 00 for medium-voltage transformers.

26 22 13	Low-Voltage Distribution Transformers
26 22 16	Low-Voltage Buck-Boost Transformers
26 22 19	Control and Signal Transformers

26 23 00 Low-Voltage Switchgear

Includes: switchgear for low-voltage applications.

316

Figure 47F Detailed breakdown of the electrical division (Sheet 6 of 13).

Number	Title	Explanation

See: 26 13 00 for medium-voltage switchgear.

26 23 13   Paralleling Low-Voltage Switchgear

26 24 00   Switchboards and Panelboards

Includes: switchboards, panelboards, and control centers.

See: 26 27 16 for electrical cabinets and enclosures.
26 29 13 for enclosed controllers.
26 29 23 for variable-frequency motor controllers.

26 24 13   Switchboards
26 24 16   Panelboards
26 24 19   Motor-Control Centers

26 25 00   Enclosed Bus Assemblies

Includes: busway, step bus, and tap boxes.

See: 33 72 26 for utility substation bus assemblies.

26 26 00   Power Distribution Units

Includes: distribution units with integral transformers, panelboards, and power conditioning components.

See: 26 24 16 for panelboards.

26 27 00   Low-Voltage Distribution Equipment

Includes: wiring devices includes receptacles, switches, dimmers, and finish plates.

See: 26 24 00 for switchboards and panelboards.
33 71 73 for utility electric meters.

26 27 13   Electricity Metering
26 27 16   Electrical Cabinets and Enclosures
26 27 19   Multi-Outlet Assemblies
26 27 23   Indoor Service Poles
26 27 26   Wiring Devices
26 27 73   Door Chimes

26 28 00   Low-Voltage Circuit Protective Devices

Includes: circuit protection devices for low-voltage applications. enclosed switches and transfer switches.

See: 26 18 00 for medium-voltage circuit protection devices.

317

Figure 47G   Detailed breakdown of the electrical division (Sheet 7 of 13).

26 28 13   Fuses
26 28 16   Enclosed Switches and Circuit Breakers
          26 28 16.13   Enclosed Circuit Breakers
          26 28 16.16   Enclosed Switches

26 29 00   Low-Voltage Controllers

Includes: contactors and motor controllers.

May Include: fuses.

Alternate Terms/Abbreviations: enclosed controllers: motor controllers.

See: 26 24 19 for motor-control centers.
26 28 13 for fuses.

26 29 13   Enclosed Controllers
          26 29 13.13   Across-the-Line Motor Controllers
          26 29 13.16   Reduced-Voltage Motor Controllers
26 29 23   Variable-Frequency Motor Controllers
26 29 33   Controllers for Fire Pump Drivers
          26 29 33.13   Full-Service Controllers for Fire Pump Electric-Motor Drivers
          26 29 33.16   Limited-Service Controllers for Fire Pump Electric-Motor Drivers
          26 29 33.19   Controllers for Fire Pump Diesel Engine Drivers

26 30 00 Facility Electrical Power Generating and Storing Equipment

Includes: equipment to generate and store electrical power for a single facility.

Notes: 48 10 00 for electrical power generation equipment.

26 31 00   Photovoltaic Collectors

Includes: solar cells to convert sunlight to electricity.

See: 07 31 00 for solar collector roof shingles.
22 33 30 for residential, collector-to-tank, solar-electric domestic water heaters.
23 56 00 for solar energy heating equipment.
42 12 23 for solar process heaters.
42 13 26 for industrial solar radiation heat exchangers.
48 14 00 for solar energy electrical power generation equipment.

26 32 00   Packaged Generator Assemblies

Includes: generators, frequency changers, and rotary converters and uninterruptible power units.

318

Figure 47H   Detailed breakdown of the electrical division (Sheet 8 of 13).

NUMBER	TITLE	EXPLANATION

See: 23 11 00 for facility fuel piping.
23 24 00 for internal-combustion engine piping.
48 11 00 for fossil fuel plant electrical power generation equipment.
48 13 00 for hydroelectric plant electrical power generation equipment.
48 15 00 for wind energy electrical power generation equipment.

26 32 13 — Engine Generators
- 26 32 13.13 — Diesel-Engine-Driven Generator Sets
- 26 32 13.16 — Gas-Engine-Driven Generator Sets
- 26 32 13.26 — Gas-Turbine Engine-Driven Generators

Alternate Terms/Abbreviations: microturbines

See: 48 11 23 Fossil Fuel Electrical Power Plant Gas Turbines

26 32 16 — Steam-Turbine Generators
26 32 19 — Hydro-Turbine Generators
26 32 23 — Wind Energy Equipment
26 32 26 — Frequency Changers
26 32 29 — Rotary Converters
26 32 33 — Rotary Uninterruptible Power Units

26 33 00 Battery Equipment

Includes: batteries, battery racks, battery chargers, static power converters, uninterruptible power supplies, and accessories.

May Include: battery-operated emergency light fixtures.

See: 25 36 23 for integrated automation battery monitors.
26 31 00 for photovoltaic collectors.
33 72 33 for electrical utility substation.
48 17 13 for electrical power generation batteries.

See Also: 26 52 00 for emergency lighting incorporating batteries.

26 33 13 — Batteries
26 33 16 — Battery Racks
26 33 19 — Battery Units
26 33 23 — Central Battery Equipment
26 33 33 — Static Power Converters
26 33 43 — Battery Chargers
26 33 46 — Battery Monitoring

319

Figure 47I Detailed breakdown of the electrical division (Sheet 9 of 13).

NUMBER	TITLE	EXPLANATION

26 33 53   Static Uninterruptible Power Supply

26 35 00   Power Filters and Conditioners

Includes: capacitors, chokes and inductors, filters, power factor controllers, and voltage regulators.

Alternate Terms/Abbreviations: EMI: electromagnetic interference. RFI: radio frequency interference. power factor correction equipment: power factor controllers.

See: 08 34 46 for RFI shielding doors.
08 56 46 for RFI shielding windows.
13 49 00 for radiation protection.
26 18 23 for medium-voltage surge arresters.
28 32 00 for radiation detection and alarm.
40 91 16 for electromagnetic process measurement devices.

26 35 13   Capacitors
26 35 16   Chokes and Inductors
26 35 23   Electromagnetic-Interference Filters
26 35 26   Harmonic Filters
26 35 33   Power Factor Correction Equipment
26 35 36   Slip Controllers
26 35 43   Static-Frequency Converters
26 35 46   Radio-Frequency-Interference Filters
26 35 53   Voltage Regulators

26 36 00   Transfer Switches

Includes: switches transfer from one source of electricity to another.

26 36 13   Manual Transfer Switches
26 36 23   Automatic Transfer Switches

26 40 00 Electrical and Cathodic Protection

26 41 00   Facility Lightning Protection

Includes: wiring and equipment for lightning protection.

See: 26 18 19 for medium-voltage lightning arresters.
33 79 00 for site grounding.
33 79 93 for site lightning protection.

26 41 13   Lightning Protection for Structures
          26 41 13.13   Lightning Protection for Buildings
26 41 16   Lightning Prevention and Dissipation
26 41 19   Early Streamer Emission Lightning Protection
26 41 23   Lightning Protection Surge Arresters and Suppressors

320

Figure 47J   Detailed breakdown of the electrical division (Sheet 10 of 13).

NUMBER	TITLE	EXPLANATION

26 42 00   Cathodic Protection

Includes: equipment, controls, and installation for cathodic protection of structures and underground metal construction and piping.

See: 40 46 42 for cathodic process corrosion protection.

26 42 13   Passive Cathodic Protection for Underground and Submerged Piping
26 42 16   Passive Cathodic Protection for Underground Storage Tank

26 43 00   Transient Voltage Suppression

Includes: devices to protect against voltage surges on electrical distribution systems.

26 43 13   Transient-Voltage Suppression for Low-Voltage Electrical Power Circuits

26 50 00 Lighting

Includes: luminaries, lighting equipment, ballasts, dimming controls, and lighting accessories. fluorescent, high intensity discharge, incandescent, mercury vapor, neon, and sodium vapor lighting.

Alternate Terms/Abbreviations: HID: high intensity discharge.

See: 10 84 00 for Gas Lighting.
25 36 26 for integrated automation lighting relays.
26 09 23 for lighting controls.
26 20 00 for low-voltage electrical transmission.

26 51 00   Interior Lighting

Includes: lighting for interior locations, except for emergency lighting, lighting in hazardous locations, and special purpose lighting. chandeliers, troffers.

See: 09 54 16 for luminous ceilings.
09 58 00 for integrated ceiling assemblies.
10 14 33 for illuminated panel signage.

26 51 13   Interior Lighting Fixtures, Lamps, And Ballasts

26 52 00   Emergency Lighting

Includes: equipment for exitway lighting and other emergency applications, including emergency battery units, fixtures with integral batter power supplies.

See: 26 53 00 for exit signs.

321

Figure 47K   Detailed breakdown of the electrical division (Sheet 11 of 13).

DIVISION **26**

NUMBER	TITLE	EXPLANATION
26 53 00	**Exit Signs**	Includes: electric exit signs. See: 26 52 00 for emergency lighting.
26 54 00	**Classified Location Lighting**	Includes: lighting for application in areas classified as hazardous. See: 26 55 33 for hazard warning lighting.
26 55 00	**Special Purpose Lighting**	Includes: lighting equipment for specialized applications. Alternate Terms/Abbreviations: healthcare lighting: medical lighting. See: 11 13 26 for loading dock lights. 11 18 00 for security equipment. 11 19 00 for detention equipment. 11 59 00 for exhibit and display equipment. 11 61 00 for theater and stage equipment. 11 70 00 for healthcare equipment. 13 10 00 for swimming pools. 13 12 00 for fountains. 13 14 00 for aquatic park structures. 13 17 00 for tubs and pools. 26 54 00 for classified location lighting. 34 40 00 for transportation signals. 35 13 13 for navigation signals. See Also: 11 61 00 for theatrical lighting.
26 55 23	Outline Lighting	
26 55 29	Underwater Lighting	
26 55 33	Hazard Warning Lighting	
26 55 36	Obstruction Lighting	
26 55 39	Helipad Lighting	See: 34 43 00 Airfield Signaling and Control Equipment
26 55 53	Security Lighting	
26 55 59	Display Lighting	
26 55 61	Theatrical Lighting	
26 55 63	Detention Lighting	
26 55 70	Healthcare Lighting	

322

Figure 47L Detailed breakdown of the electrical division (Sheet 12 of 13).

NUMBER	TITLE	EXPLANATION

26 56 00   **Exterior Lighting**

Includes: lighting equipment for exterior locations, except for special purpose and signal lighting. airfield general exterior lighting.

Alternate Terms/Abbreviations: athletic lighting: sports lighting.

See: 10 14 33 for illuminated panel signage.
11 13 26 for loading dock lights.
11 68 23 for exterior court athletic equipment.
32 94 00 for planting accessories.
34 41 13 for traffic signals.
34 42 13 for railway signals.
34 43 13 for airfield signals.
34 43 16 for airfield landing equipment.
34 71 00 for roadway construction.
34 72 00 for railway construction.
34 73 00 for airfield construction.
34 75 00 for roadway equipment.

26 56 13	Lighting Poles and Standards
26 56 16	Parking Lighting
26 56 19	Roadway Lighting
26 56 23	Area Lighting
26 56 26	Landscape Lighting
26 56 29	Site Lighting
26 56 33	Walkway Lighting
26 56 36	Flood Lighting
26 56 68	Exterior Athletic Lighting

Figure 47M   Detailed breakdown of the electrical division (Sheet 13 of 13).

Additional Resources

Electronics Fundamentals: Circuits, Devices, and Applications, Thomas L. Floyd. New York: Prentice Hall.

Principles of Electric Circuits, Thomas L. Floyd. New York: Prentice Hall.

3.0.0 Section Review

1. The letter designation SL on an electrical drawing indicates a _____.

   a. limit switch
   b. low-voltage switch
   c. load switch
   d. light switch

2. Existing power lines on an electrical site plan may be indicated by the letter _____.

   a. E
   b. I
   c. P
   d. V

3. The specific types of luminaires required in an installation can be found on the _____.

   a. power plan
   b. electrical site plan
   c. schematic diagram
   d. lighting fixture schedule

4. A power-riser diagram is a type of _____.

   a. block diagram
   b. detail view
   c. sectional view
   d. schematic

5. In the MasterFormat® specification, electronic safety and security can be found under Division _____.

   a. 25
   b. 26
   c. 27
   d. 28

SUMMARY

This module presents the symbols and conventions used on architectural and engineering drawings. As an electrician, you need to know how to recognize the basic symbols used on electrical drawings and other drawings used in the building construction industry. You should also know where to find the meaning of symbols that you do not immediately recognize. Schedules, diagrams, and specifications often provide detailed information that is not included on the working drawings.

Building projects require detailed specifications. These written specifications are complex and detailed and need a unified format to be easily usable by the trades. The specification format most commonly used is the *MasterFormat®* developed by CSI and CSC. The *MasterFormat®* was updated in 2010 with changes to the division numbering system.

Reading architectural and engineering drawings takes practice and study. Now that you have the basic skills, take the time to master them.

1. A section line on a drawing shows ______.
   a. the north orientation
   b. the location of the section on the plan
   c. where to locate receptacles in that section
   d. the section scale

2. In a drawing with dimension lines, the values shown inside the lines are ______.
   a. for general reference only
   b. reduced to scale
   c. always shown in decimal values
   d. the actual dimensions

3. Dashed lines used to represent a branch circuit on a drawing mean that the wiring is to be ______.
   a. concealed in the ceiling or wall
   b. concealed in the floor
   c. exposed
   d. turned up

4. An electrical drafting line with a double arrowhead represents ______.
   a. wiring concealed in the floor
   b. wiring turned down
   c. a branch circuit homerun
   d. wiring concealed in a ceiling or wall

Figure RQ01

5. The symbol shown in *Figure RQ01* represents a ______.
   a. head guy
   b. receptacle outlet
   c. sidewalk guy
   d. manhole

Figure RQ02

6. The symbol shown in *Figure RQ02* represents a ______.
   a. single receptacle outlet
   b. duplex receptacle outlet
   c. duplex receptacle outlet split wired
   d. triplex receptacle outlet

Figure RQ03

7. The symbol shown in *Figure RQ03* represents a ______.
   a. temperature control relay
   b. transformer
   c. incandescent fixture (surface or pendant)
   d. incandescent fixture with pull chain (surface or pendant)

Figure RQ04

8. The symbol shown in *Figure RQ04* represents a ______.
   a. single receptacle outlet
   b. duplex receptacle outlet
   c. duplex receptacle outlet split wired
   d. triplex receptacle outlet

Figure RQ05

9. The symbol shown in *Figure RQ05* represents a ______.

    a. single receptacle outlet
    b. duplex receptacle outlet
    c. duplex receptacle outlet split wired
    d. quadruplex receptacle outlet

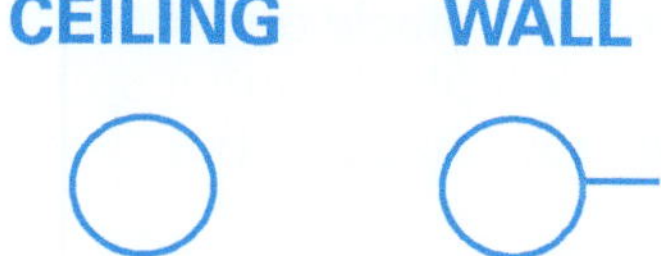

Figure RQ06

10. The symbol shown in *Figure RQ06* represents a ______.

    a. temperature control relay
    b. transformer
    c. incandescent fixture (surface or pendant)
    d. incandescent fixture with pull chain (surface or pendant)

Figure RQ07

11. What electrical symbol is shown in *Figure RQ07*?

    a. Wiring turned up
    b. Sidewalk guy
    c. Service weatherhead
    d. Head guy

12. To meet general recommendations, a residential branch circuit rated for 2,400VA should have a connected load of no more than ______.

    a. 1,680VA
    b. 1,920VA
    c. 2,040VA
    d. 2,160VA

13. Power-riser diagrams are used to show the ______.

    a. arrangement of electric service equipment
    b. branch circuit layout for power
    c. branch circuit layout for lighting
    d. panelboard schedule

14. The symbols T_1, T_2, and T_3 in a typical motor starter schematic represent ______.

    a. voltage supply lines
    a. auxiliary contacts
    b. motor terminals
    c. line contacts

15. The current *MasterFormat*® standard covering communications systems is under ______.

    a. Division 16
    b. Division 27
    c. Division 37
    d. Division 48

Fill in the blank with the correct term that you learned from your study of this module.

1. __________ typically include the following information: a site plan, floor plans, elevations of all exterior faces of the building, and large-scale detail drawings.

2. A(n) __________ is an exact copy or reproduction of an original drawing.

3. A simple, single-line diagram used to show electrical equipment and related connections is a(n) __________.

4. A(n) __________ shows the path of an electrical circuit or system of circuits, along with the circuit components.

5. To convey a substantial amount of detailed information to installation electricians, an engineer will use a(n) __________.

6. Shown in a separate view, a(n) __________ view is an enlarged, detailed view taken from an area of a drawing.

7. A cutaway drawing that shows the inside of an object or building is a(n) __________ drawing.

8. The sizes or measurements that are printed on a drawing are called __________.

9. The relationship between an object's size in a drawing and the object's actual size is the __________.

10. The height of the front, rear, or sides of a building is shown in a(n) __________.

11. A building's location on the site is shown in a(n) __________.

12. A drawing that has a top-down view of a building is a(n) __________.

13. A drawing that has a top-down view of a single object is a(n) __________.

14. A(n) __________ is a single-line block diagram used to indicate the electric service equipment, service conductors and feeders, and subpanels.

15. Owners, architects, and engineers use __________ to specify material and workmanship requirements.

16. A(n) __________ is a systematic way of presenting equipment lists on a drawing in tabular form.

17. Complicated circuits, such as control circuits, are shown in a(n) __________.

18. Usually developed by manufacturers, fabricators, or contractors, a(n) __________ shows specific dimensions and other information about a piece of equipment and its installation methods.

19. A special ruler with increments in inches is the __________.

20. A special ruler with increments in decimal units is the __________.

Trade Terms

Architect's scale
Architectural drawings
Block diagram
Blueprint
Detail drawing

Dimensions
Electrical drawing
Elevation drawing
Engineer's scale
Floor plan

One-line diagram
Plan view
Power-riser diagram
Scale
Schedule

Schematic diagram
Sectional view
Shop drawing
Site plan
Written specifications

1. A(n) ___________ indicates the location of the building on the property.

2. The ___________ show the walls and partitions for each floor or level.

3. What are the three main functions of electrical drawings?

   ____

4. The title block of an electrical drawing should contain the following ten items:

   ____

5. Match the following names to their corresponding electrical drafting lines.

   (A) ——————— E ———————   _____ WIRING TURNED UP

   (B) ——————————————   _____ BRANCH CIRCUIT HOMERUN TO PANELBOARD

   (C) — — — — — — — — — —   _____ WIRING TURNED DOWN

   (D) ——————————————○   _____ EXPOSED WIRING

   (E) ——————————————●   _____ WIRING CONCEALED IN FLOOR

   (F) ——————————————▶▶   _____ WIRING CONCEALED IN CEILING OR WALL

   or
   1   2
   ——————————◢◤◢◤

6. What does the letter F stand for in reference to safety switches? ___________

7. On a floor plan with a scale of ½" = 1'–0", what would be the equivalent distance if you measured 3¾" on the drawing? ___________

8. The purpose of a(n) ___________ is to identify that part of the project to which the sheet applies.

9. Single-line block diagrams are also known as ___________.

10. Division ___________ of the current CSI specification covers electrical work.

Wayne Stratton
Associated Builders and Contractors

How did you choose a career in the electrical field?

Three events in my childhood created the desire to learn the electrical trade. At age six, the farmhouse we lived in was totally destroyed by fire. The cause was electrical. As a young teen, a local electrician had incorrectly wired a heating element and electrocuted several pigs. In 1973, my father hired this electrician to install a motor starter on a grain conveyor. He could not figure it out. I wanted to learn how to do this type of work and do it safely.

Tell us about your apprenticeship experience.

My education is from a technical school. I have attended several manufacturers' training sessions.

I had to gain the hands-on experience after learning the trade. My observation of the apprenticeship programs is this: you get hands-on experience while you learn.

What positions have you held in the industry?

I worked as a plant industrial electrician responsible for motor control, DC motors, co-generation, and medium voltage distribution. Later, I began working for an electrical contractor who wanted to expand his business into the industrial field. I worked as a PLC technician designing and installing control systems. In 1987, I began teaching apprenticeship classes.

What would you say is the primary factor in achieving success?

The desire to learn all that I can learn, the ability to think outside the box, and the opportunities to gain a variety of experiences. All this helps me continue to learn and share with trainees.

What does your current job involve?

I teach electrical apprenticeship levels one through four at two different locations in Iowa. My other responsibilities involve task training for electrical licensing, fire alarm, and code updates.

Do you have any advice for someone just entering the trade?

Continue to learn. Completing an apprenticeship program or acquiring an electrician's license is not the end of learning. With code changes every 3 years, there is always more to learn. If you don't understand something, ask! Observe and learn from experienced individuals.

METRIC CONVERSION CHART

METRIC CONVERSION CHART

| INCHES | | METRIC | INCHES | | METRIC | INCHES | | METRIC |
Fractional	Decimal	mm	Fractional	Decimal	mm	Fractional	Decimal	mm
.	0.0039	0.1000	.	0.5512	14.0000	.	1.8898	48.0000
.	0.0079	0.2000	9/16	0.5625	14.2875	.	1.9291	49.0000
.	0.0118	0.3000	.	0.5709	14.5000	.	1.9685	50.0000
1/64	0.0156	0.3969	37/64	0.5781	14.6844	2	2.0000	50.8000
.	0.0157	0.4000	.	0.5906	15.0000	.	2.0079	51.0000
.	0.0197	0.5000	19/32	0.5938	15.0813	.	2.0472	52.0000
.	0.0236	0.6000	39/64	0.6094	15.4781	.	2.0866	53.0000
.	0.0276	0.7000	.	0.6102	15.5000	.	2.1260	54.0000
1/32	0.0313	0.7938	5/8	0.6250	15.8750	.	2.1654	55.0000
.	0.0315	0.8000	.	0.6299	16.0000	.	2.2047	56.0000
.	0.0354	0.9000	41/64	0.6406	16.2719	.	2.2441	57.0000
.	0.0394	1.0000	.	0.6496	16.5000	2 1/4	2.2500	57.1500
.	0.0433	1.1000	21/32	0.6563	16.6688	.	2.2835	58.0000
3/64	0.0469	1.1906	.	0.6693	17.0000	.	2.3228	59.0000
.	0.0472	1.2000	43/64	0.6719	17.0656	.	2.3622	60.0000
.	0.0512	1.3000	11/16	0.6875	17.4625	.	2.4016	61.0000
.	0.0551	1.4000	.	0.6890	17.5000	.	2.4409	62.0000
.	0.0591	1.5000	45/64	0.7031	17.8594	.	2.4803	63.0000
1/16	0.0625	1.5875	.	0.7087	18.0000	2 1/2	2.5000	63.5000
.	0.0630	1.6000	23/32	0.7188	18.2563	.	2.5197	64.0000
.	0.0669	1.7000	.	0.7283	18.5000	.	2.5591	65.0000
.	0.0709	1.8000	47/64	0.7344	18.6531	.	2.5984	66.0000
.	0.0748	1.9000	.	0.7480	19.0000	.	2.6378	67.0000
5/64	0.0781	1.9844	3/4	0.7500	19.0500	.	2.6772	68.0000
.	0.0787	2.0000	49/64	0.7656	19.4469	.	2.7165	69.0000
.	0.0827	2.1000	.	0.7677	19.5000	2 3/4	2.7500	69.8500
.	0.0866	2.2000	25/32	0.7813	19.8438	.	2.7559	70.0000
.	0.0906	2.3000	.	0.7874	20.0000	.	2.7953	71.0000
3/32	0.0938	2.3813	51/64	0.7969	20.2406	.	2.8346	72.0000
.	0.0945	2.4000	.	0.8071	20.5000	.	2.8740	73.0000
.	0.0984	2.5000	13/16	0.8125	20.6375	.	2.9134	74.0000
7/64	0.1094	2.7781	.	0.8268	21.0000	.	2.9528	75.0000
.	0.1181	3.0000	53/64	0.8281	21.0344	.	2.9921	76.0000
1/8	0.1250	3.1750	27/32	0.8438	21.4313	3	3.0000	76.2000
.	0.1378	3.5000	.	0.8465	21.5000	.	3.0315	77.0000
9/64	0.1406	3.5719	55/64	0.8594	21.8281	.	3.0709	78.0000
5/32	0.1563	3.9688	.	0.8661	22.0000	.	3.1102	79.0000
.	0.1575	4.0000	7/8	0.8750	22.2250	.	3.1496	80.0000
11/64	0.1719	4.3656	.	.8858	22.5000	.	3.1890	81.0000
.	0.1772	4.5000	57/64	.89063	22.6219	.	3.2283	82.0000
3/16	0.1875	4.7625	.	.9055	23.0000	.	3.2677	83.0000
.	0.1969	5.0000	29/32	.90625	23.0188	.	3.3071	84.0000
13/64	0.2031	5.1594	59/64	.92188	23.4156	.	3.3465	85.0000
.	0.2165	5.5000	.	.9252	23.5000	.	3.3858	86.0000
7/32	0.2188	5.5563	15/16	.93750	23.8125	.	3.4252	87.0000
15/64	0.2344	5.9531	.	.9449	24.0000	.	3.4646	88.0000
.	0.2362	6.0000	61/64	.95313	24.2094	3 1/2	3.5000	88.9000
1/4	0.2500	6.3500	.	.9646	24.5000	.	3.5039	89.0000
.	0.2559	6.5000	31/32	.96875	24.6063	.	3.5433	90.0000
17/64	0.2656	6.7469	.	.9843	25.0000	.	3.5827	91.0000
.	0.2756	7.0000	63/64	.98438	25.0031	.	3.6220	92.0000
9/32	0.2813	7.1438	1	1.000	25.40	.	3.6614	93.0000
.	0.2953	7.5000	.	1.0039	25.5000	.	3.7008	94.0000
19/64	0.2969	7.5406	.	1.0236	26.0000	.	3.7402	95.0000
5/16	0.3125	7.9375	.	1.0433	26.5000	.	3.7795	96.0000
.	0.3150	8.0000	.	1.0630	27.0000	.	3.8189	97.0000
21/64	0.3281	8.3344	.	1.0827	27.5000	.	3.8583	98.0000
.	0.3346	8.5000	.	1.1024	28.0000	.	3.8976	99.0000
11/32	0.3438	8.7313	.	1.1220	28.5000	.	3.9370	100.0000
.	0.3543	9.0000	.	1.1417	29.0000	4	4.0000	101.6000
23/64	0.3594	9.1281	.	1.1614	29.5000	.	4.3307	110.0000
.	0.3740	9.5000	.	1.1811	30.0000	4 1/2	4.5000	114.3000
3/8	0.3750	9.5250	.	1.2205	31.0000	.	4.7244	120.0000
25/64	0.3906	9.9219	1 1/4	1.2500	31.7500	5	5.0000	127.0000
.	0.3937	10.0000	.	1.2598	32.0000	.	5.1181	130.0000
13/32	0.4063	10.3188	.	1.2992	33.0000	.	5.5118	140.0000
.	0.4134	10.5000	.	1.3386	34.0000	.	5.9055	150.0000
27/64	0.4219	10.7156	.	1.3780	35.0000	6	6.0000	152.4000
.	0.4331	11.0000	.	1.4173	36.0000	.	6.2992	160.0000
7/16	0.4375	11.1125	.	1.4567	37.0000	.	6.6929	170.0000
.	0.4528	11.5000	.	1.4961	38.0000	.	7.0866	180.0000
29/64	0.4531	11.5094	1 1/2	1.5000	38.1000	.	7.4803	190.0000
15/32	0.4688	11.9063	.	1.5354	39.0000	.	7.8740	200.0000
.	0.4724	12.0000	.	1.5748	40.0000	8	8.0000	203.2000
31/64	0.4844	12.3031	.	1.6142	41.0000	.	9.8425	250.0000
.	0.4921	12.5000	.	1.6535	42.0000	10	10.0000	254.0000
1/2	0.5000	12.7000	.	1.6929	43.0000	20	20.0000	508.0000
.	0.5118	13.0000	.	1.7323	44.0000	30	30.0000	762.0000
33/64	0.5156	13.0969	1 3/4	1.7500	44.4500	40	40.0000	1016.000
17/32	0.5313	13.4938	.	1.7717	45.0000	60	60.0000	1524.000
.	0.5315	13.5000	.	1.8110	46.0000	80	80.0000	2032.000
35/64	0.5469	13.8906	.	1.8504	47.0000	100	100.0000	2540.000

TO CONVERT TO MILLIMETERS, MULTIPLY INCHES × 25.4
TO CONVERT TO INCHES, MULTIPLY MILLIMETERS × 0.03937*
*FOR SLIGHTLY GREATER ACCURACY WHEN CONVERTING TO INCHES, DIVIDE MILLIMETERS BY 25.4

Architect's scale: A special ruler with various measurement scales that can be used when drafting or making measurements on architectural drawings. Architect's scales measure distances in inches.

Architectural drawings: Working drawings consisting of site plans, floor plans, elevations, sectional views, details, and other information necessary for the construction of a building.

Block diagram: A single-line diagram used to show electrical equipment and related connections. See power-riser diagram.

Blueprint: An exact copy or reproduction of an original drawing.

Detail drawing: An enlarged, detailed view taken from an area of a drawing and shown in a separate view.

Dimensions: Sizes or measurements printed on a drawing.

Electrical drawing: A means of conveying a large amount of exact, detailed information in an abbreviated language. Consists of lines, symbols, dimensions, and notations to accurately convey an engineer's designs to electricians who install the electrical system on a job.

Elevation drawing: An architectural drawing showing height, but not depth; usually the front, rear, and sides of a building or object.

Engineer's scale: A special ruler with various measurement scales that can be used when drafting or making measurements on architectural drawings. Engineer's scales measure distances in decimal units.

Floor plan: A drawing of a building as if a horizontal cut were made through a building at about window level, and the top portion removed. The floor plan is what would appear if the remaining structure were viewed from above.

One-line diagram: A drawing that shows, by means of lines and symbols, the path of an electrical circuit or system of circuits along with the various circuit components. Also called a single-line diagram.

Plan view: A drawing made as though the viewer were looking straight down (from above) on an object.

Power-riser diagram: A single-line block diagram used to indicate the electric service equipment, service conductors and feeders, and subpanels. Notes are used on power-riser diagrams to identify the equipment; indicate the size of conduit; show the number, size, and type of conductors; and list related materials. A panelboard schedule is usually included with power-riser diagrams to indicate the exact components (panel type and size), along with fuses, circuit breakers, etc., contained in each panelboard.

Scale: On a drawing, the size relationship between an object's actual size and the size it is drawn. Scale also refers to the measuring tool used to determine this relationship.

Schedule: A systematic method of presenting equipment lists on a drawing in tabular form.

Schematic diagram: A detailed diagram showing complicated circuits, such as control circuits.

Sectional view: A cutaway drawing that shows the inside of an object or building.

Shop drawing: A drawing that is usually developed by manufacturers, fabricators, or contractors to show specific dimensions and other pertinent information concerning a particular piece of equipment and its installation methods.

Site plan: A drawing showing the location of a building or buildings on the building site. Such drawings frequently show topographical lines, electrical and communication lines, water and sewer lines, sidewalks, driveways, and similar information.

Written specifications: A written description of what is required by the owner, architect, and engineer in the way of materials and workmanship. Together with working drawings, the specifications form the basis of the contract requirements for construction.

Additional Resources

This module presents thorough resources for task training. The following resource material is suggested for further study.

Electronics Fundamentals: Circuits, Devices, and Applications, Thomas L. Floyd. New York, NY: Pearson Education, Inc.

Principles of Electric Circuits, Thomas L. Floyd. New York: Prentice Hall.

Figure Credits

MSA The Safety Company, Module Opener

John Traister, Figures 6–18, 24–26, 28–31, 33, 34, 36–40, 43–45

Mike Powers, Figures 20, 22

2016 *MasterFormat®*, Figures 46 and 47.

MasterFormat® Numbers and Titles used in this book are from MasterFormat®, published by CSI and Construction Specifications Canada (CSC), and are used with permission from CSI. For those interested in a more in-depth explanation of MasterFormat® and its use in the construction industry visit **www.masterformat.com** or contact:

CSI
110 South Union Street, Suite 100
Alexandria, VA 22314
800-689-2900; 703-684-0300
www.csinet.org

Section Review Answer Key

Answer	Section Reference	Objective
Section One		
1. c	1.1.0	1a
2. b	1.2.0	1b
3. d	1.3.0	1c
4. a	1.4.0	1d
5. c	1.5.2	1e
6. a	1.6.0	1f
Section Two		
1. c	2.0.0	2a
2. d	2.2.0	2b
3. d	2.3.0	2c
Section Three		
1. b	3.1.0; Figure 24	3a
2. a	3.2.1; Figure 29	3b
3. d	3.3.0	3c
4. a	3.4.1	3d
5. d	3.5.2	3e

NCCER CURRICULA — USER UPDATE

NCCER makes every effort to keep its textbooks up-to-date and free of technical errors. We appreciate your help in this process. If you find an error, a typographical mistake, or an inaccuracy in NCCER's curricula, please fill out this form (or a photocopy), or complete the online form at **www.nccer.org/olf**. Be sure to include the exact module ID number, page number, a detailed description, and your recommended correction. Your input will be brought to the attention of the Authoring Team. Thank you for your assistance.

Instructors – If you have an idea for improving this textbook, or have found that additional materials were necessary to teach this module effectively, please let us know so that we may present your suggestions to the Authoring Team.

NCCER Product Development and Revision

13614 Progress Blvd., Alachua, FL 32615

Email: curriculum@nccer.org
Online: www.nccer.org/olf

❏ Trainee Guide   ❏ Lesson Plans   ❏ Exam   ❏ PowerPoints   Other ___________________

Craft / Level: Copyright Date:

Module ID Number / Title:

Section Number(s):

Description:

____

____

____

____

Recommended Correction:

____

____

____

____

Your Name:

Address:

Email: Phone:

Electrical Test Equipment

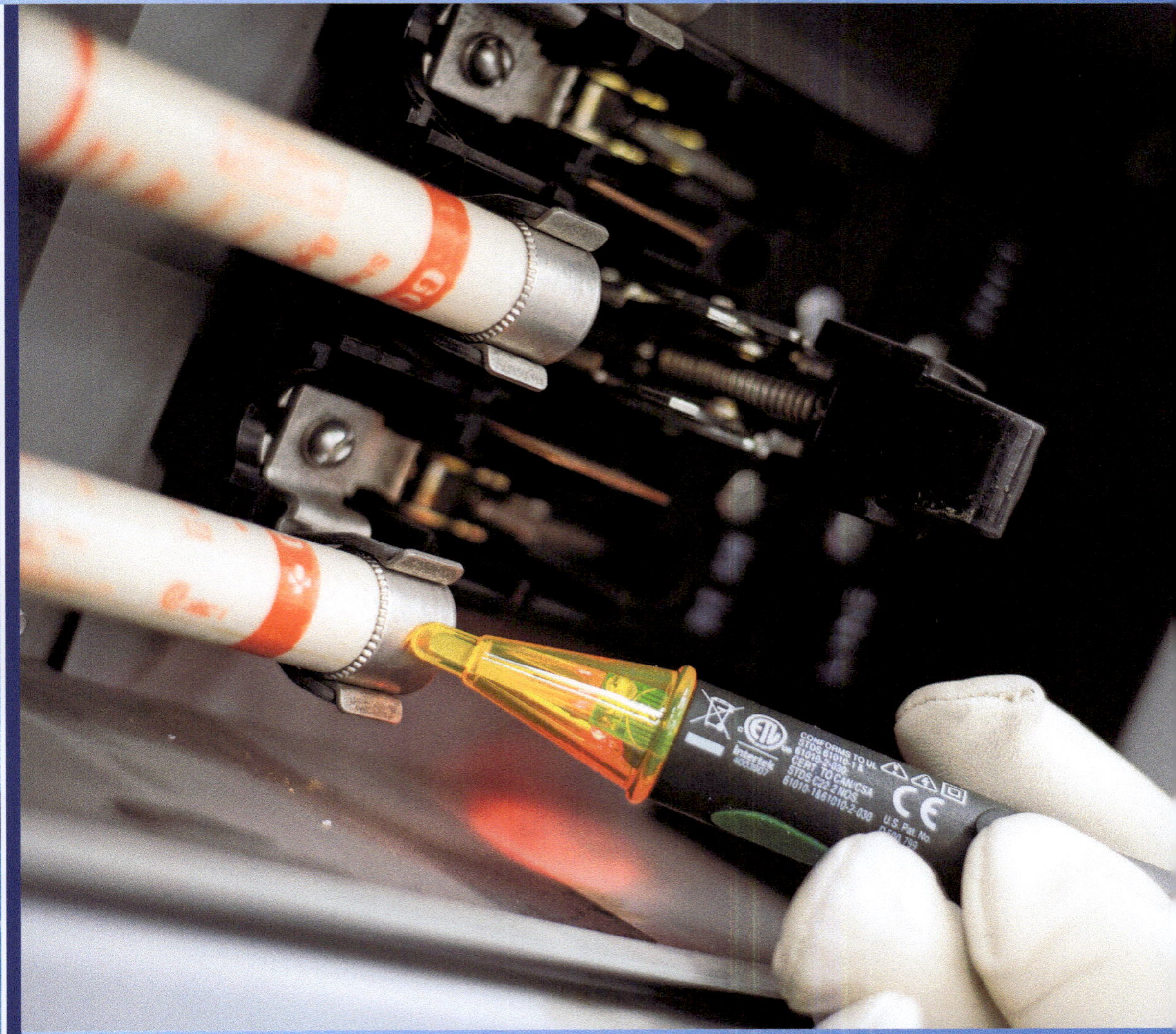

OVERVIEW

The test equipment selected for a specific task depends on the type of measurement and the level of accuracy required. This module covers the applications of various types of electrical test equipment. It also describes meter safety precautions and category ratings.

Module 26112-17

Objectives

When you have completed this module, you will be able to do the following:

1. Identify various types of electrical test equipment.
   a. Identify the applications of a voltmeter.
   b. Identify the applications of an ohmmeter.
   c. Identify the applications of a clamp-on ammeter.
   d. Identify the applications of a multimeter.
   e. Identify the applications of other meters.
2. Select a meter with the correct category rating for an application.
   a. Identify electrical test equipment safety hazards.

Performance Tasks

Under the supervision of the instructor, you should be able to do the following:

1. Measure the voltage in the classroom from line to neutral and neutral to ground.
2. Use an ohmmeter to measure the value of various resistors.

Trade Terms

Backfeed
Coil
Continuity

d'Arsonval meter movement
Frequency

Industry Recognized Credentials

If you are training through an NCCER-accredited sponsor, you may be eligible for credentials from NCCER's Registry. The ID number for this module is 26112-17. Note that this module may have been used in other NCCER curricula and may apply to other level completions. Contact NCCER's Registry at 888.622.3720 or go to **www.nccer.org** for more information.

Note

NFPA 70®, *National Electrical Code*® and *NEC*® are registered trademarks of the National Fire Protection Association, Quincy, MA.

Contents

Figures and Tables

1.0.0 METERS

Objective

Identify various types of electrical test equipment.

a. Identify the applications of a voltmeter.
b. Identify the applications of an ohmmeter.
c. Identify the applications of a clamp-on ammeter.
d. Identify the applications of a multimeter.
e. Identify the applications of other meters.

Trade Terms

Coil: A number of turns of wire, especially in spiral form, used for electromagnetic effects or for providing electrical resistance.

Continuity: An electrical term used to describe a complete (unbroken) circuit that is capable of conducting current. Such a circuit is also said to be closed.

d'Arsonval meter movement: A meter movement that uses a permanent magnet and moving coil arrangement to move a pointer across a scale.

Frequency: The number of cycles completed each second by a given AC voltage; usually expressed in hertz. One hertz equals one cycle per second.

Electronic test instruments and meters are generally used for the following list of tasks:

- Troubleshooting electrical/electronic circuits and equipment
- Verifying proper operation of instruments and associated equipment

The test equipment selected for a specific task depends on the type of measurement and the level of accuracy required. This module will focus on some of the test equipment used by electricians. Upon completion of this module, you should be able to select the appropriate test equipment for a specific application and identify the applicable safety hazards.

In 1882, a Frenchman named Arsene d'Arsonval invented the galvanometer. This meter used a stationary permanent magnet and a coil that moved to indicate current flow on a calibrated scale. The early galvanometer was very accurate but could measure only very small currents. Over the following years, many improvements were made that extended the range of the meter and increased its ruggedness. The d'Arsonval meter movement (*Figure 1*) is the basis for analog meters.

A moving-coil meter movement operates on the electromagnetic principle. In its simplest form, the moving-coil meter uses a coil of very fine wire wound on a light aluminum frame. A permanent magnet surrounds the coil. The aluminum frame is mounted on pivots to allow it and the coil to rotate freely between the poles of the permanent magnet. When current flows through the coil, it becomes magnetized, and the polarity of the coil is repelled by the field of the permanent magnet. This causes the coil frame to overcome the force of a spring and rotate on a pivot. The distance it rotates is determined by the amount of current that flows through the coil. By attaching a pointer to the coil frame and adding a calibrated scale, the amount of current flowing through the meter can be measured. Multiplier resistors are used to extend the range of the meter movement for voltage measurements, while shunt resistors are used to extend the range of the meter movement for current measurements.

Today, most meters are solid-state digital systems; they are easier to read than mechanical (analog) meters and have no meter movement or moving parts.

1.1.0 Voltmeter

A voltmeter is used to measure voltage, also known as potential difference or electromotive force (emf). It is connected in parallel with the circuit or component being measured. An analog meter uses the basic d'Arsonval meter movement with internally switched resistors to measure different voltage ranges. A digital meter uses an analog-to-digital converter chip to convert the sensed values into a digital or graphic display.

Phantom Readings

The sensitivity of a digital meter can sometimes produce a low reading known as a phantom or ghost reading. This is due to the induction from the electrical field around the energized conductors in close proximity to the meter.

Many digital voltmeters are autoranging, which means that the meter will automatically search for the correct scale. When using a voltmeter that is not autoranging, always start with the highest voltage range and work down until the indication reads somewhere between half and three-quarter scale. This will provide a more accurate reading and prevent damage to the meter. On many meters, a DC value is indicated by a straight line with three dashes beneath it, while an AC value is indicated by a sine wave.

> **WARNING!**
>
> Measuring voltages above 50V exposes the technician to potentially life-threatening hazards. Follow all applicable safety procedures as found in *NFPA 70E®*, OSHA standards, and company/institutional policies and standards.

A voltmeter is used when the exact value of the voltage is required. However, electricians are often concerned with identifying only whether voltage is present, and if so, the general range of the voltage. In other words, is it energized, and if so, is it at 120V, 240V, or 480V? In these cases, a voltage tester is used. The range of voltage and the type of current (AC and/or DC) that a voltage tester is capable of measuring are usually

Figure 2 Voltage tester.

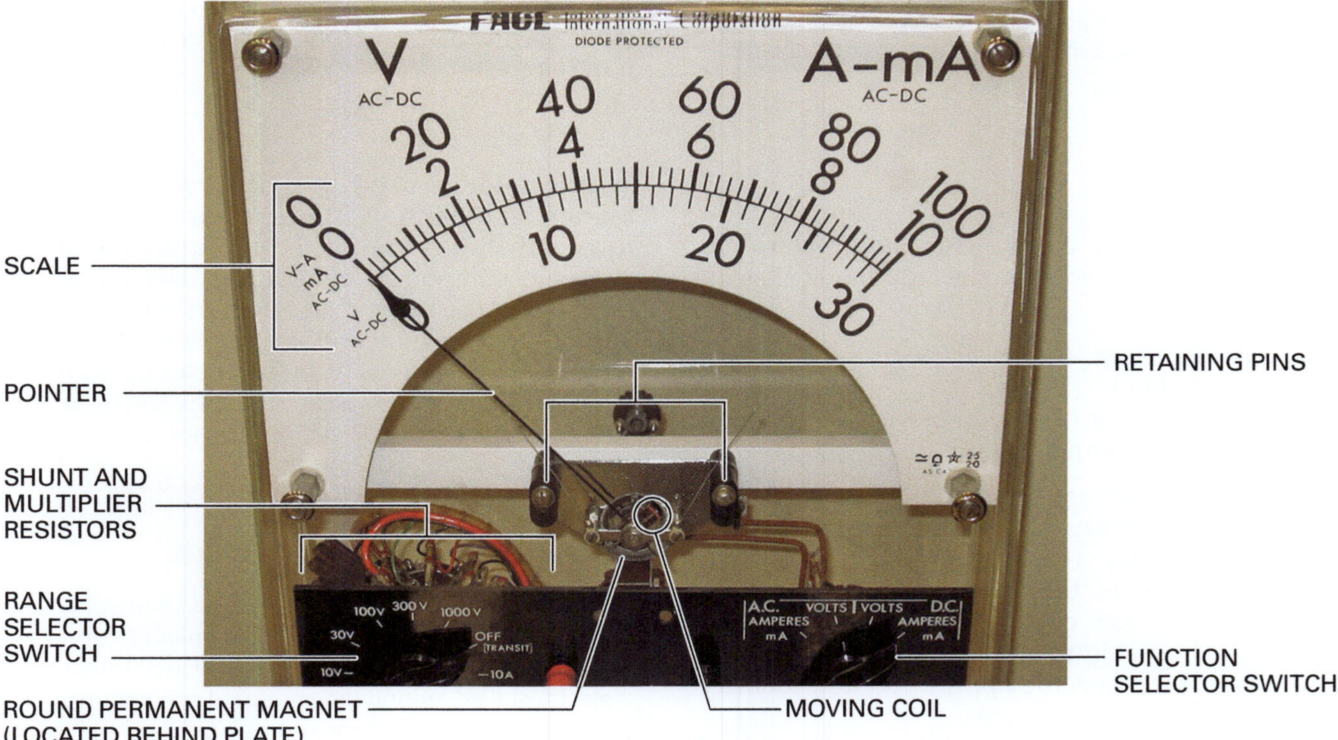

Figure 1 d'Arsonval meter movement.

NCCER – *Maritime Electrical*

indicated on the scales that display the reading (*Figure 2*).

Advanced voltage testers offer additional features, such as a digital readout, GFCI test capability, and even the ability to switch between use as a contact and noncontact detector (*Figure 3*).

A voltage tester must be checked before each use to make sure that it is in good condition and is operating correctly. The external check of the tester should include a careful inspection of the insulation on the leads for cracks or frayed areas. Faulty leads constitute a safety hazard, so they must be replaced. As a check to make sure that the voltage tester is operating correctly, the probes of the tester are first connected to a known energized source. The voltage indicated on the tester should match the voltage of the source. If there is no indication, the voltage tester is not operating correctly, and it must be repaired or replaced. It must also be repaired or replaced if it indicates a voltage different from the known voltage of the source.

Voltage testers are used to make sure that voltage is available when it is needed and to ensure that power has been cut off when it should have been. In a troubleshooting situation, it might be necessary to verify that power is available in order to be sure that lack of power is not the problem. For example, if there were a problem with a power tool, such as a drill, a voltage tester might be used to make sure that power is available to run the drill. A voltage tester might also be used to verify that there is power available to a three-phase motor that will not start.

> **CAUTION**
>
> Care should be taken when placing the probes of the tester across the voltage source. Some voltage testers are designed to take quick readings and may be damaged if left in contact with the voltage source for too long.

> **WARNING!**
>
> When testing for voltage to verify that a circuit is de-energized as part of an electrical lockout, always perform a live-dead-live test. This involves first verifying operation of the test equipment on a known energized (live) source, then de-energizing and testing the target circuit to ensure it is dead, then again testing the known energized (live) source before making contact with the de-energized circuit. This is an OSHA requirement for systems above 600V but is a good practice at all voltage levels.

> **WARNING!**
>
> Make sure to select a meter with the correct category rating for the environment. Always check test leads for wear, cracks in the insulation, or signs of damage before using.

Voltage Detectors

Simple noncontact (proximity) voltage detectors can also be used to indicate the presence of voltage within their specified range rating, but do not discriminate between ranges of values in the same way as a voltage tester. They are handy for quickly scanning for the presence of voltage in junction boxes or termination cabinets, and can even be used to trace circuits through walls. The voltage detector shown here glows in the presence of voltages between 50V and 1,000V.

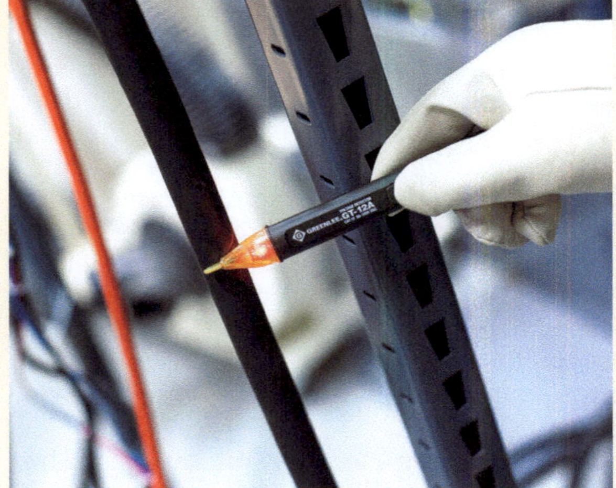

Figure Credit: Greenlee / A Textron Company

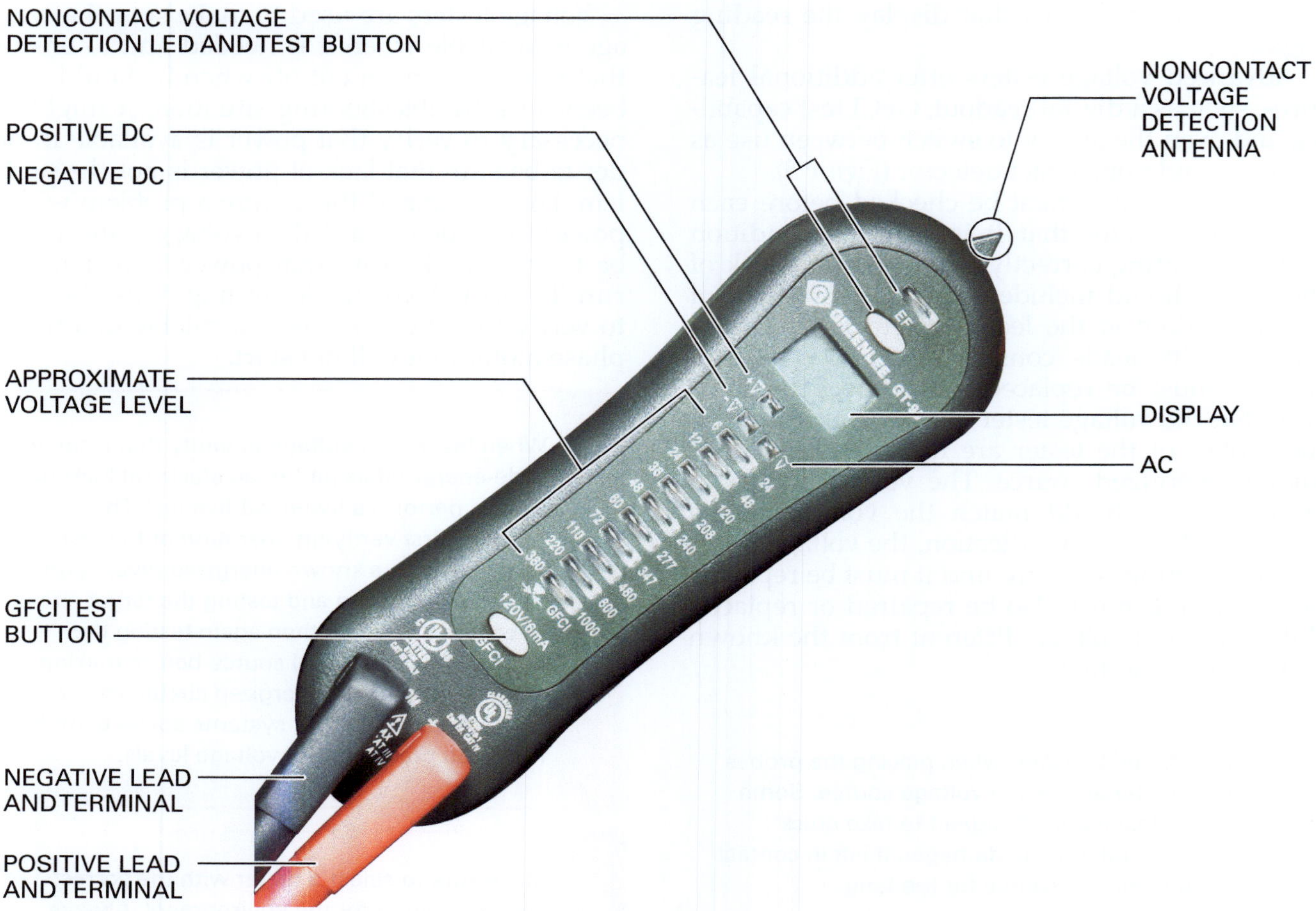

Figure 3 Multifunction voltage tester.

1.2.0 Ohmmeter

An ohmmeter measures the resistance of a circuit or component. It can also be used to locate open circuits or shorted circuits. An ohmmeter consists of a DC current meter movement, a low-voltage DC power source (usually a battery), and current-limiting resistors, all of which are connected in series with the meter (*Figure 4*).

Before measuring the resistance of an unknown resistor or electrical circuit, connect the test leads together. This zeroes out or nulls the resistance of the leads. Some analog ohmmeters have a zero adjustment knob. With the leads connected together, turn the adjustment knob until the meter registers zero ohms. This adjustment must be made each time a different range is selected.

Many digital ohmmeters are autoranging. The correct scale is internally selected and the reading will indicate the range (ohms, K-ohms, or M-ohms). Analog ohmmeters require that you

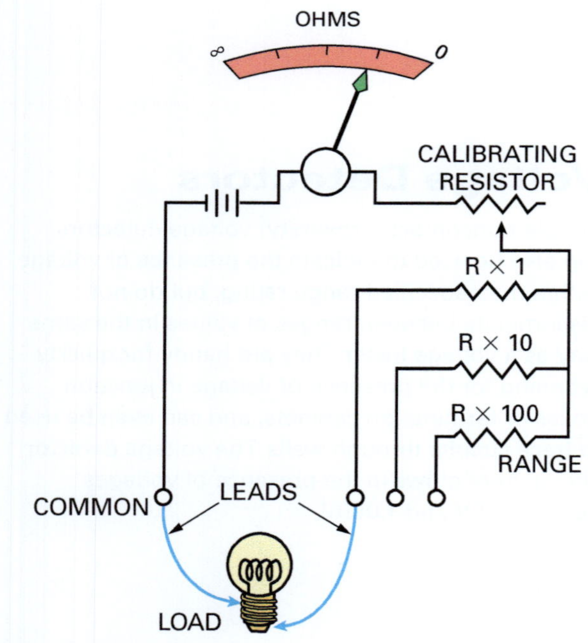

Figure 4 Ohmmeter schematic.

select the desired range. Most digital meters also have an audible tone when the measured value is very low or at zero ohms. This indicates a closed circuit and is useful when using the meter as a continuity tester. A continuity test is used to determine if a circuit is complete.

> **WARNING!**
>
> Prior to taking a reading with an ohmmeter, verify that both sides of the circuit are de-energized by using a voltmeter. If the circuit were energized, its voltage could cause a current to flow through the meter. This can damage the meter and/or circuit and cause personal injury.

When making resistance measurements in circuits, each component in the circuit can be tested individually by removing the component from the circuit and connecting the ohmmeter leads across it. However, the component does not have to be totally removed from the circuit. Usually, the part can be effectively isolated by disconnecting one of its leads from the circuit. Note that this method can still be somewhat time consuming.

1.3.0 Ammeter

A clamp-on ammeter, also known as a clamp meter, can measure current without having to make contact with uninsulated wires (*Figure 5*). This type of meter operates by sensing the strength of the electromagnetic field around the wire(s).

Clamp-on ammeters measure current by using simple transformer principles. The conductor(s) being measured would be the primary and the jaws (clamp) of the meter would be the secondary. The current in the primary winding induces a current in the secondary winding. If the ratio of the primary winding to the secondary winding is 1,000, then the secondary current is $^1/_{1000}$ of the current flowing in the primary. The smaller secondary current is connected to the meter's input. For example, a 1A current in the conductor will produce 0.001A (1mA) in the meter.

To measure the current, open the jaws of the meter and close them around the conductor(s) to be measured. Make sure that the jaws are clean and close tightly. Then read the magnitude of the current on the meter display.

Many meters have a Hold function. This is useful in tight locations when it is hard to read the meter while it is clamped around the conductor(s). Just press the Hold button when the value is measured, then remove and read the meter.

> **CAUTION**
>
> When using a clamp-on ammeter, make sure that the range of the meter is at least as high as the current to be measured. If the meter is digital and the current is too high, the display will read OL (overload). This means that the meter has been overloaded. If the meter is an analog meter, the indicating needle will peg (move) above the maximum limit on the scale, which might damage the meter.

Some meters have a Min/Max or peak function. This allows the technician to record the maximum inrush, as with a motor start.

> **WARNING!**
>
> Using a clamp-on ammeter may expose you to energized systems and equipment. Never use this type of meter unless you are qualified and are following NFPA, OSHA, and company/institutional safety procedures.

Think About It

Resistance

Why does the resistance vary when holding a resistor by pinching the meter leads against the resistor with your fingers when measuring it, versus measuring it while holding it in clips?

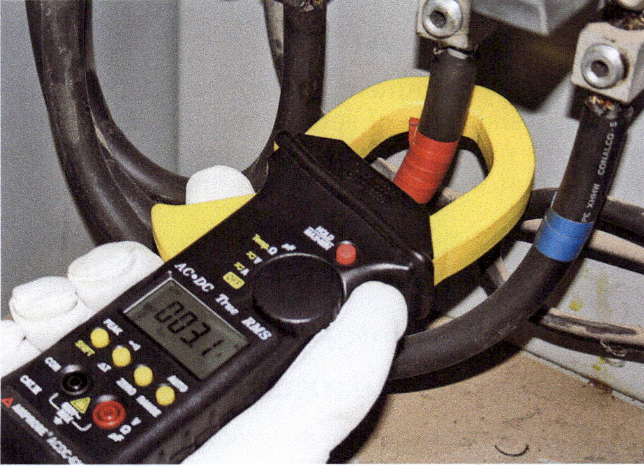

Figure 5 Clamp-on ammeter.

1.4.0 Multimeter

The multimeter is also known as a volt-ohm-milliammeter (VOM). An analog VOM is shown in *Figure 6*. It is a multipurpose instrument that combines the three previous meters discussed. When using an analog meter, you must select the proper voltage (DC or AC) and range (in volts, amps, or ohms). When using a digital VOM (*Figure 7*), you must select the proper voltage. Most have an autoranging feature for the magnitude.

Before use, a VOM must be checked on a known power source.

Use of a VOM is the same as using the individual voltmeter, ohmmeter, and clamp-on ammeter. Current clamps can be used with most multimeters for measuring AC and DC currents above the milliamp level. These can be plugged into either the amp or voltage test lead connections, depending on the current clamp. Always refer to the manufacturer's instructions for the device in use. Current clamps are available in various current ranges, from 50A to several thousand amps. The jaws are available in different shapes and sizes to suit various applications, from round to rectangular, and even flexible.

Some multimeters can also measure the **frequency** of an AC waveform—this function is useful for diagnosing harmonic problems in an electrical distribution system.

Figure 6 Analog VOM.

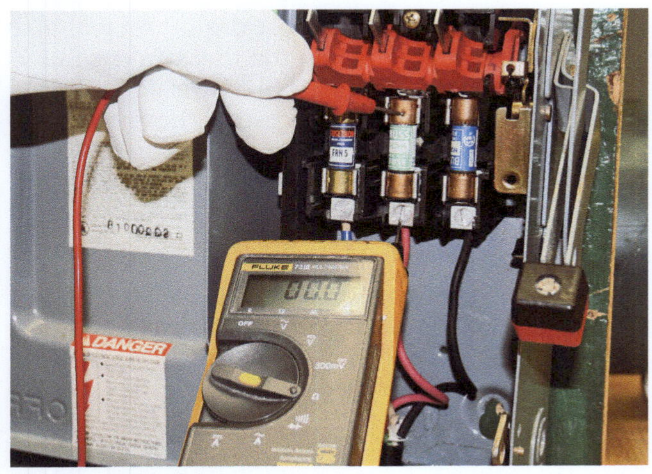

Figure 7 Digital VOM.

Another additional feature is the Min/Max memory function. It will record the minimum and maximum readings over the time period selected. Other common multimeter functions include capacitance measurement, diode and transistor testers, temperature measurement, and true RMS measurement for accurate voltage and current readings at different frequencies. Refer to the manufacturer's instructions for the meter in use. In addition to the current clamps used with standard multimeters, clamp-on multimeters are also available (*Figure 8*). They are used to measure AC current, AC and DC voltage, resistance, and other values.

1.5.0 Specialty Meters

In addition to the more common meters, a variety of special meters are available to complete various electrical measurements. Some of these meters are megohmmeters, motor and phase rotation testers, and recording instruments.

1.5.1 Megohmmeters

An ordinary ohmmeter cannot be used for measuring resistances of several million ohms, such as those found in conductor insulation or between motor or transformer windings. The in-

Think About It

Current Measurements

What happens if you loop the conductor so that the ammeter measures two turns instead of one?

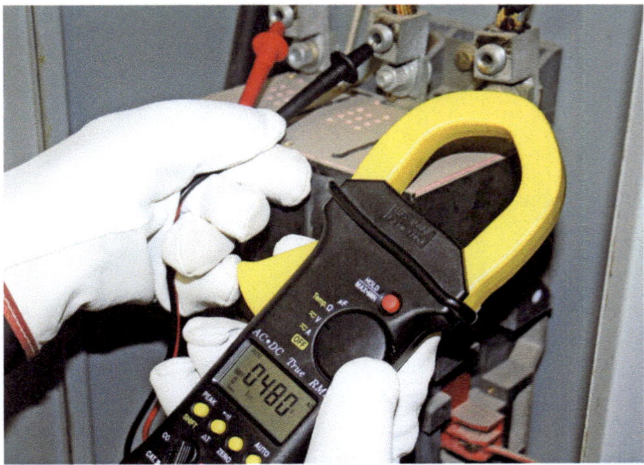

Figure 8 Clamp-on multimeter.

strument used to measure very high resistances is known as a megohmmeter. Megohmmeters are also called Meggers®, or insulation resistance testers. They can be powered by alternating current, battery (*Figure 9*), or hand cranking (*Figure 10*). When using a megohmmeter, you could be injured or cause damage to the equipment you are working on if the following minimum safety precautions are not followed:

- High voltages are present when using a megohmmeter. For example, in 600V class systems, applied megohmmeter voltages are typically 500V and 1,000V. Only qualified individuals may use this equipment. Always wear appropriate personal protective equipment when approaching energized parts.

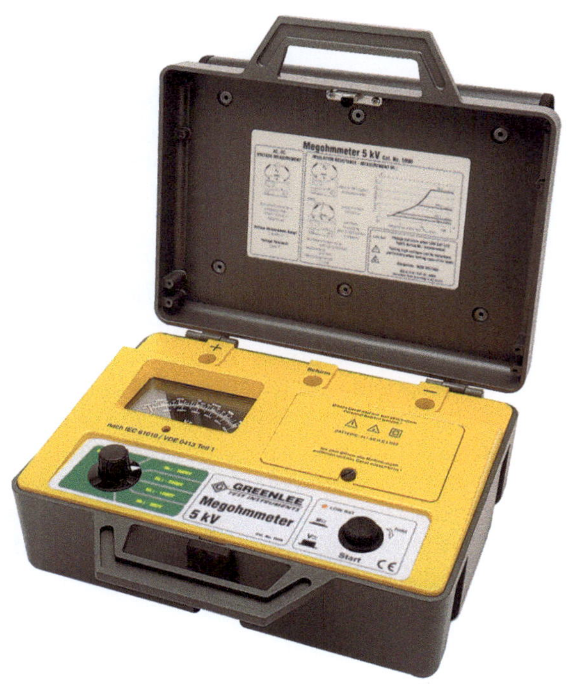

Figure 9 Battery-operated megohmmeter.

- De-energize and verify the de-energization of the circuit before connecting the meter. Make sure all capacitors are discharged.
- If possible, disconnect the item being checked from the other circuit components before using the meter.
- Do not exceed the manufacturer's recommended voltage test levels for the cable or equipment under test. Many manufacturers have different test levels based on the age of the cable or equipment being tested.
- Never touch the test leads when the meter is energized or powered. Meggers generate high voltage, and touching the leads could result in injury or electrical shock.
- After the test, discharge any energy that may be left in the circuit by grounding the conductor or equipment for a period of time equal to the duration of the test.
- When megging cables or bus ducts where you have exposed parts that are remote from your testing position, safely secure or barricade the exposed end to protect others from inadvertent contact with the test voltage.

If a megohmmeter is used to test switchgear, all of the electronics must be disconnected prior to testing the switchgear. The voltage produced by the megohmmeter may damage electronic equipment.

Meter manufacturers supply detailed manuals for testing various devices and equipment. Always follow these instructions.

1.5.2 Motor and Phase Rotation Testers

Before connecting a three-phase motor to a circuit, you must first match the legs or windings of the motor (T1, T2, and T3) to the phases of the circuit (L1, L2, and L3). This ensures that the motor rotates in the proper direction. A motor rotation tester can be used to identify the legs of the motor (*Figure 11*), while a phase rotation tester can be used to identify the phases of the circuit (*Figure 12*).

Do not connect a motor rotation tester to energized equipment. This can result in injury and equipment damage.

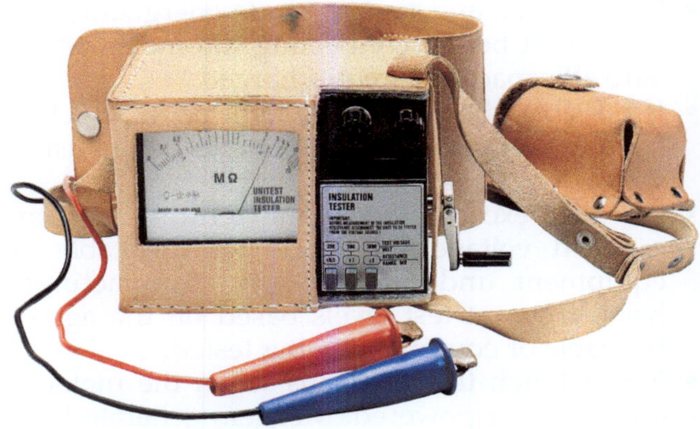

Figure 10 Hand-crank megohmmeter.

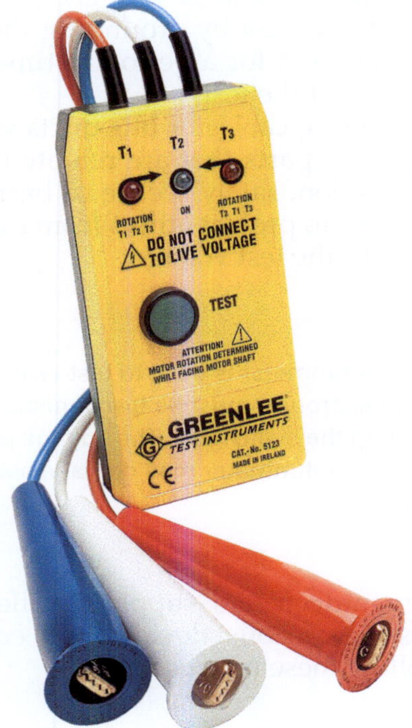

Figure 11 Motor rotation tester.

A motor rotation tester is a passive device; it operates on residual magnetism present in the motor after it has been run or tested by the manufacturer prior to shipping. To use a motor rotation tester, connect the three motor wires to the T1, T2, and T3 leads on the tester, then rotate the motor shaft a half-turn while pressing the Test button (the direction of rotation depends on the tester in use; always follow the manufacturer's instructions). Either the clockwise or counterclockwise LED will light up. If the required rotation is clockwise and the clockwise LED lights up, tag the motor

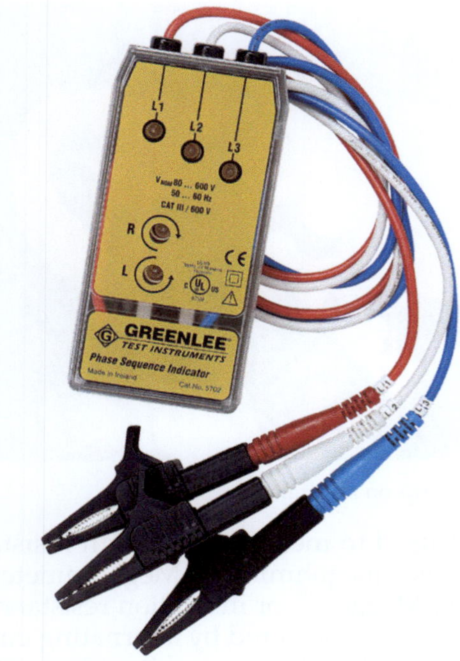

Figure 12 Phase rotation tester.

wires to correspond to the motor rotation leads. If the required rotation is clockwise and the counterclockwise LED lights up, switch a pair of leads and retest.

A phase rotation tester, also called a phase sequence indicator, is used on three-phase electrical systems to indicate the phase sequence rotation of the voltages. These testers typically have LEDs to indicate the phase rotation. A phase sequence is measured as clockwise or counterclockwise rotation.

Phase rotation testers may only be used by qualified individuals. Always wear appropriate personal protective equipment when approaching energized parts.

A phase rotation tester is used when it is necessary to ensure the same phase rotation throughout a facility. To test phase rotation, de-energize and lock out power to the circuit, then connect the three leads of the tester to the phase conductors in the circuit. Next, safely energize the circuit and observe the meter. Make note of the color scheme of the connected leads to the system, along with the phase sequences as indicated on

Meter Care

Like all meters, a megohmmeter is a sensitive instrument. Treat it with care and keep it in its case when not in use.

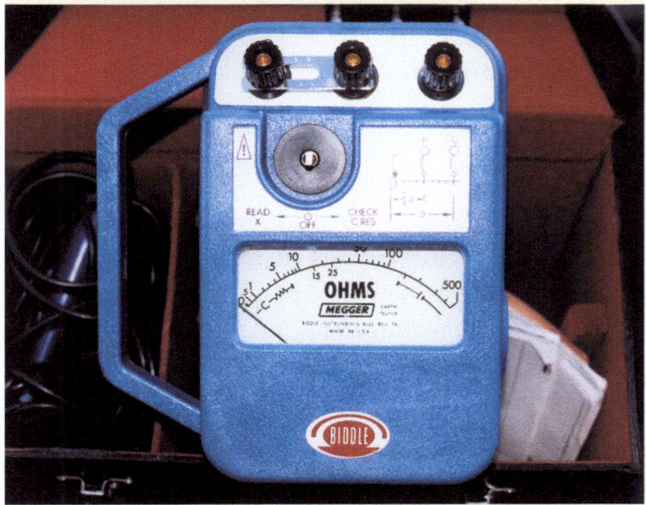

the meter. This is necessary to ensure that the added equipment follows the same phase rotation. De-energize and lock out the circuit before disconnecting the leads.

1.5.3 Recording Instruments

The term *recording instrument* describes many instruments that make a permanent record of measured quantities over a period of time. Recording instruments use either a paper strip or electronic accessible memory. Those using electronic memory are usually called data loggers. These instruments record electrical quantities, including potential difference, current, power, resistance,

Phase Rotation Tester

The leads for a phase rotation tester may not correspond to typical circuit color coding. A good idea is to put phase tape on your meter leads to correspond to the circuit color coding. This will help to ensure correct connections.

and frequency. They can also record nonelectrical quantities by electrical means, such as a temperature recorder that uses a potentiometer system to record thermocouple output.

It is often necessary to know the conditions that exist in an electrical circuit over a period of time to determine such things as peak loads, voltage fluctuations, and so on. An automatic recording instrument can be connected to take readings at specified intervals for later review and analysis. Some meters can upload data to a PC for real-time data logging and graphing (*Figure 13*).

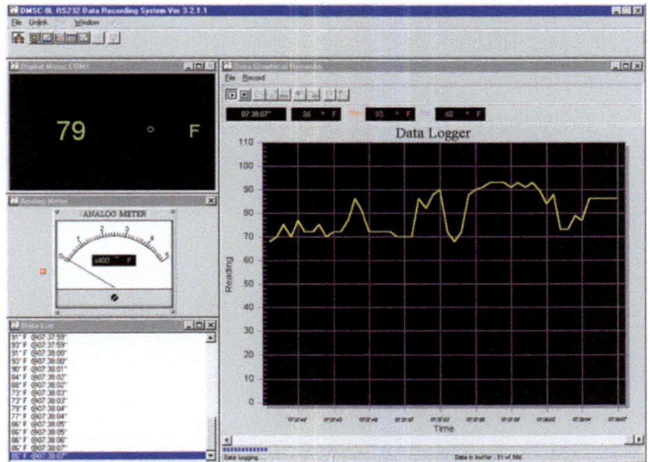

Figure 13 Data recording system.

Additional Resources

ABCs of DMMs, Multimeter Features and Functions Explained. Everett, WA: Fluke Corporation.

ABCs of Multimeter Safety. Everett, WA: Fluke Corporation.

Clamp Meter ABCs. Everett, WA: Fluke Corporation.

Electronics Fundamentals: Circuits, Devices, and Applications, Thomas L. Floyd. New York: Prentice Hall.

Power Quality Analyzer Uses for Electricians. Everett, WA: Fluke Corporation.

Principles of Electric Circuits, Thomas L. Floyd. New York: Prentice Hall.

1.0.0 Section Review

1. A voltmeter is used to measure the value of a circuit's ______.

   a. electromagnetic field
   b. resistance
   c. electromotive force
   d. current

2. A closed circuit is indicated by an ohmmeter reading of ______.

   a. zero
   b. infinity
   c. ERR
   d. ten

3. A meter that can take measurements without making conductor contact is a(n) ______.

   a. clamp-on ammeter
   b. megger
   c. phase rotation tester
   d. ohmmeter

4. Current measurements in milliamps are typically read using a(n) ______.

   a. clamp-on ammeter
   b. VOM
   c. megger
   d. insulation resistance tester

5. An expected resistance of 1,000,000 ohms would be measured using a(n) ______.

   a. recording instrument
   b. clamp-on ammeter
   c. multimeter
   d. megger

2.0.0 Category Ratings

Objective

Select a meter with the correct category rating for an application.

a. Identify electrical test equipment safety hazards.

Performance Tasks

1. Measure the voltage in the classroom from line to neutral and neutral to ground.
2. Use an ohmmeter to measure the value of various resistors.

Trade Terms

Backfeed: The reverse flow of electrical power in a distribution system caused by power induced into the system from outside sources, such as when using a generator during outages or when using a solar photovoltaic (PV) system. Backfeeds can also be caused by equipment overloads.

Distribution systems and loads are becoming more complex, increasing the risk of transient power spikes. Lightning strikes on outdoor transmission lines and switching surges from normal switching operations can also produce dangerous high-energy transients. Motors, capacitors, variable speed drives, and power conversion equipment can also generate power spikes.

Safety systems are built into test equipment to protect electricians from transient power spikes. The International Electrotechnical Commission (IEC) developed a safety standard, *IEC 1010*, for test equipment that was adapted as *ULStandardUL3111-1*. These standards define four overvoltage installation categories, often abbreviated as CAT I, CAT II, CAT III, and CAT IV (*Table 1*). These categories identify the hazards posed by transients; the higher the category number, the greater the risk to the electrician.

A higher category number refers to an installation with higher power available and higher-energy transients.

When selecting a meter, choose a meter rated for the highest category you will be working in. Then select the appropriate voltage level. In addition, make sure that your test leads are rated as high as your meter. Choose meters that are independently tested and certified by UL, CSA, or another recognized testing organization. Certified meters are marked with the category rating on the meter housing (*Figure 14*).

2.1.0 Safety

Safety must be the primary responsibility of all personnel on a job site. The safe installation, maintenance, and operation of electrical equipment requires strict adherence to local and national codes and safety standards, as well as facility and company safety policies. Carelessness can result in serious injury or death due to electrical shock, burns, falls, flying objects, etc. After an accident has occurred, investigation almost invariably shows that it could have been prevented by the exercise of simple safety precautions and procedures. It is your personal responsibility to identify and eliminate unsafe conditions and unsafe acts that cause accidents.

You must bear in mind that de-energizing main supply circuits by opening supply switches will not necessarily de-energize all circuits in a given piece of equipment. A source of danger that has often been neglected or ignored, sometimes with tragic results, is the input to electrical equipment from other sources, such as a backfeed. The rescue of a victim shocked by the power input from a backfeed is often hampered because of the time required to determine the source of power and isolate it. Always turn off all power inputs before working on equipment and lock out and tag, then check with an operating voltage tester to be sure that the equipment is safe to work on.

> **WARNING!**
>
> When performing lockout/tagout procedures, remember that other forms of energy may be present and must also be locked out and tagged. These include water pressure, steam, springs, gravity, and other forms of energy.

Table 1 Overvoltage Installation Categories

Overvoltage Category	Installation Examples
CAT I	Electronic equipment and circuitry
CAT II	Single-phase loads such as small appliances and tools, outlets at more than 30 feet from a CAT III source or 60 feet from a CAT IV source
CAT III	Three-phase motors, single-phase commercial or industrial lighting, switchgear, busduct and feeders in industrial plants
CAT IV	Three-phase power at meter, service-entrance, or utility connection, any outdoor conductors

Safety can never be stressed enough. There are times when your life literally depends on it. Always observe the following precautions:

- Remember that the common 120V power supply is not a low, relatively harmless voltage but is a voltage that has caused more deaths than any other.
- Thoroughly inspect all test equipment before each use. Check for broken leads or knobs, damaged plugs, or frayed cords. Do not use equipment that is wet or damaged.
- Make sure the rating of any leads or accessories meets or exceeds the rating of the meter.
- Do not work with energized equipment unless you are both qualified and approved by your supervisor.
- Never shortcut safety; strictly adhere to all energized work policies and procedures.
- When testing circuits, test at higher ranges first, then work your way down to lower ranges.
- Always have a standby person present during hot work. This person should know whom to contact in case of emergency and how to disconnect the power.

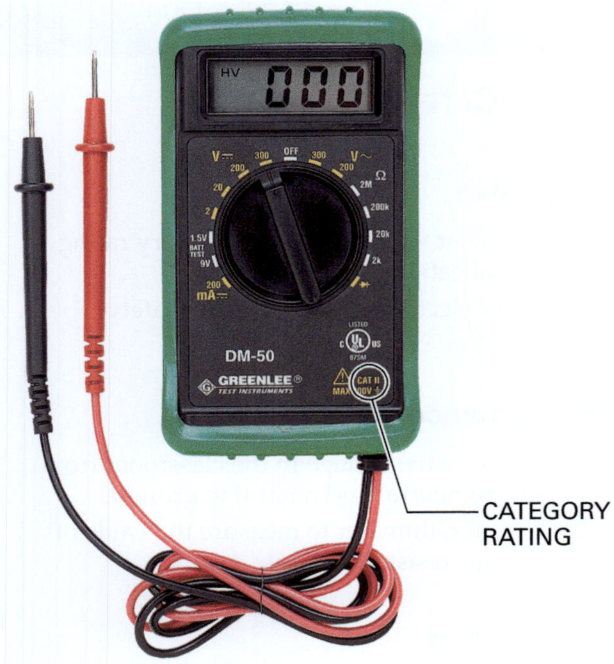

Figure 14 Category rating on a typical meter.

Think About It

Putting It All Together

Which test instrument would be required for each of the following applications?

- Identify a short circuit in house wiring
- Measure the secondary voltage of an AC transformer
- Identify a blown fuse in a circuit
- Identify the contact configuration of a three-way switch or multi-pole relay

2.0.0 Section Review

1. Which of the following is true with regard to meter safety?

   a. 120V power is relatively harmless.
   b. All test leads carry the same category rating.
   c. When testing circuits, test at the lowest ranges first, then work your way up.
   d. Thoroughly inspect all test equipment before each use.

SUMMARY

Meters and other devices are used to test and troubleshoot circuits and electrical equipment. One of the most important tests you will perform is verifying the absence of voltage before working on a device or circuit. This is often done using a voltage tester. Other common test equipment includes multimeters, clamp-on ammeters, megohmmeters, and motor/phase rotation testers.

You must understand the operation of and safety precautions for each piece of test equipment. In addition, you must be able to select the appropriate test equipment based on the task and category rating of the environment in which the work is to be performed. Always inspect and verify the operation of all test equipment before using it. Your life may depend on it.

1. Which of the following is true regarding analog meters?
   a. Analog meters are solid-state instruments.
   b. Analog meters provide a digital readout.
   c. Analog meters rely on a moving-coil meter movement.
   d. Analog meters provide an autoranging setting.

2. A voltmeter is used to test ______.
   a. exact voltages
   b. voltage ranges
   c. power
   d. sine waves

3. In order to ensure safety, before measuring low voltages, you should first test for ______.
   a. resistance
   b. current
   c. vibration
   d. higher voltages

4. An ammeter is used to measure ______.
   a. current
   b. voltage
   c. resistance
   d. insulation value

5. Clamp-on ammeters operate by ______.
   a. using d'Arsonval meter movement
   b. sensing the strength of the electromagnetic field around the wire
   c. measuring the high resistance end of a power transformer
   d. using a resistive shunt

6. An insulation tester is another name for a(n) ______.
   a. megohmmeter
   b. ammeter
   c. multimeter
   d. continuity tester

7. A motor rotation tester ______.
   a. tests an energized motor
   b. gets connected to the motor supply conductors
   c. works on residual magnetism
   d. works only on clockwise rotation

8. The highest level of protection is provided by instruments rated as ______.
   a. CAT I
   b. CAT II
   c. CAT III
   d. CAT IV

9. The International Electrotechnical Commission (IEC) developed a safety standard for overvoltage installation categories of CAT I, CAT II, CAT III, and CAT IV for ______.
   a. wiring
   b. electrical equipment
   c. test equipment
   d. signaling circuits

10. When a supply circuit is de-energized, the associated circuits ______.
    a. are always de-energized
    b. remain unaffected
    c. may still be energized due to backfeeds
    d. do not require lockout/tagout

Fill in the blank with the correct term that you learned from your study of this module.

1. Used for electromagnetic effects or for providing electrical resistance, a __________ is a number of turns of wire.

2. A __________ uses a permanent magnet and moving coil arrangement to move a pointer across a scale.

3. Usually expressed in hertz, __________ is the number of cycles completed each second by a given AC voltage.

4. __________ is an uninterrupted electrical path for current flow.

5. Power from a solar PV system can cause a dangerous __________ in equipment assumed to be de-energized.

Trade Terms

Backfeed
Coil
Continuity

d'Arsonval meter movement
Frequency

1. The measurement of the electromotive force of a circuit is accomplished using a(n) ___________
   a. ammeter
   b. wattmeter
   c. voltmeter
   d. ohmmeter

2. When using a voltmeter that is not autoranging, start with the highest setting and work down until the meter reads somewhere between

   ____

3. A(n) ___________ is used to extend the range of a meter movement for current measurements.

4. True or False? Always connect an ohmmeter in parallel with a load.

5. The voltage range of a meter movement can be extended by adding a(n) ___________ in series.

6. Short circuits can be detected by using a(n) ___________.

7. What type of test equipment would you use to check the resistance between motor windings?

   ____

8. The phase sequence of a circuit is identified using a(n) ___________.

9. True or False? A voltage tester is used for precise voltage measurements.

10. A(n) ___________ is used to take electrical readings at specified intervals.

Clarence "Ed" Cockrell

HR/Safety Manager
Vector Electric & Controls, Inc.

How did you get started in the construction industry?

I worked as a summer electrical helper during high school and college and found it very rewarding. I was looking for a job that would hold my interest for more than a year. I studied electrical engineering at Louisiana State University for two years.

Who inspired you to enter the industry? Why?

My brother-in-law and father-in-law inspired me to enter the industry.

What do you enjoy most about your job?

I started off as an electrician's apprentice, which offered many potential job opportunities. I enjoyed seeing the work progress from dirt to a functional building and finding new challenges as new technology changes the way we install the electrical components. My electrical training also opened the door to the possibility of being a field superintendent, project manager, senior office manager, human resource/training manager, and then human resource/safety/training manager.

Do you think training and education are important in construction? If so, why?

Training gives an apprentice the opportunity to become a great electrician and not an electrical laborer, by that I mean not just a conduit or cable tray installer or wire puller, but a well-rounded electrician. It also allows the apprentice to advance beyond the limits of being an electrician. Almost all of our supervisors, estimators, and project managers have been trained by an apprenticeship program and followed it up with more NCCER training.

How important are NCCER credentials to your career?

I went through a state certified apprenticeship and then completed the CSST training. Without these certifications, I believe I would be no more than a second-rate electrician.

How has training/construction impacted your life and you career?

It has allowed me to raise a family, buy a house and raise a child in a comfortable fashion. I have advanced many times at work, which has led to increased wages. Through my job, I have met and had dealings with many interesting people from all walks of life.

Would you suggest construction as a career to others? If so, why?

Yes. It offers a rewarding career opportunity to anyone willing to take pride and ownership in their learning and work.

How do you define craftsmanship?

I believe that craftsmanship is always delivering a great quality job in whatever you do. It takes pride and self-esteem to deliver work deemed to meet this definition of craftsmanship. This pride in their work helps build life qualities that become the building blocks of a truly honorable life. They walk through their community as a positive contributor as well as helping to build a secure and prosperous business. They teach their children through their actions how those who build contribute to the well-being of their community and country.

Trade Terms Introduced in This Module

Backfeed: The reverse flow of electrical power in a distribution system caused by power induced into the system from outside sources, such as when using a generator during outages or when using a solar photovoltaic (PV) system. Backfeeds can also be caused by equipment overloads.

Coil : A number of turns of wire, especially in spiral form, used for electromagnetic effects or for providing electrical resistance.

Continuity: An electrical term used to describe a complete (unbroken) circuit that is capable of conducting current. Such a circuit is also said to be closed.

d'Arsonval meter movement : A meter movement that uses a permanent magnet and moving coil arrangement to move a pointer across a scale.

Frequency : The number of cycles completed each second by a given AC voltage; usually expressed in hertz. One hertz equals one cycle per second.

Additional Resources

This module presents thorough resources for task training. The following reference material is recommended for further study:

ABCs of DMMs, Multimeter Features and Functions Explained. Everett, WA: Fluke Corporation.

ABCs of Multimeter Safety. Everett, WA: Fluke Corporation.

Clamp Meter ABCs. Everett, WA: Fluke Corporation.

Electronics Fundamentals: Circuits, Devices, and Applications, Thomas L. Floyd. New York: Prentice Hall.

Power Quality Analyzer Uses for Electricians. Everett, WA: Fluke Corporation.

Principles of Electric Circuits, Thomas L. Floyd. New York: Prentice Hall.

Figure Credits

Greenlee / A Textron Company, Module Opener, Figures 2, 3, 6, 9–14,
Tim Dean, Figures 1, 5, 7, 8

Section Review Answer Key

Answer	Section Reference	Objective
Section One		
1. c	1.1.0	1a
2. a	1.2.0	1b
3. a	1.3.0	1c
4. b	1.4.0	1d
5. d	1.5.1	1e
Section Two		
1. d	2.1.0	2a

Raceways and Fittings

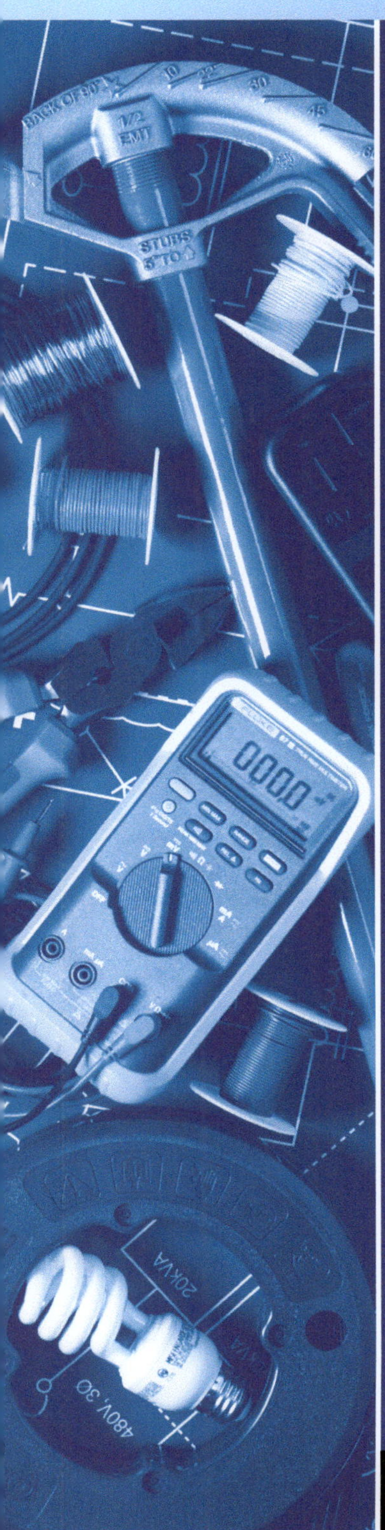

OVERVIEW

Electrical raceways present challenges and requirements involving proper installation techniques, general understanding of raceway systems, and applications of the *NEC*® to raceway systems. Acquiring quality installation skills for raceway systems requires practice, knowledge, and training. This module describes various types of raceway systems, along with their installation and *NEC*® requirements. It also describes the use of various conduit bodies.

Module 26108-17

RACEWAYS AND FITTINGS

Objectives

When you have completed this module, you will be able to do the following:

1. Select and install raceway systems.
   a. Identify types of conduit and their applications.
   b. Properly bond conduit for use as a ground path.
   c. Install metal conduit fittings.
   d. Make conduit-to-box connections.
   e. Identify raceway supports.
   f. Identify installation requirements for various construction methods.
2. Select fasteners and anchors for the installation of raceway systems.
   a. Select and install tie wraps.
   b. Select and install screws.
   c. Select and install hammer-driven pins and studs.
   d. Identify the safety requirements for stud-type guns.
   e. Select and install masonry anchors.
   f. Select and install hollow-wall anchors.
   g. Select and install epoxy anchoring systems.
3. Select and install wireways and other specialty raceways.
   a. Identify types of wireways and their components.
   b. Install wireway supports.
   c. Identify and install specialty raceways.
4. Select and install cable trays.
   a. Identify cable tray types and fittings.
   b. Install cable tray supports.
5. Handle and store raceways.
   a. Handle raceways.
   b. Store raceways.

Performance Tasks

Under the supervision of the instructor, you should be able to do the following:

1. Identify the appropriate conduit body for a given application.
2. Identify and select various types and sizes of raceways, fittings, and fasteners for a given application.
3. Demonstrate how to install a raceway system.
4. Terminate a selected raceway system.

Trade Terms

Accessible
Approved
Bonding wire
Cable trays
Conduit
Exposed location
Kick

Raceways
Splice
Tap
Trough
Underwriters Laboratories, Inc. (UL)
Wireways

Industry Recognized Credentials

If you are training through an NCCER-accredited sponsor, you may be eligible for credentials from NCCER's Registry. The ID number for this module is 26108. Note that this module may have been used in other NCCER curricula and may apply to other level completions. Contact NCCER's Registry at 888.622.3720 or go to **www.nccer.org** for more information.

Note

NFPA 70®, *National Electrical Code*® and *NEC*® are registered trademarks of the National Fire Protection Association, Quincy, MA.

Contents

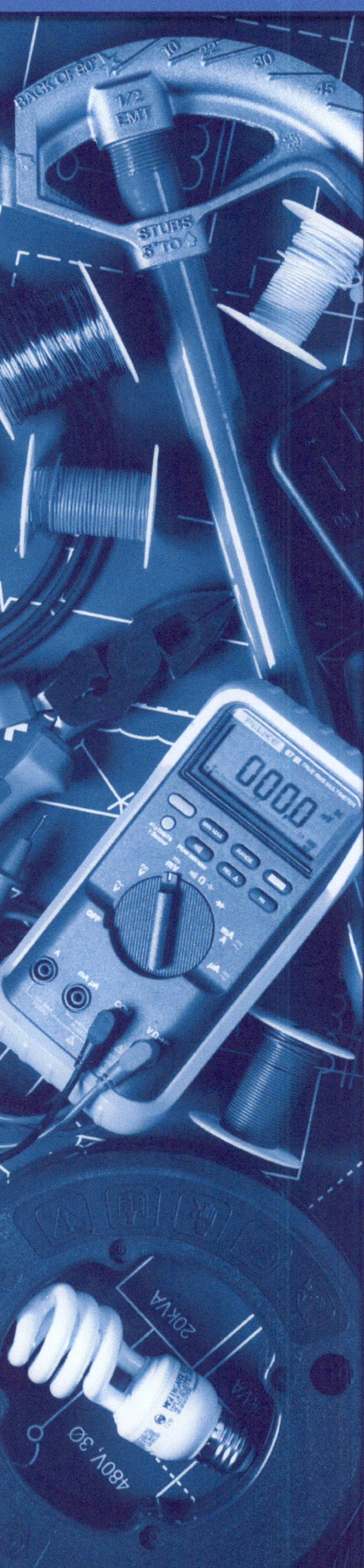

1.0.0 RACEWAY SYSTEMS

Objective

Select and install raceway systems.
a. Identify types of conduit and their applications.
b. Properly bond conduit for use as a ground path.
c. Install metal conduit fittings.
d. Make conduit-to-box connections.
e. Identify raceway supports.
f. Identify installation requirements for various construction methods.

Performance Task

1. Identify the appropriate conduit body for a given application.

Trade Terms

Accessible: Able to be reached, as for service or repair.

Approved: Meeting the requirements of an appropriate regulatory agency.

Bonding wire: A wire used to make a continuous grounding path between equipment and ground.

Conduit: A round raceway, similar to pipe, that houses conductors.

Exposed location: Not permanently closed in by the structure or finish of a building; able to be installed or removed without damage to the structure.

Kick: A bend in a piece of conduit, usually less than 45°, made to change the direction of the conduit.

Raceways: Enclosed channels designed expressly for holding wires, cables, or busbars, with additional functions as permitted in the *NEC*®.

Splice: Connection of two or more conductors.

Tap: Intermediate point on a main circuit where another wire is connected to supply electrical current to another circuit.

Underwriters Laboratories, Inc. (UL): An agency that evaluates and approves electrical components and equipment.

Wireways: Steel troughs designed to carry electrical wire and cable.

The term raceways refers to a wide range of circular and rectangular enclosed channels used to house electrical wiring. Raceways can be metallic or nonmetallic and come in different shapes. Depending on the particular purpose for which they are intended, raceways include enclosures such as underfloor raceways, flexible metal conduit, tubing, wireways, surface metal raceways, surface nonmetallic raceways, and support systems such as cable trays.

1.1.0 Types of Conduit and Their Applications

Conduit is a raceway with a circular cross section, similar to pipe, that contains wires or cables. Conduit is used to provide protection for conductors and route them from one place to another. Metal conduit also provides a permanent electrical path to ground. This equipment must be listed in accordance with the *NEC*®. There are many types of conduit used in the construction industry. The size of conduit to be used is determined by engineering specifications, local codes, and the *NEC*®. Refer to *NEC Chapter 9, Tables 1 through 4* and *Informative Annex C* for conduit fill with various conductors.

1.1.1 Electrical Metallic Tubing

Electrical metallic tubing (EMT) is the lightest duty tubing available for enclosing and protecting electrical wiring. EMT is widely used for residential, commercial, and industrial wiring systems. It is lightweight, easily bent and/or cut to shape, and is the least costly type of metallic conduit. Because the wall thickness of EMT is less than that of rigid conduit, it is often referred to as thinwall conduit. A comparison of inside and outside diameters of EMT to rigid metal conduit (RMC) and intermediate metal conduit (IMC) is shown in *Figure 1*.

NEC Section 358.10(A) permits the installation of EMT for either exposed or concealed work. Per *NEC Section 358.10(B)(1)*, galvanized steel and stainless steel EMT, elbows, and fittings shall be permitted to be installed in concrete, in direct contact with the earth, or in areas subject to severe corrosive influences where protected by corrosion protection and approved as suitable for the condition.

According to *NEC Section 358.10(D)*, where EMT is installed in wet locations, all supports, bolts, straps, screws, and so forth shall be of corrosion-resistant materials or protected against corrosion by corrosion-resistant materials.

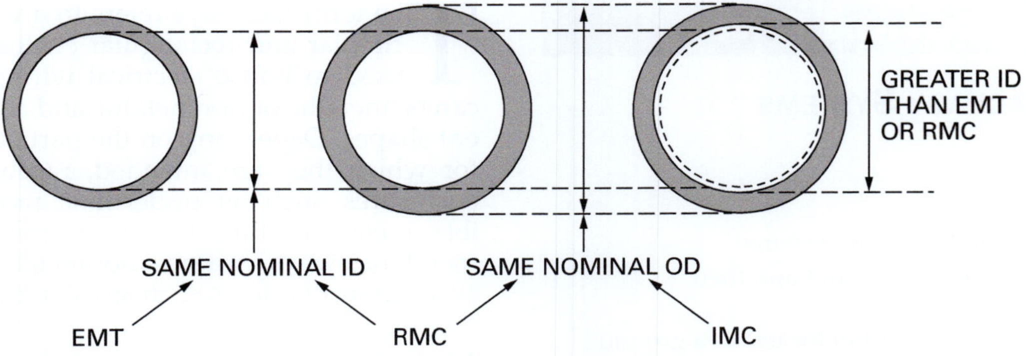

Figure 1 Conduit comparison.

According to *NEC Section 358.12*, EMT shall not be used where subject to severe physical damage or for the support of luminaires or other equipment, except conduit bodies no larger than the largest trade size of the tubing.

In a wet area, EMT and other metallic conduit must be installed to prevent water from entering the conduit system. In locations where walls are subject to regular wash-down [see *NEC Section 300.6(D)*], the entire conduit system must be installed to provide a $\frac{1}{4}$" (6 mm) air space between it and the wall or supporting surface. The entire conduit system is considered to include conduit, boxes, and fittings. To ensure resistance to corrosion caused by wet environments, EMT is galvanized. The term galvanized is used to describe the procedure in which the interior and exterior of the conduit are coated with a corrosion-resistant zinc compound.

EMT, being a good conductor of electricity, may be used as an equipment grounding conductor [see *NEC Section 250.118(4)*]. The conduit system must be tightly connected at each joint and provide a continuous grounding path from each electrical load to the service equipment. The connectors used in an EMT system ensure electrical and mechanical continuity throughout the system (see *NEC Sections 250.96, 300.10, and 358.42*).

> **NOTE**
>
> Support requirements for EMT are also covered in *NEC Section 358.30*. The types of supports will be discussed later in this module.

Because EMT is too thin for threads, fittings listed for EMT must be used. For wet or damp locations, listed compression fittings such as those shown in *Figure 2* are used. These fittings contain a compression ring made of metal that forms a raintight seal.

> **Think About It**
>
> # EMT Use
>
> Where would you use EMT? Are there any circumstances where EMT cannot be run through a suspended ceiling? What are some differences between EMT and rigid conduit?

When EMT compression couplings are used, they must be securely tightened, and when installed in masonry concrete, they must be of the concrete-tight type. If installed in a wet location, they must be the raintight type. Refer to *NEC Section 358.42*.

EMT fittings for dry locations can be either the setscrew type or the indenting type. To use the setscrew type, the ends of the EMT are inserted into the sleeve and the setscrews are tightened to make the connection. Various types of setscrew fittings are shown in *Figure 3*.

EMT sizes of $2\frac{1}{2}$" (MD 63) and larger have the same outside diameter as corresponding sizes of galvanized RMC. RMC threadless connectors may be used to connect EMT.

> **NOTE**
>
> EMT connectors smaller than $2\frac{1}{2}$" (MD 63), although they are the same size as RMC threadless connectors, may not be used to connect RMC.

Both setscrew and compression couplings are available in die-cast or steel construction. Steel couplings are stronger, but may not seal as well.

Support requirements for EMT are presented in *NEC Section 358.30*. As with most other metal conduit, EMT must be supported at intervals not to exceed 10' (3 m) and within 3' (900 mm) of

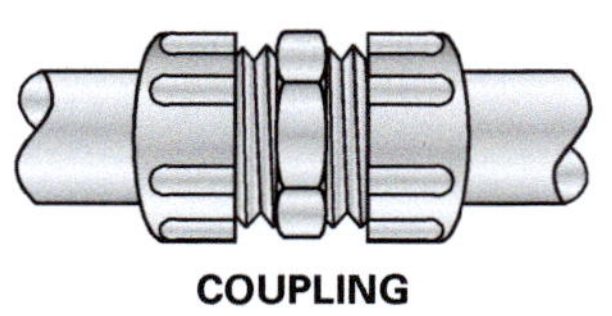

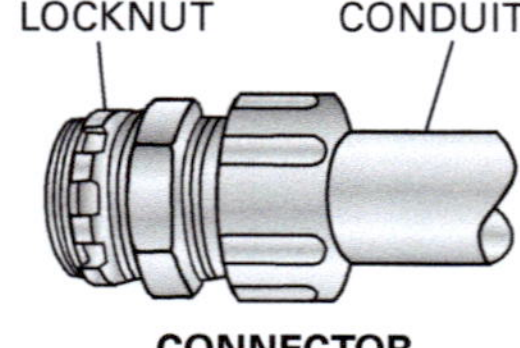

Figure 2 Compression fittings.

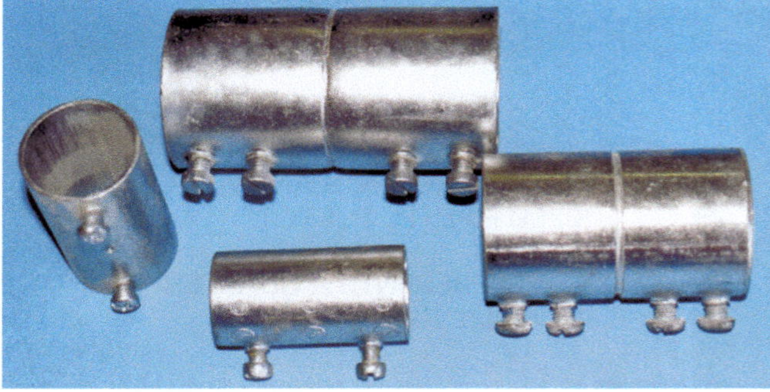

Figure 3 Setscrew fittings.

each outlet box, junction box, cabinet, fitting, or terminating end of the conduit. An exception to *NEC Section 358.30(A), Exception 1* allows the fastening of unbroken lengths of EMT to be increased to a distance of 5' (1.5 m) where structural members do not readily permit fastening within 3' (900 mm).

Electrical nonmetallic tubing (ENT) is also available. It provides an economical alternative to EMT, but it can only be used in certain applications. See *NEC Article 362*.

1.1.2 Rigid Metal Conduit

Rigid metal conduit (RMC) is conduit that is constructed of metal of sufficient thickness to permit the cutting of pipe threads at each end. Specific information on RMC may be found in *NEC Article 344*. RMC provides the best physical protection for conductors of any of the various types of conduit. RMC is supplied in 10' (3 m) lengths, including a threaded coupling on one end.

RMC may be made from steel or aluminum. Rigid metal steel conduit may be stainless, galvanized, or enamel-coated inside and out. Because of its threaded fittings, RMC provides an excellent equipment grounding conductor as defined in *NEC Section 250.118(2)*. A piece of RMC is shown in *Figure 4 (A)*. The support requirements for RMC are presented in *NEC Section 344.30(A) and (B) and NEC Table 344.30(B)(2)*.

RMC is mostly used in industrial applications. RMC is heavier than EMT and IMC. It is more

EMT Installation

EMT is easily cut to size using a conduit cutter, such as the one shown here.

Figure Credit: Greenlee / A Textron Company

difficult to cut and bend, usually requires threading of cut ends, and has a higher purchase price than EMT and IMC. As a result, the cost of installing RMC is generally higher than the cost of installing EMT and IMC.

1.1.3 Plastic-Coated RMC

Plastic-coated RMC has a thin coating of polyvinyl chloride (PVC) over the RMC. See *Figure 4 (B)*. This combination is useful when an environment calls for the ruggedness of RMC along with the corrosion resistance of rigid nonmetallic conduit. Plastic-coated RMC requires special threading and bending techniques. Typical installations where plastic-coated RMC may be required are:

- Chemical plants
- Food plants
- Refineries
- Fertilizer plants
- Paper mills
- Wastewater treatment plants

1.1.4 Aluminum Conduit

Aluminum conduit has several characteristics that distinguish it from steel conduit. Because it has better resistance to wet environments and some chemical environments, aluminum conduit generally requires less maintenance in installations such as sewage treatment plants.

NEC Section 300.6(B) states that aluminum conduit used in concrete or in direct contact with soil requires supplementary corrosion protection. According to Underwriters Laboratories *Electrical Construction Equipment Directory* (UL Green Book), examples of supplementary protection are paints

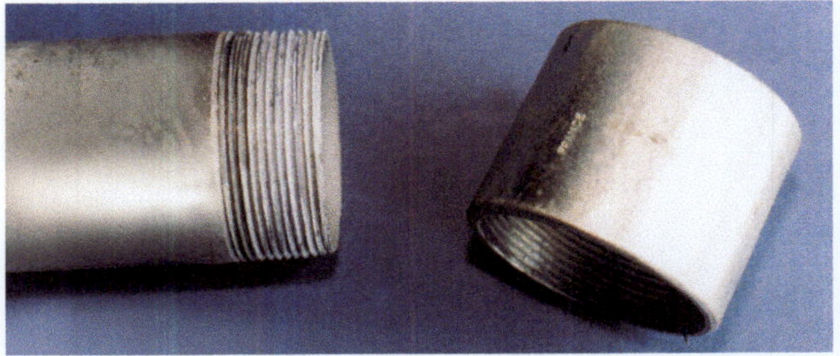

(A) RIGID METAL CONDUIT (RMC)

(B) PLASTIC-COATED RMC

Figure 4 Types of rigid metal conduit (RMC).

approved for the purpose (such as bitumastic paint), tape wraps approved for the purpose, or PVC-coated conduit.

> **NOTE**
>
> Caution must be exercised to avoid burial of aluminum conduit in soil or concrete that contains calcium chloride. Calcium chloride may interfere with the corrosion resistance of aluminum conduit. Calcium chloride and similar materials are often added to concrete to speed concrete setting. It is important to determine if chlorides are to be used in the concrete prior to installing aluminum conduit. If chlorides are to be used, aluminum conduit must be avoided. Check with local authorities regarding this type of usage.

1.1.5 Black Enamel Steel Conduit

Rigid black enamel steel conduit (often called black conduit) is steel conduit that is coated with a black enamel. In the past, this type of conduit was used exclusively for indoor wiring. Black enamel steel conduit is no longer manufactured for sale in the United States. It is mentioned only because it may still be found in existing installations.

1.1.6 Intermediate Metal Conduit

Intermediate metal conduit (IMC) is a type of rigid steel conduit. It has a wall thickness that is less than that of RMC but greater than that of EMT. The weight of IMC is approximately $\frac{2}{3}$ that of RMC. Because of its lower purchase price, lighter weight, and thinner walls, IMC installations are generally less expensive than comparable RMC installations. However, IMC installations still have high strength ratings.

> **NOTE**
>
> Additional information on IMC may be found in *NEC Article 342*.

The outside diameter of a given size of IMC is the same as that of the comparable size of RMC. Therefore, RMC fittings may be used with IMC. Because the threads on IMC and RMC are the same size, no special threading tools are needed to thread IMC. Some electricians feel that threading IMC is more difficult than threading RMC because IMC is somewhat harder.

The internal diameter of a given size of IMC is somewhat larger than the internal diameter of the same size of RMC because of the difference in wall thickness. Bending IMC requires the use of special shoes to support the conduit so it does not collapse.

The *NEC*® requires that IMC be identified along its length at 5' (1.5 m) intervals with the letters IMC. *NEC Sections 110.21(A)(1) and 342.120* describe this marking requirement.

Like RMC, IMC is permitted to act as an equipment grounding conductor, as defined in *NEC Section 250.118(3)*. The use of IMC may be restricted in some jurisdictions. It is important to investigate the requirements of each jurisdiction before selecting any materials.

1.1.7 Rigid Polyvinyl Chloride Conduit

The most common type of rigid nonmetallic conduit is manufactured from polyvinyl chloride (PVC). See *NEC Article 352*. Because PVC is noncorrosive, chemically inert, and non-aging, it is often used for installation in wet or corrosive environments. Corrosion problems found with steel and aluminum RMC do not occur with PVC. However, PVC may deteriorate under some conditions, such as extreme sunlight, unless marked sunlight resistant.

All PVC is marked according to standards established by the National Electrical Manufacturers Association (NEMA) or **Underwriters Laboratories, Inc. (UL)**. A section of PVC is shown in *Figure 5*.

Since PVC is lighter than steel or aluminum rigid conduit, IMC, or EMT, it is considered easier to handle. PVC can usually be installed much faster than other types of conduit because the joints are made up with cement and require no threading.

PVC contains no metal. This characteristic reduces the voltage drop of conductors carrying alternating current in PVC compared to identical conductors in steel conduit.

Because PVC is nonconducting, it cannot be used as an equipment grounding conductor. An

<table>
<tr><td>

Use of Aluminum Conduit

Aluminum conduit is used for special purposes such as high-cycle lines (400 cycles or above); around cooling towers, food service areas, and other applications in which corrosion is a factor; or where magnetic induction is a concern, such as near magnetic resonance imaging (MRI) equipment in hospitals.

</td></tr>
</table>

equipment grounding conductor sized in accordance with *NEC Table 250.122* must be pulled in each PVC conductor run (except for underground service-entrance conductors).

PVC is available in a variety of lengths. However, some jurisdictions require it to be cut to 10' (3 m) prior to installation. PVC is subject to expansion and contraction directly related to the difference in temperature, plus any radiating effects on the conduit. In moderate climates, even a 10' (3 m) installation of PVC would require an expansion joint per the *NEC*®. Each straight section of conduit run must be treated independently from other sections when connected by elbows. To avoid damage to PVC caused by temperature changes, expansion couplings are used. *Figure 6* shows various PVC fittings. The inside of the coupling is sealed with one or more O-rings. This type of coupling may allow up to 6" (150 mm) of movement. Check the requirements of the local jurisdiction prior to installing PVC.

PVC is manufactured in the following two types:

- *Type EB* – Thin wall for underground use only when encased in concrete. Also referred to as Type I.
- *Type DB* – Thick wall for underground use without encasement in concrete. Also referred to as Type II.

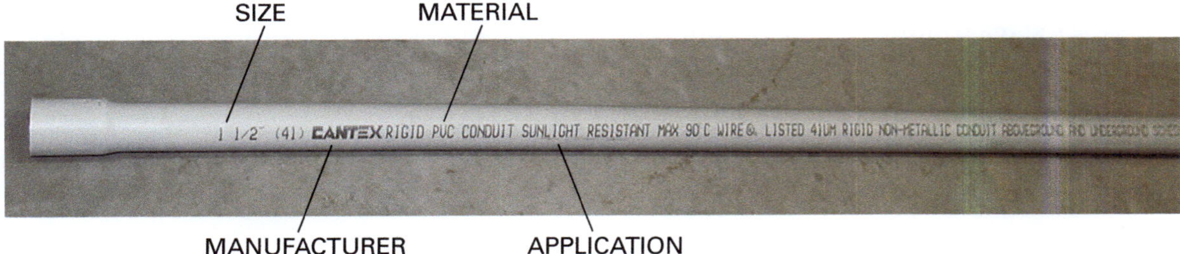

Figure 5 Rigid nonmetallic conduit.

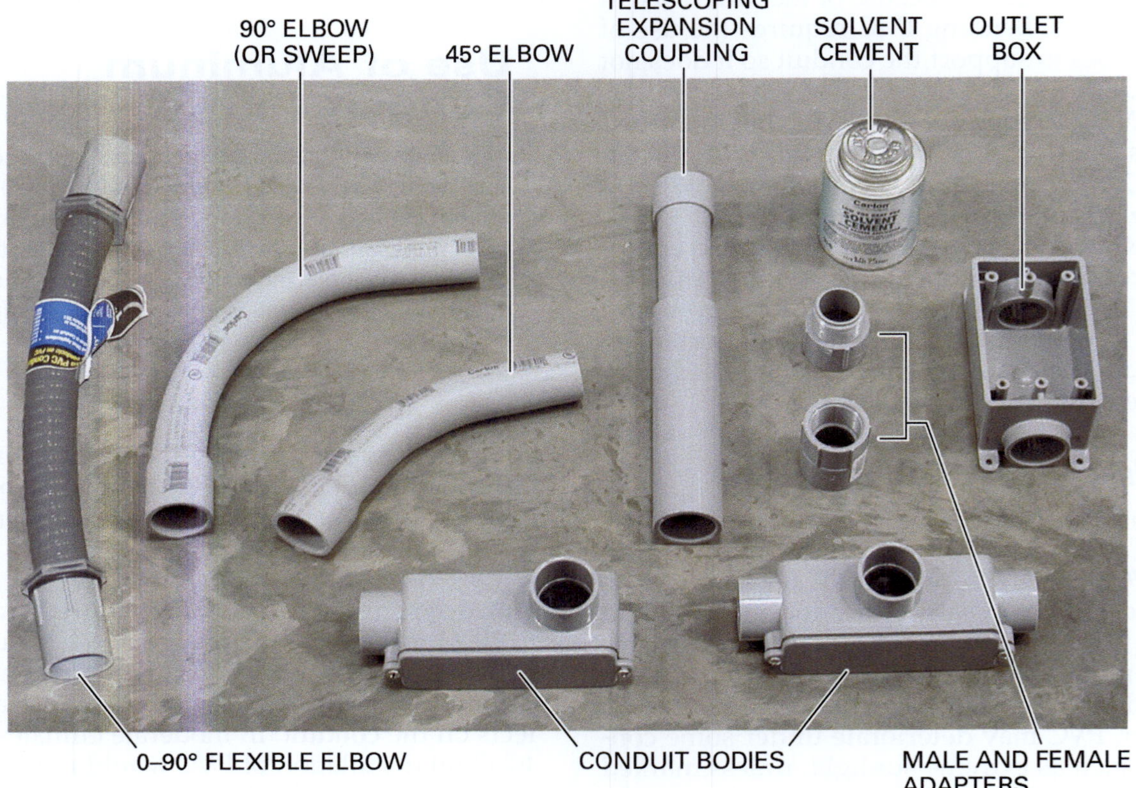

Figure 6 PVC fittings.

Type DB is available in the following two wall thicknesses:

- Schedule 40 is heavy wall for direct burial in the earth and aboveground installations.
- Schedule 80 is extra heavy wall for direct burial in the earth, aboveground installations for general applications, and installations where the conduit is subject to physical damage.

PVC is affected by higher-than-usual ambient temperatures. Support requirements for PVC are found in *NEC Section 352.30(B) and Table 352.30*. As with other conduit, it must be supported within 3' (900 mm) of each device or outlet box, junction box, conduit body, or other termination, but the maximum spacing between supports depends upon the size of the conduit. Some of the regulations for the maximum spacing of supports are:

- $\frac{1}{2}$" to 1" conduit (MD 16 to MD 27): every 3' (900 mm)
- $1\frac{1}{4}$" to 2" conduit (MD 35 to MD 53): every 5' (1.5 m)
- $2\frac{1}{2}$" to 3" conduit (MD 63 to MD 78): every 6' (1.8 m)
- $3\frac{1}{2}$" to 5" conduit (MD 91 to MD 129): every 7' (2.1 m)
- 6" conduit (MD 155): every 8' (2.5 m)

Liquidtight Conduit

Liquidtight conduit protects conductors from vapors, liquids, and solids. Liquidtight conduit that includes an inner metal core is widely used in commercial and industrial construction.

NCCER – *Maritime Electrical*

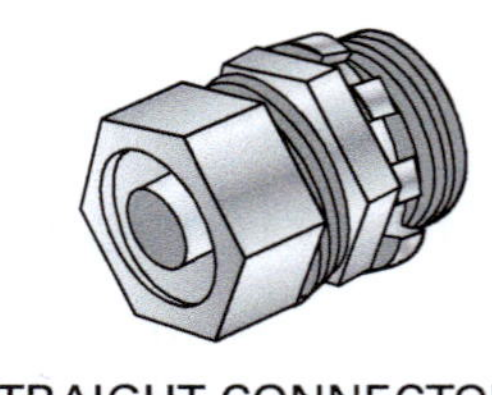
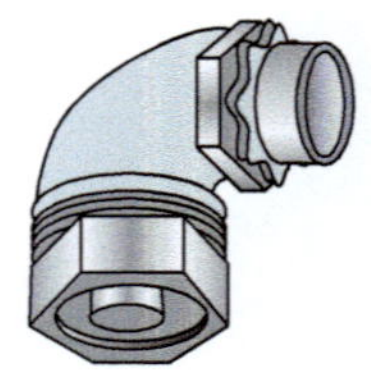

Figure 7 Liquidtight flex connectors.

1.1.8 High-Density Polyethylene Conduit

High-density polyethylene conduit (HDPE) is a rigid nonmetallic conduit listed for underground installations. (It is not listed for aboveground use.) See *NEC Article 353*. It is suitable for direct burial or where encased in concrete. In many signaling and communications applications, it is provided on reels with conductors pre-installed and may be laid in a trench or plowed into the earth.

1.1.9 Liquidtight Flexible Nonmetallic Conduit

Liquidtight flexible nonmetallic conduit (LFNC) was developed as a raceway for industrial equipment where flexibility was required and protection of conductors from liquids was also necessary. LFNC is covered in *NEC Article 356*. Usage of LFNC has been expanded from industrial applications to outside and direct burial usage where listed.

Several varieties of LFNC have been introduced. The first product (LFNC-A) is commonly referred to as hose. It consists of an inner and outer layer of neoprene with a nylon reinforcing web between the layers. A second-generation product (LFNC-B), and most widely used, consists of a smooth wall, flexible PVC with a rigid PVC integral reinforcement rod. The third product (LFNC-C) is a nylon corrugated shape without any integral reinforcements. These three permitted LFNC raceway designs must be flame resistant with fittings **approved** for installation of electrical conductors. Nonmetallic connectors are listed for use and some liquidtight metallic flexible conduit connectors are dual-listed for both metallic and nonmetallic liquidtight flexible conduit.

LFNC is sunlight-resistant and suitable for use at conduit temperatures of 80°C dry and 60°C wet. It is available in $\frac{3}{8}$" (MD 12) through 4" (MD 103) sizes. *NEC Section 356.12* states that LFNC cannot be used where subject to physical damage. LFNC is also limited in length to no longer than 6' (1.8 m), except where properly secured, where flexibility is required, or as permitted by *NEC Section 356.10*. Also, it cannot be used in any hazardous (classified) locations except as specified in other articles of the *NEC®*.

Liquidtight flexible metal conduit is a raceway of circular cross section having an outer liquidtight, nonmetallic, sunlight-resistant jacket over an inner flexible metal core with associated couplings and connectors covered by *NEC Article 350*.

Compression connectors are used to connect liquidtight flexible conduit to boxes or equipment. They are available in straight, 45°, and 90° configurations (*Figure 7*).

1.1.10 Flexible Metal Conduit

Flexible metal conduit (FMC), also called flex, may be used for many kinds of wiring systems. FMC is covered in *NEC Article 348*. Flexible metal conduit is made from a single strip of steel or aluminum, wound and interlocked. It is typically available in diameters from $\frac{3}{8}$" (MD 12) through 4" (MD 103). An illustration of flexible metal conduit is shown in *Figure 8*.

Flexible metal conduit is often used to connect equipment or machines that vibrate or move slightly during operation. Also, final connection to equipment having an electrical connection point that is marginally **accessible** is often accomplished with flexible metal conduit.

Flexible metal conduit is easily bent, but the minimum bending radius is the same as for other types of conduit. It should not be bent more than the equivalent of four quarter bends (360° total) between pull points (e.g., conduit bodies and boxes). It can be connected to boxes with a flexible conduit connector and to rigid conduit or EMT by using a combination coupling. A flexible-to-rigid combination coupling is shown in *Figure 9*.

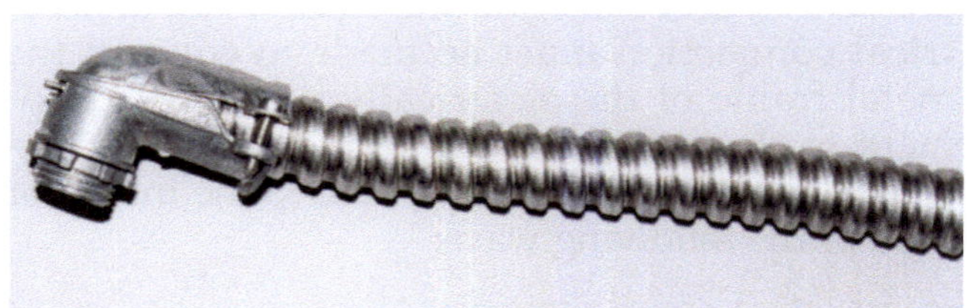
Figure 8 Flexible metal conduit.

Figure 9 Combination coupling.

Flexible metal conduit is generally available in two types: nonliquidtight and liquidtight. *NEC Articles 348 and 350* cover the uses of flexible metal conduit.

Liquidtight flexible metal conduit has an outer covering of liquidtight, sunlight-resistant flexible material that acts as a moisture seal. It is intended for use in wet locations. It is used primarily for equipment and motor connections when movement of the equipment is likely to occur. The number of bends, size, and support requirements for liquidtight conduit are the same as for all flexible conduit. Fittings used with liquidtight conduit must also be of the liquidtight type.

Support requirements for flexible metal conduit are found in *NEC Sections 348.30 and 350.30*. Straps or other means of securing the flexible metal conduit must be spaced every 4 $\frac{1}{2}$' (1.4 m) and within 12" (300 mm) of each end. (This spacing is closer together than for rigid conduit.) However, at terminals where flexibility is necessary, longer lengths are permitted per *NEC Sections 348.30 and 350.30, Exceptions*. For example, where flexibility is required for conduit sizes $\frac{1}{2}$" through $1\frac{1}{4}$" (MD 16 through MD 35), lengths of up to 36" (900 mm) without support are permitted. Failure to provide proper support for flexible conduit can make pulling conductors difficult.

1.2.0 Bonding Conduit

For safety reasons, most equipment that receives electrical power and has a metallic frame is bonded. In order to bond the equipment, an electrical connection must be made to connect the metal frame of the electrically powered equipment to the grounding point at the service-entrance equipment. This is usually done in one or both of the following ways:

- The frame of the equipment is connected to a wire (equipment grounding conductor), which is directly connected to the ground point at the grounding terminal.
- The frame of the equipment is connected (bonded) to a metal conduit or other type of raceway system, which provides an uninterrupted and low-impedance circuit to the ground point at the service-entrance equipment. The metal raceway or conduit acts as the equipment grounding conductor.

According to *NEC Section 250.96*, metal raceways, cable trays, cable armor, cable sheath, enclosures, frames, fittings, and other metal noncurrent-carrying parts that are to serve as equipment grounding conductors, with or without the use of supplementary equipment grounding conductors, shall be bonded where necessary to ensure electrical continuity and safely conduct any fault current likely to be imposed on them. Any nonconductive paint, enamel, or similar coating shall be removed at threads, contact points, and contact surfaces or be connected by means of fittings designed so as to make such removal unnecessary.

The purpose of the equipment grounding conductor is to provide a low-resistance path to ground for all equipment that receives power. This is done so that if an ungrounded conductor comes in contact with the frame of a piece of equipment, the circuit overcurrent device immediately acts to open the circuit. It also reduces the voltage to ground that would be present on the faulted equipment if a person came in contact with the equipment frame.

1.3.0 Metal Conduit Fittings

A large variety of conduit fittings are available to do electrical work. Manufacturers design and construct fittings to permit a multitude of applications. The type of conduit fitting used in a particular application depends upon the size and type of conduit, the type of fitting needed for the application, the location of the fitting, and the installation method. The requirements and proper applications of boxes, conduit bodies, or fittings are found in *NEC Section 300.15*. Some of the more common types of fittings are examined in the following sections.

1.3.1 Couplings

Couplings are sleeve-like fittings that are threaded inside to join two male threaded pieces of rigid conduit or IMC. A piece of conduit with a coupling is shown in *Figure 10*.

Other types of couplings may be used depending upon the location and type of conduit. Several types are shown in *Figure 11*.

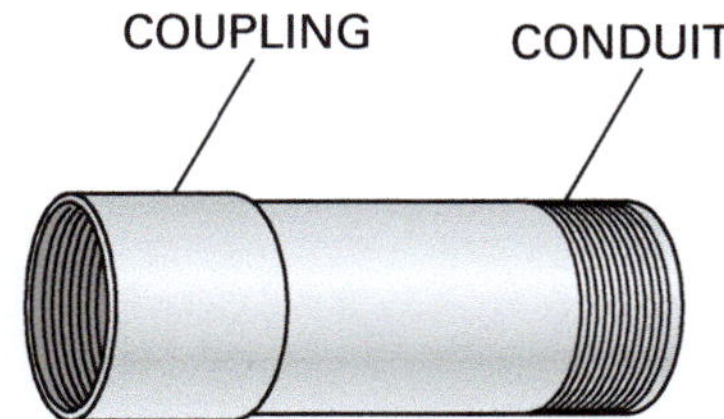

Figure 10 Conduit and coupling.

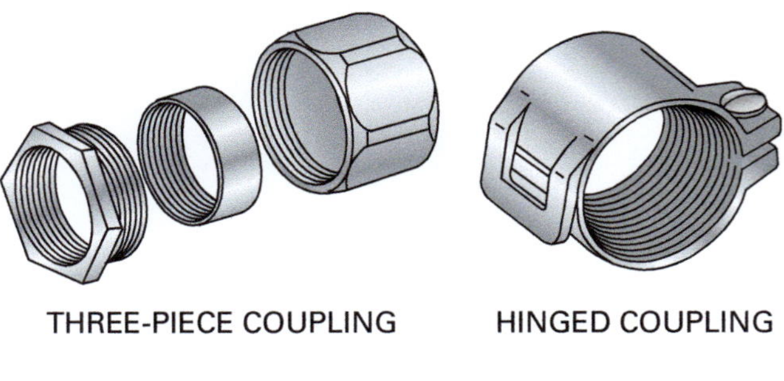
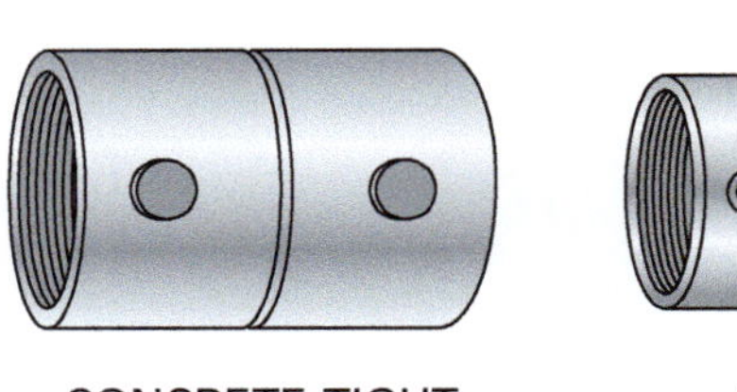
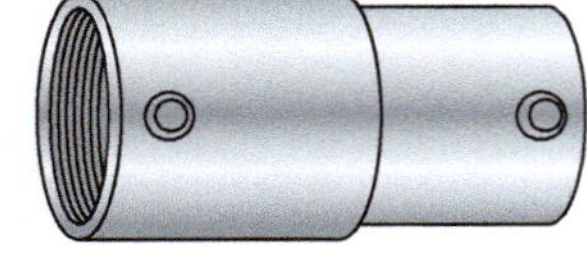

Figure 11 Metal conduit couplings.

1.3.2 Insulating Bushings

An insulating bushing is either nonmetallic or has an insulated throat. Insulating bushings are installed on the threaded end of conduit that enters a sheet metal enclosure. Bushings can be grounding or nongrounding.

The purpose of a nongrounding insulating bushing is to protect the conductors from being damaged by the sharp edges of the threaded conduit end. *NEC Section 300.15(C)* states that where a conduit enters a box, fitting, enclosure, or conduit termination, a fitting must be provided to protect the wire from abrasion. *NEC Section 312.6(C)* references *NEC Section 300.4(G)*, which states that where ungrounded conductors of No. 4 AWG or larger enter a raceway in a cabinet or box enclosure, the conductors shall be protected by a substantial fitting providing a smoothly rounded insulating surface, unless the conductors are separated from the raceway fitting by substantial insulating material securely fastened in place. An exception is where threaded hubs or bosses that are an integral part of a cabinet, box, enclosure, or raceway provide a smoothly rounded or flared entry for conductors. Insulating bushings are shown in *Figure 12*.

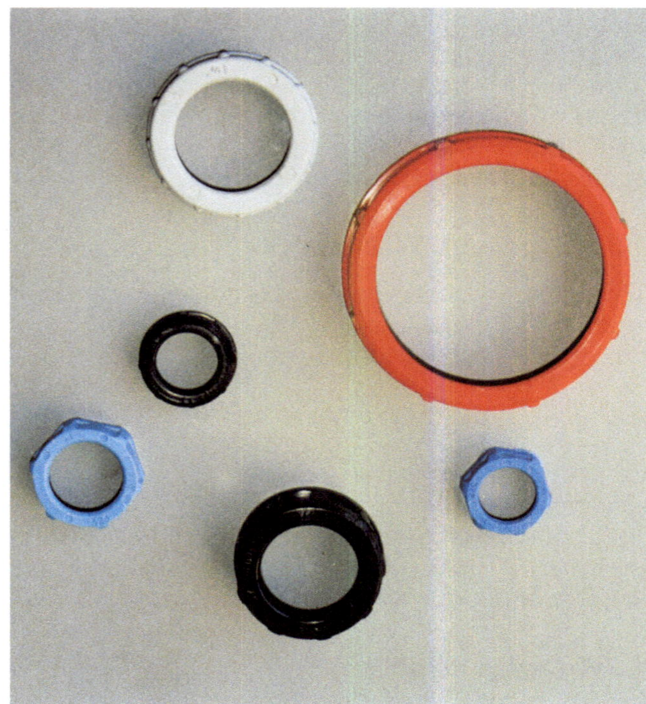

Figure 12 Insulating bushings.

Installation of Conduit Bodies

It will be much easier to identify conduit bodies once you begin to see them in use. Here you see liquidtight nonmetallic conduit entering a Type T conduit body (A) and a Type LB conduit body in an outdoor commercial application (B).

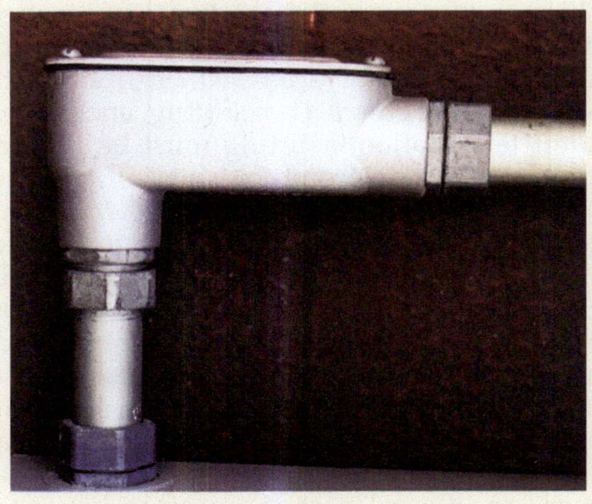

(B)

(A)

Grounded insulating bushings, usually called grounding bushings, are used to protect conductors and also have provisions for connection of an equipment grounding conductor. The ground wire, once connected to the grounding bushing, may be connected to the enclosure to which the conduit is connected. Grounding insulating bushings are shown in *Figure 13*.

1.3.3 Threaded Weatherproof Hubs

Threaded weatherproof hubs are used for conduit entering a box in a wet location. *Figure 14* shows typical threaded weatherproof hubs.

1.3.4 Offset Nipples

Offset nipples are used to connect two pieces of electrical equipment in close proximity where a slight offset is required. They come in sizes ranging from $\frac{1}{2}$" to 2" (MD 16 to MD 53) in diameter. See *Figure 15*.

1.3.5 Conduit Bodies

Conduit bodies, also called condulets, are a separate portion of a conduit or tubing system that provide access through a removable cover(s) to the interior of the system at a junction of two or more sections of the system, a pull point, or at a terminal point of the system. They are usually cast and

Figure 13 Grounding insulating bushings.

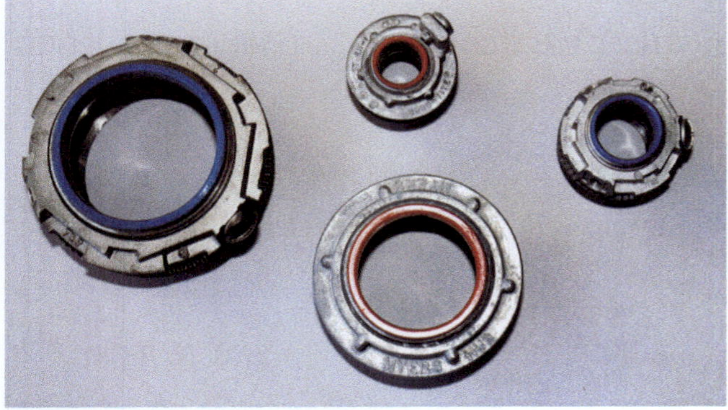

Figure 14 Threaded weatherproof hubs.

are significantly higher in cost than the stamped steel boxes permitted with EMT. However, there are situations in which conduit bodies are preferable, such as in outdoor locations, for appear-

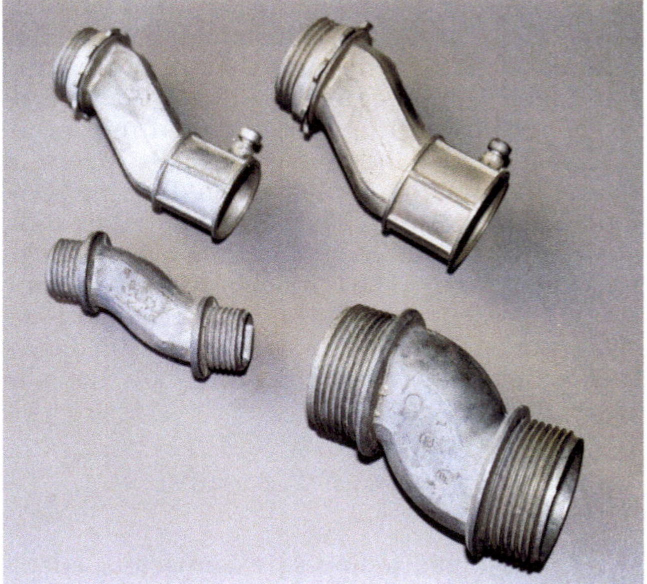

Figure 15 Offset nipples.

ance's sake in an **exposed location**, or to change types or sizes of raceways. Also, conduit bodies do not have to be supported, as do stamped steel pull and junction boxes. They are also used when elbows or bends would not be appropriate.

NEC Section 314.16(C)(2) states that conduit bodies cannot contain a **splice**, **tap**, or device unless they are durably and legibly marked by the manufacturer with their volume capacity. The maximum number of conductors permitted in a conduit body is found using *NEC Table 314.16(B)*. (This information is shown in *Table 1*.)

Type C conduit bodies may be used to provide a pull point in a long conduit run or a conduit run that has bends totaling more than 360°. A Type C conduit body is shown in *Figure 16*.

When referring to conduit bodies, the letter L represents an elbow. A Type L conduit body is used as a pulling point for conduit that requires a 90° change in direction. The cover is removed, then the wire is pulled out, coiled on the ground or floor, reinserted into the other conduit body's opening, and pulled. The cover and its associated gasket are then replaced. Type L conduit bodies are available with the cover on the back (Type LB), on the sides (Type LL or LR), or on both sides (Type LRL). Type L conduit bodies are shown in *Figure 17*.

<table>
<tr><td rowspan="2" style="border:1px solid #000; padding:4px;">NOTE</td><td style="border:1px solid #000; padding:4px;">The cover and gasket must be ordered separately. Do not assume that these parts come with conduit bodies when they are ordered.</td></tr>
</table>

Table 1 Volume Required per Conductor [Data from NEC Table 314.16(B)]

Size of Conductor (AWG)	Free Space Within Box for Each Conductor
No. 18	1.5 cu in
No. 16	1.75 cu in
No. 14	2.0 cu in
No. 12	2.25 cu in
No. 10	2.5 cu in
No. 8	3.0 cu in
No. 6	5.0 cu in

Reprinted with permission from NFPA 70-2017, *National Electrical Code®*, Copyright © 2016, National Fire Protection Association, Quincy, MA. This reprinted material is not the complete and official position of the NFPA on the referenced subject, which is represented only by the standard in its entirety which may be obtained through the NFPA website at **www.nfpa.org**.

Figure 16 Type C conduit body.

To identify Type L conduit bodies, use the following method:

Step 1   Hold the body like a pistol.

Step 2   Locate the opening on the body:

- If the opening is to the left, it is a Type LL.
- If the opening is to the right, it is a Type LR.
- If the opening is on top (back), it is a Type LB.
- If there are openings on both the left and the right, it is a Type LRL.

Type T conduit bodies are used to provide a junction point for three intersecting conduits and are used extensively in conduit systems. A Type T conduit body is shown in *Figure 18*.

Type X conduit bodies are used to provide a junction point for four intersecting conduits. The removable cover provides access to the interior of the X so that wire pulling and splicing may be performed. A Type X conduit body is shown in *Figure 19*.

1.3.6 Sealing Fittings

Hazardous locations in manufacturing plants and other industrial facilities involve a wide vari-

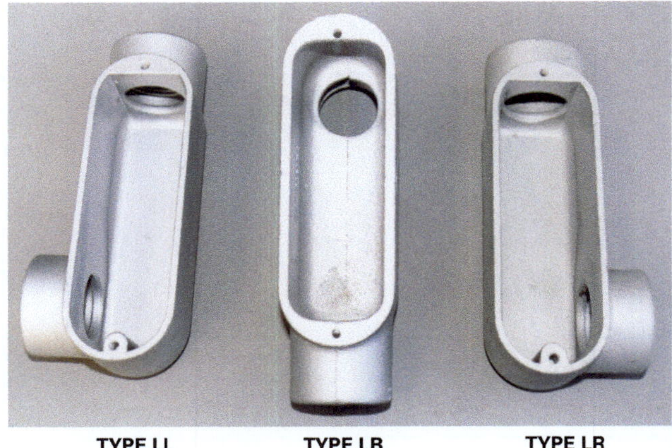

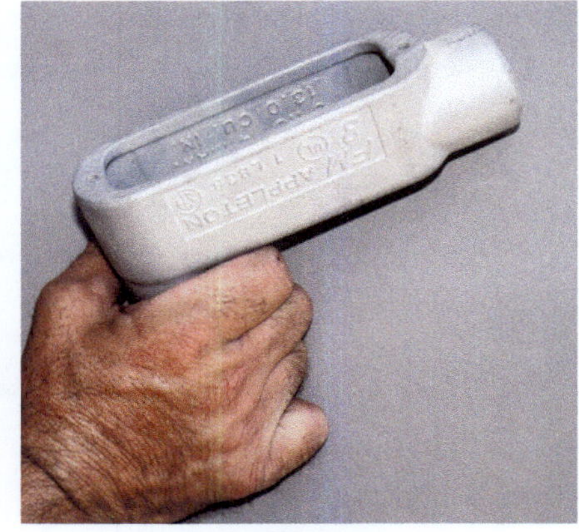

Figure 17 Type L conduit bodies and how to identify them.

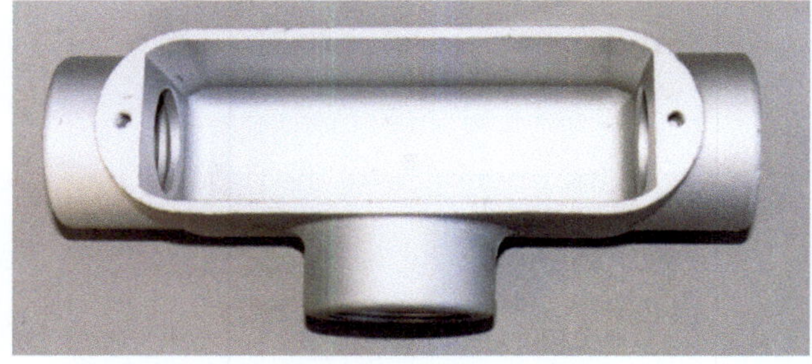

Figure 18 Type T conduit body.

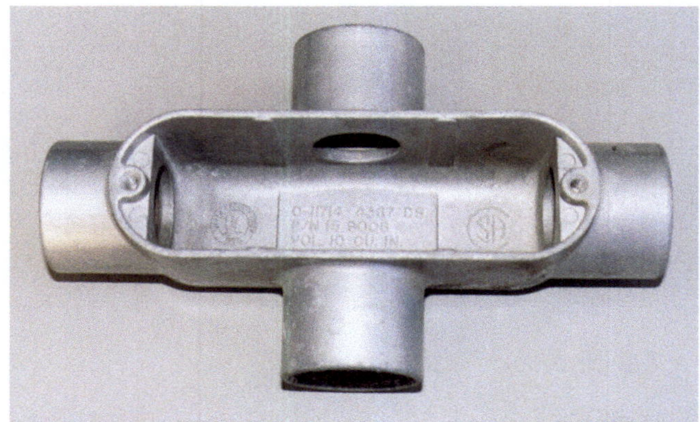

Figure 19 Type X conduit body.

ety of flammable gases and vapors and ignitable dusts. These hazardous substances have widely different flash points, ignition temperatures, and flammable limits requiring fittings that can be sealed. Sealing fittings are installed in conduit runs to minimize the passage of gases, vapors, or flames through the conduit and reduce the accumulation of moisture. They are required by *NEC Article 500* in hazardous locations where explosions may occur. They are also required where conduit passes from a hazardous location of one classification to another or to an unclassified location. Several types of sealing fittings are shown in *Figure 20*.

1.4.0 Making a Conduit-to-Box Connection

Conduit is joined to boxes by connectors, adapters, threaded hubs, or locknuts.

Bushings protect the wires from the sharp edges of the conduit. As previously discussed, bushings are usually made of plastic or metal. Some metal bushings have a grounding screw to permit a bonding wire to be installed.

Locknuts (*Figure 21*) are used on the inside and outside walls of the box to which the conduit is connected. A grounding locknut may be needed if a bonding wire is to be installed. Special sealing locknuts are also used in wet locations.

When joining metal conduit to metal boxes, a means must be provided in each metal box for the connection of an equipment grounding conductor. The means shall be permitted to be a tapped hole or equivalent per *NEC Section 314.40(D)*. A proper conduit-to-box connection is shown in *Figure 22*.

In order to make a good connection, use the following procedure:

Step 1 Thread the external locknut onto the conduit. Run the locknut to the bottom of the threads.

Step 2 Insert the conduit into the box opening.

Step 3 If an inside locknut or grounding locknut is required, screw it onto the conduit inside the box opening.

Step 4 Screw the bushing onto the threads projecting into the box opening. Make sure the bushing is tightened as much as possible.

Step 5 Tighten the external locknut to secure the conduit to the box.

It is important that the bushings and locknuts fit tightly. For this reason, the conduit must enter

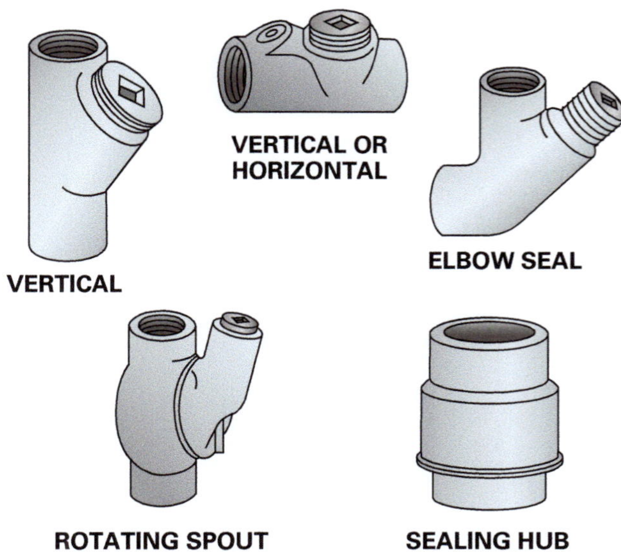

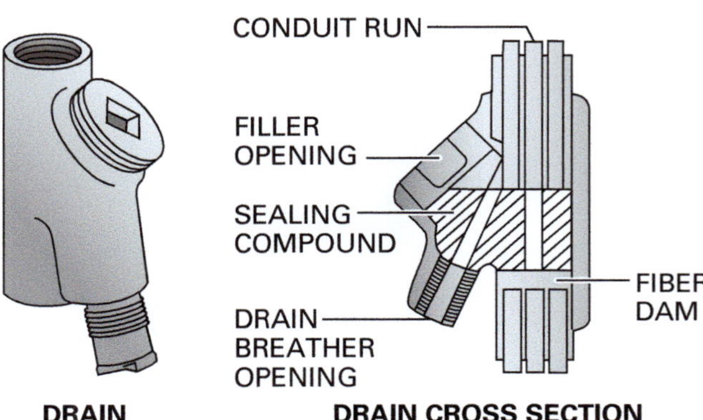

Figure 20 Sealing fittings.

straight into the box. This may require that a box offset or **kick** be made in the conduit.

1.5.0 Raceway Supports

Raceway supports are available in many types and configurations. This section discusses common conduit supports found in electrical installations. *NEC Section 300.11(B)* discusses the requirements for branch circuit wiring that is supported from above suspended ceilings. Electrical equipment and raceways must have their own supporting methods and may not be supported by the supporting hardware of a fire-rated roof/ceiling assembly.

1.5.1 Straps

Straps are used to support conduit to a surface (see *Figure 23*). The spacing of these supports must conform to the minimum support spacing requirements for each type of conduit. One- and two-hole straps are used for all types of conduit: EMT, RMC,

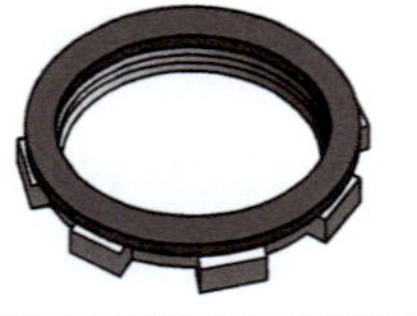

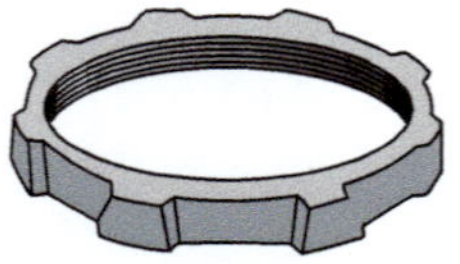

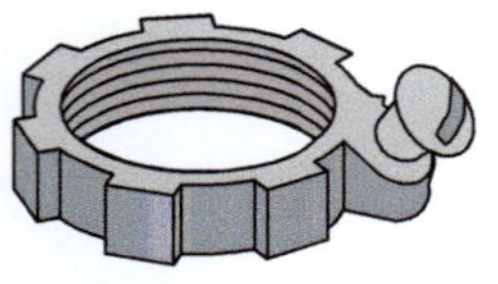

Figure 21 Locknuts.

IMC, PVC, and flex. The straps can be flexible or rigid. Two-part straps are used to secure conduit to electrical framing channels (struts). Parallel and right angle beam clamps are also used to support conduit from structural members.

Clamp back straps can also be used with a backplate to maintain the $\frac{1}{4}$" (6 mm) spacing from the surface required for installations in wet locations.

1.5.2 Standoff Supports

The standoff support, often referred to as a Minerallac® (the name of a manufacturer of this type of support), is used to support conduit away from the supporting structure. In the case of the one-hole and two-hole straps, the conduit must be offset wherever a fitting occurs. If standoff supports are used, the conduit is held away from the supporting surface, and no offsets are required in the conduit at the fittings. Standoff supports may be used to support all types of conduit including RMC, IMC, EMT, PVC, and flex, as well as tubing installations. A standoff support is shown in *Figure 24*.

1.5.3 Electrical Framing Channels

Electrical framing channels or other similar framing materials are used together with Unistrut®-type conduit clamps to support conduit (*Figure 25*). They may be attached to a ceiling, wall, or other surface or be supported from a trapeze hanger.

1.5.4 Beam Clamps

Beam clamps are used with suspended hangers. The raceway is attached to or laid in the hanger. The hanger is suspended by a threaded rod. One end of the threaded rod is attached to the hanger and the other end is attached to a beam clamp.

Installing Sealing Fittings

These fittings must be sealed after the wires are pulled. A fiber dam is first packed into the base of the fitting between and around the conductors, then the liquid sealing compound is poured into the fitting. Speed sealing materials are also available that eliminate the need to insert a fiber dam.

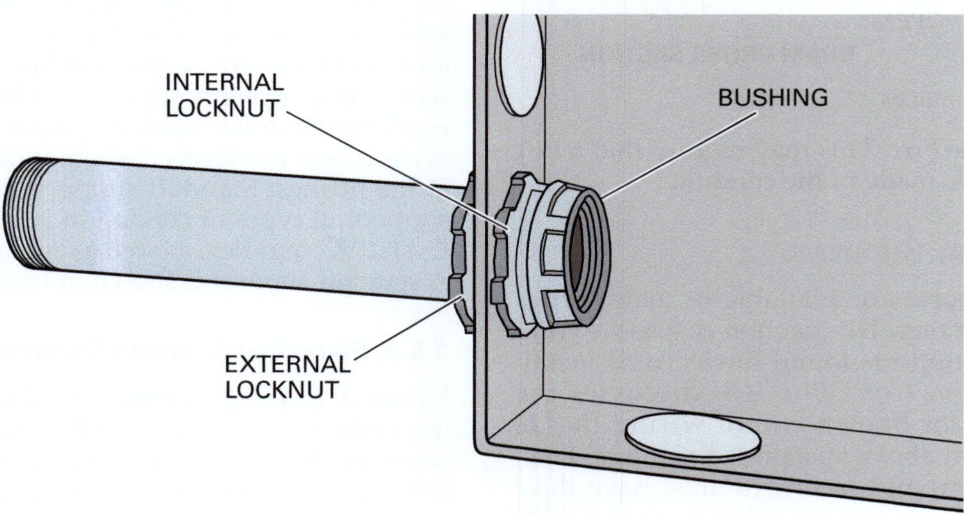

Figure 22   Conduit-to-box connection.

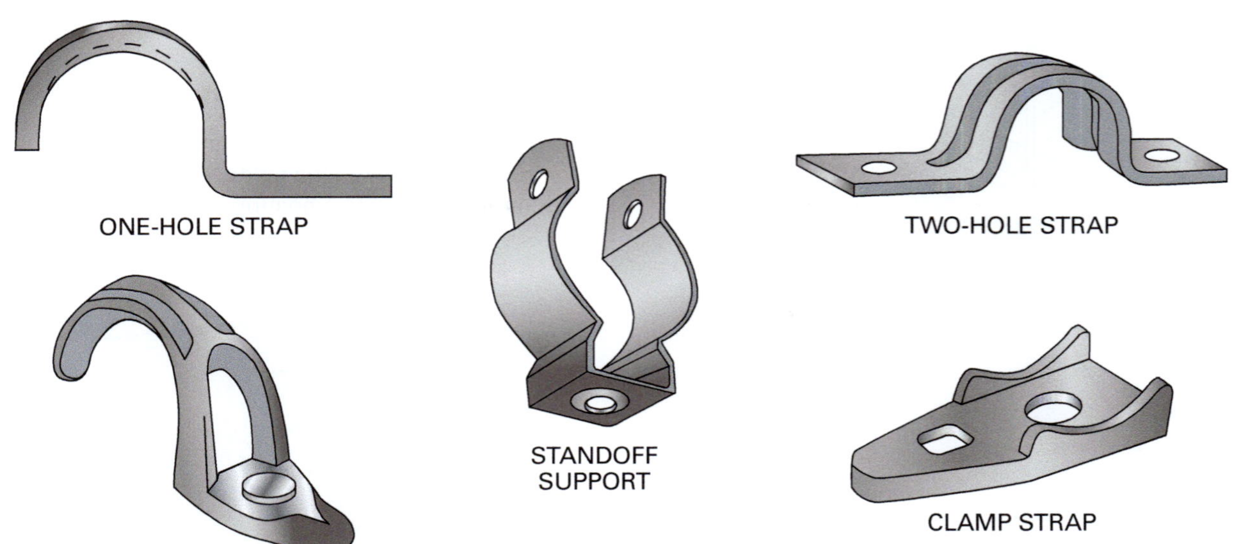

Figure 23 Straps.

Figure 24 Standoff support.

Wall-Mounted Supports

This wall-mounted support has been fabricated to hold the conduit away from the metal building. While it is shown with only one raceway, additional raceways can be added to the framing channel.

Figure Credit: Tim Dean

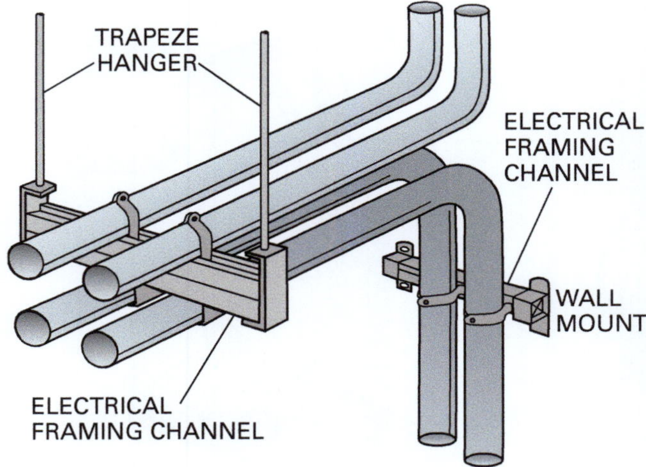

Figure 25 Electrical framing channels.

The beam clamp is then attached to a beam. A beam clamp with wireway support assembly is shown in *Figure 26*.

1.6.0 Installation Requirements for Various Construction Methods

Conduit and box installation varies with the type of construction. This section discusses some special requirements for masonry and concrete, metal framing, wood, and structural steel construction.

1.6.1 Masonry and Concrete Flush-Mount Construction

In a reinforced concrete construction environment, the conduit and boxes must be embedded in the concrete to achieve a flush surface. Ordinary boxes may be used, but special concrete boxes are preferred and are available in depths up to 6" (150 mm). These boxes have special ears by which they are nailed to the wooden forms for the concrete. When installing them, stuff the boxes tightly with paper to prevent concrete from seeping in. *Figure 27* shows an installed box.

Flush construction can also be done on existing concrete walls, but this requires chiseling a channel and box opening, anchoring the box and conduit, and then resealing the wall.

To achieve flush construction with masonry walls, the most acceptable method is for the electrician to work closely with the mason laying the blocks. When the construction blocks reach the convenience outlet elevation, boxes are made up as shown in *Figure 28*. The figure shows a raised tile ring or box device cover.

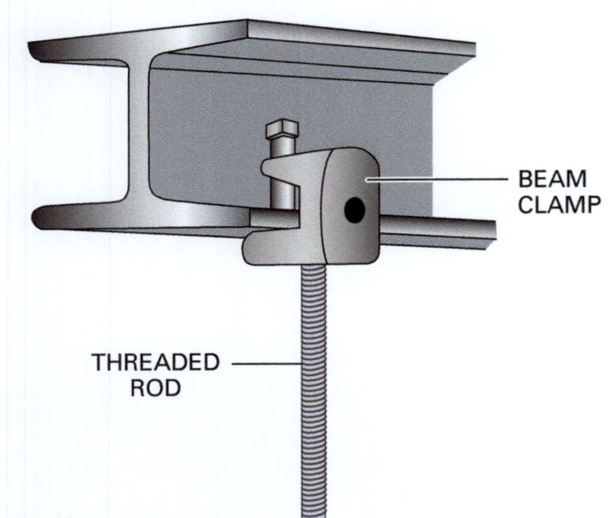

Bundling Conductors

When conductors are bundled together in a wireway their magnetic fields tend to cancel, thus minimizing inductive heating in the conductors.

Figure 26 Beam clamp.

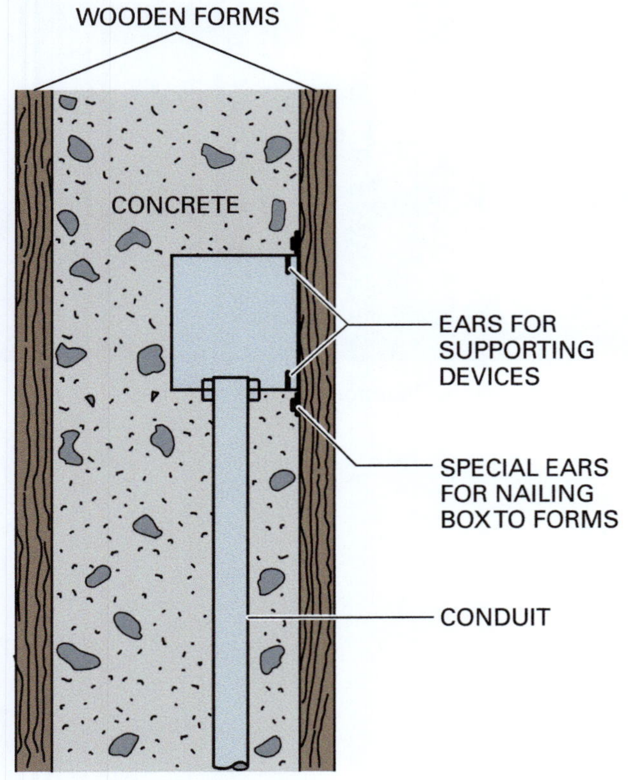

Figure 27 Concrete flush-mount installation.

Figure 29 shows a masonry box that needs no extension or deep plaster ring to bring it to the surface.

Sections of conduit are then coupled in short (approximately 3' to 5' or 900 mm to 1. 5 m) lengths. This is done because it is impractical for the mason to maneuver blocks over 10' (3 m) sections of conduit.

The electrician must work with the mason to ensure the box is properly grouted and sealed.

1.6.2 Metal Stud Environment

Metal stud walls are a popular method of construction for the interior walls of commercial buildings. Metal stud framing consists of rela-

Figure 28 Box with raised ring.

Figure 29 Three-gang concrete box.

tively thin metal channel studs, usually constructed of galvanized steel and with an overall dimension the same as standard 2 x 4 wooden studs. Wiring in this type of construction is relatively easy when compared to masonry.

EMT conduit and MC cable are the most common type of wiring methods for metal stud environments. Metal studs usually have some number of pre-punched holes that can be used to route the conduit. If a pre-punched hole is not located where it needs to be, holes can be easily punched in the metal stud with a hole cutter or knockout punch (*Figure 30*).

Boxes can be secured to the metal stud using self-tapping screws or one of the many types of box supports available. EMT conduit is supported by the metal studs using conduit straps or

Cutting or punching metal studs can create sharp edges. Avoid contact that can result in cuts.

other approved methods. It is important that the conduit be properly supported to facilitate pulling the conductors through the tubing. Boxes are mounted on the metal studs so that the box will be flush with the finished walls. You must know what the finished wall thickness is going to be to properly secure the boxes to the metal studs. For example, if the finished wall will be $\frac{5}{8}$" (16 mm) drywall, then the box must be fastened so that it protrudes $\frac{5}{8}$" (16 mm) from the metal stud.

According to *NEC Section 300.4(B)(1)*, NM cable run through metal studs must be protected by listed bushings or listed grommets (*Figure 31*). This protects the cables from the friction of pull-

When using a screw gun or cordless drill to mount boxes to studs, keep the hand holding the box away from the gun/drill to avoid injury.

ing during installation and from the weight of the cable and vibrations following the installation.

1.6.3 Wood Frame Environment

At one time, the use of rigid conduit in partitions and ceilings was a time-consuming

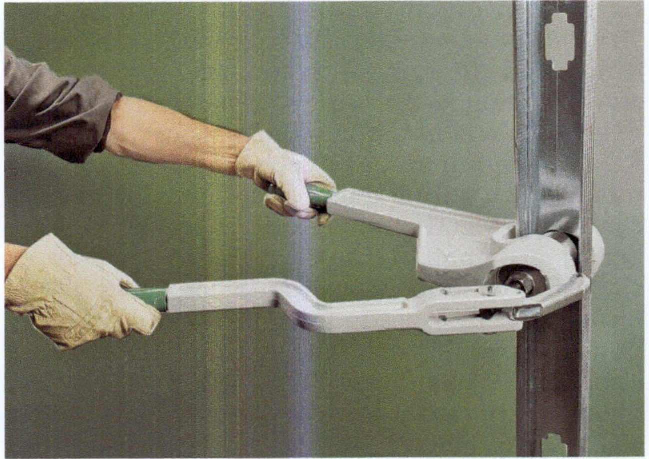

Figure 30 Metal stud punch.

operation. Thinwall conduit makes an easier and quicker job, largely because of the types of fittings that are specially adapted to it.

Figure 32 shows two methods of running thinwall conduit in these locations: boring timbers and notching them. When boring, holes must be drilled large enough for the tubing to be inserted between the studs. The tubing is cut rather short, calling for multiple couplings. EMT can be bowed quite a bit while threading through holes in studs. Boring is the preferred method.

NEC Section 300.4 addresses the requirements to prevent physical damage to conductors and cabling in wood members. By keeping the edge of

the drilled hole $1\frac{1}{4}"$ (32 mm) from the closest edge of the stud, nails are not likely to penetrate the stud far enough to damage the cables. The building codes provide maximum requirements for bored or notched holes in studs.

NEC Section 300.4(A)(1) requires the use of a steel plate or bushing at least $\frac{1}{16}"$ (1.6 mm) thick or a listed steel nail plate where wiring is installed through bored wooden members less than $1\frac{1}{4}"$ (32 mm) from the nearest edge (*Figure 33*). Nail plates are also required to protect the conductors in all notched wooden members per *NEC Section 300.4(A)(2)*.

The exception in the *NEC*® permits IMC, RMC, PVC, and EMT to be installed through bored holes or laid in notches less than $1\frac{1}{4}"$ (32 mm)

Figure 31 NM cable protected by grommets.

from the nearest edge without a steel plate or bushing.

Because of its weakening effect upon the structure, notching should be resorted to only where absolutely necessary. Notches should be as narrow as possible and in no case deeper than $\frac{1}{16}$ the stock of a bearing timber. A bearing timber supports floor joists or other weight.

Some wood I-beams are manufactured with perforated knockouts in their web, approximately 12" (300 mm) apart. Never notch or drill through

the beam flange or cut other openings in the web without checking the manufacturer's specification sheet. Also, do not drill or notch other types of engineered lumber without first checking the specification sheets.

1.6.4 Metal Buildings

Many commercial and industrial buildings are prefabricated structures with steel structural supports, and roofing and siding made of light-gauge metal sheets (*Figure 34*). Conduit can be routed across the structural members that support the roof. *NEC Section 300.4(E)* states that a cable, raceway, or box in exposed or concealed locations under metal-corrugated sheet roof decking must be installed and supported so the nearest outside

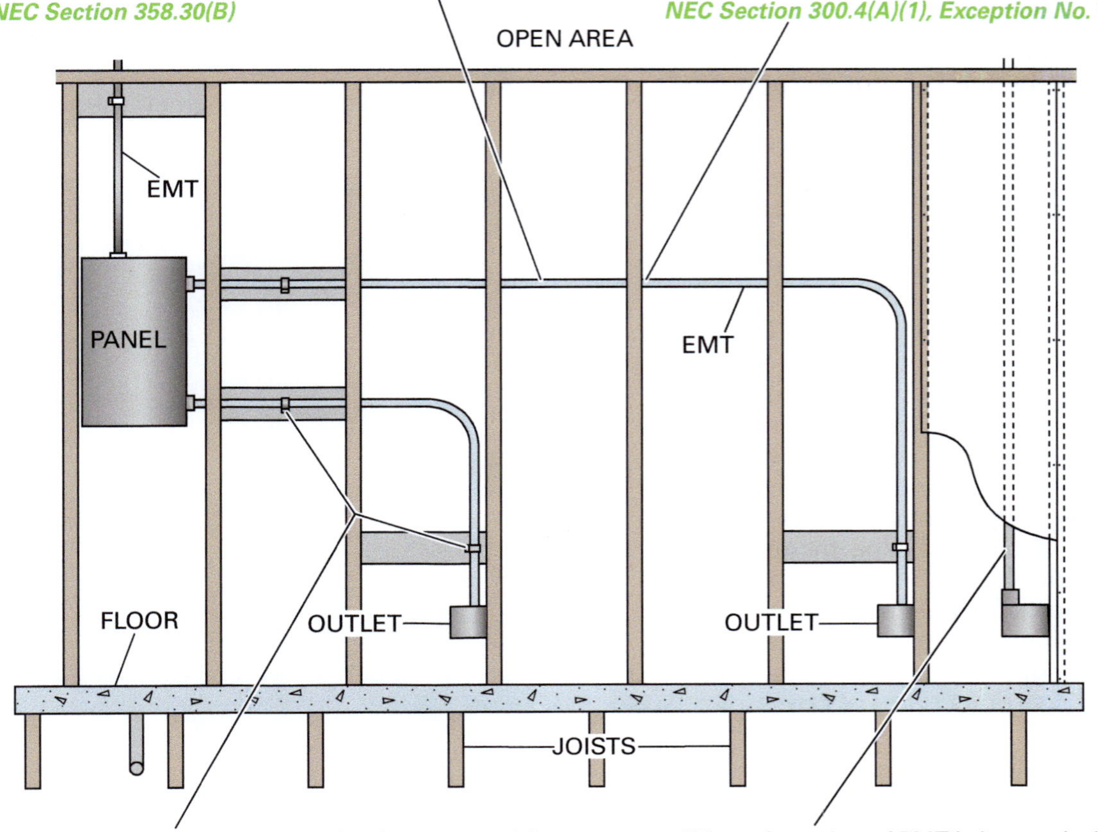

Figure 32 Installing wire or conduit in a wood-frame building.

Figure 33 Steel nail plate.

Figure 34 Metal building.

surface of the cable or raceway is not less than $1\frac{1}{2}$" (38 mm) from the nearest surface of the roof decking. The roof structure can consist of beams and purlins (*Figure 35*) or open-web steel joists (*Figure 36*).

Beams and purlins should not be drilled through; consequently, the conduit is supported from the metal beams by anchoring devices designed especially for that purpose. The supports attach to the beams or supports and have clamps to secure the conduit to the structure. All conduit runs should be plumb since they are exposed. Bends should be correct and have a neat and orderly appearance.

Rigid metal conduit is often required in metal buildings. If a large number of conduits are run along the same path, strut-type systems are used. These systems are sometimes referred to as Unistrut® systems (Unistrut® is a manufacturer of these systems). Another manufacturer of strut systems is B-Line systems. Both are very similar. These systems use a channel-type member that can support conduits from the ceiling by using threaded rod supports for the channel, as shown in *Figure 37*. Strut channel can also be secured to masonry walls to support vertical runs of conduit, wireways, and various types of boxes.

Additional Resources

Benfield Conduit Bending Manual, 2nd Edition. Overland

Figure 36 Open-web steel joist roof supports.

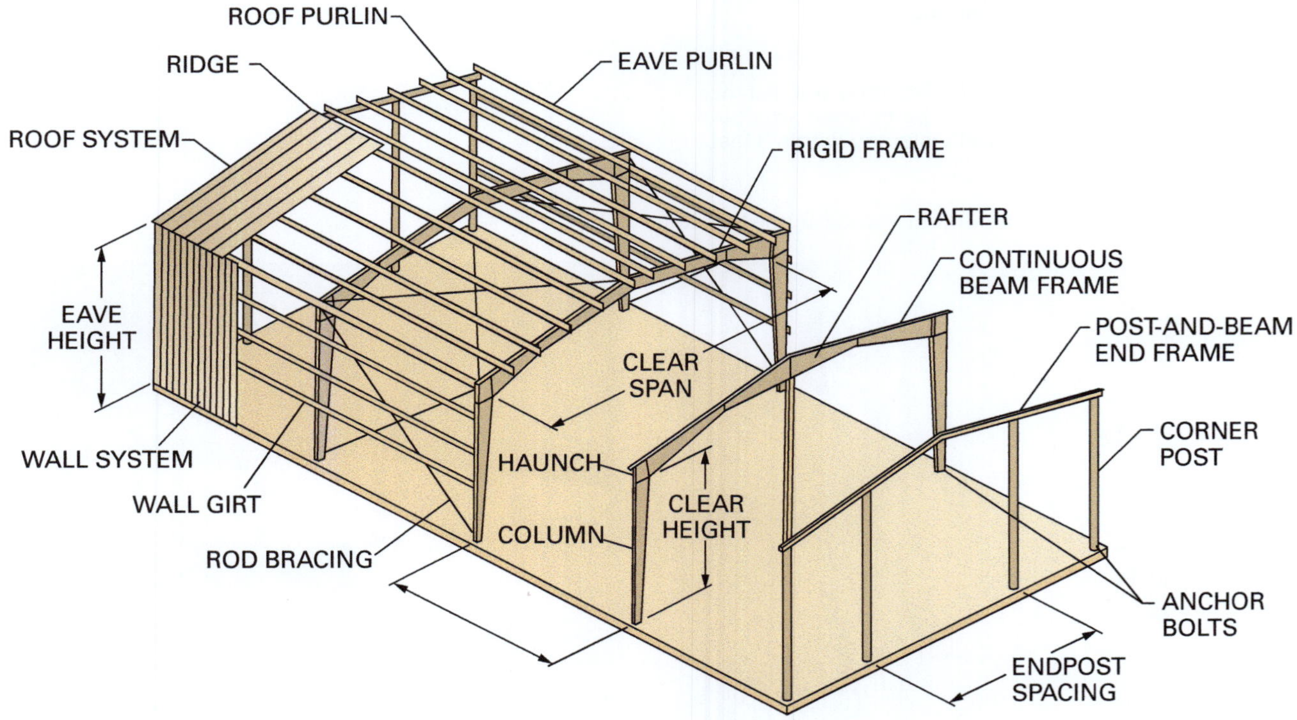

Figure 35 Beam and purlin roof system.

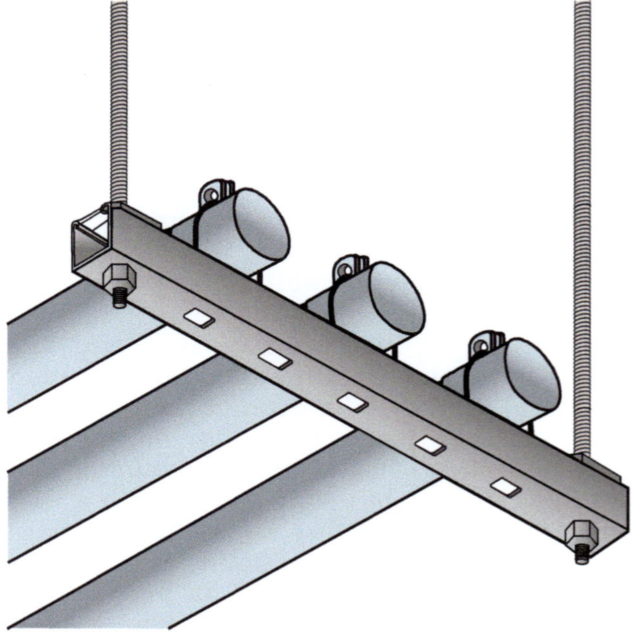

Figure 37 Steel strut system.

What's wrong with this picture?

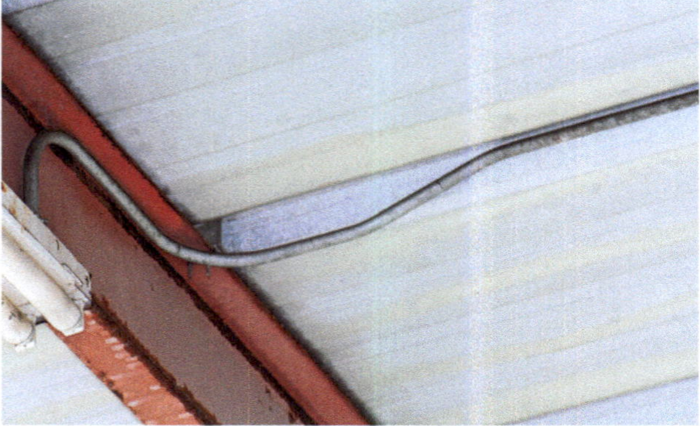

Figure Credit: Tim Dean

Park, KS: EC&M Books.

National Electrical Code® Handbook, Latest Edition. Quincy, MA: National Fire Protection Association.

1.0.0 Section Review

1. Information on conduit fill for various conductors can be found in *NEC®* ______.

   a. Informative Annex A
   b. Informative Annex B
   c. Informative Annex C
   d. Informative Annex D

2. Metal raceways and other enclosures are bonded to ______.

   a. prevent leakage onto any electrical systems below
   b. ensure electrical continuity and safely conduct any fault current
   c. ensure electrical continuity and prevent leakage
   d. ensure mechanical continuity

3. A fitting that does not require support is a(n) ______.

   a. outlet box
   b. pull box
   c. junction box
   d. conduit body

4. If the conduit does *not* enter straight into the box, ______.

   a. it may result in a voltage drop
   b. the fittings will be too tight
   c. a kick is required
   d. cut the conduit short

5. Beam clamps are used with ______.

   a. straps
   b. standoff supports
   c. framing channels
   d. suspended hangers

6. When installing conduit in a masonry environment, it is best to use ______.

   a. 3' to 5' (900 mm to 1.5 m) lengths of conduit
   b. 6' (1.8 m) lengths of conduit
   c. 8' to 10' (2.4 m to 3 m) lengths of conduit
   d. 20' (6 m) lengths of conduit

2.0.0 FASTENERS AND ANCHORS FOR RACEWAY SYSTEMS

Objective

Select fasteners and anchors for the installation of raceway systems.

a. Select and install tie wraps.
b. Select and install screws.
c. Select and install hammer-driven pins and studs.
d. Identify the safety requirements for stud-type guns.
e. Select and install masonry anchors.
f. Select and install hollow-wall anchors.
g. Select and install epoxy anchoring systems.

Performance Tasks

2. Identify and select various types and sizes of raceways, fittings, and fasteners for a given application.
3. Demonstrate how to install a raceway system.
4. Terminate a selected raceway system.

Conduit and other types of raceways used to carry wiring and cables must be properly supported. This generally means attaching the raceway to the building structure. Depending on the type of construction, the raceways may have to be attached to wood, concrete, or metal. Each of these materials requires the use of fasteners designed for the specific use. Using the wrong fastener, or installing the right fastener incorrectly, can lead to a failure of the raceway support.

The project specifications and manufacturer's installation instructions may specify the type and size of fasteners to use and how to install them. In other instances, the electrician will be expected to select the right type of fastener for a given application. It is therefore important that every electrician be familiar with the different types of fasteners, their uses, and their limitations.

2.1.0 Tie Wraps

A tie wrap is a one-piece, self-locking cable tie, usually made of nylon, that is used to fasten a bundle of wires and cables together. Tie wraps can be quickly installed either manually or using a special installation tool. Black tie wraps resist ultraviolet light and are recommended for outdoor use. Special rated tie wraps may be required for use in return air ceilings; consult local requirements.

Tie wraps are made in standard, cable strap and clamp, and identification configurations (*Figure 38*). All types function to clamp bundled wires or cables together. In addition, the cable strap and clamp has a molded mounting hole in the head used to secure the tie with a rivet, screw, or bolt after the tie wrap has been installed around the wires or cable. Identification tie wraps have a large flat area provided for imprinting or writing cable identification information. A releasable tie wrap version is also available. It is a non-permanent tie used for bundling wires or cables that may require frequent additions or deletions. Cable ties are made in various lengths and colors. Tie wraps can also be attached to a variety of adhesive mounting bases made for that purpose.

2.2.0 Bolts and Screws

Bolts and screws are made in a variety of shapes and sizes for different fastening jobs. The finish or coating used on a bolt or screw determines whether it is for interior or exterior use, corrosion resistant, etc. Bolts and screws of all types have heads with different shapes and slots. Some have machine threads and are self-drilling. The size or diameter of the body or shank is given in gauge numbers ranging from No. 0 to No. 24, and in fractions of an inch for screws with diameters larger than $\frac{1}{4}$" (metric size M6). The higher the gauge number, the larger the diameter of the shank. Lengths are measured from the tip to the part of the head that is flush to the surface when driven in. When choosing a fastener for an application, you must consider the type and thickness of the materials to be fastened, the size of the fastener, the material it is made of, the shape of its head, and the type of driver. Because of the wide diversity in the types of fasteners and their applications, always follow the manufacturer's recommendations to select the right fastener for the job. To prevent damage to the fastener head or the ma-

Tie Wraps

Tie wraps are available in a wide variety of colors that can be used to color code different cable bundles.

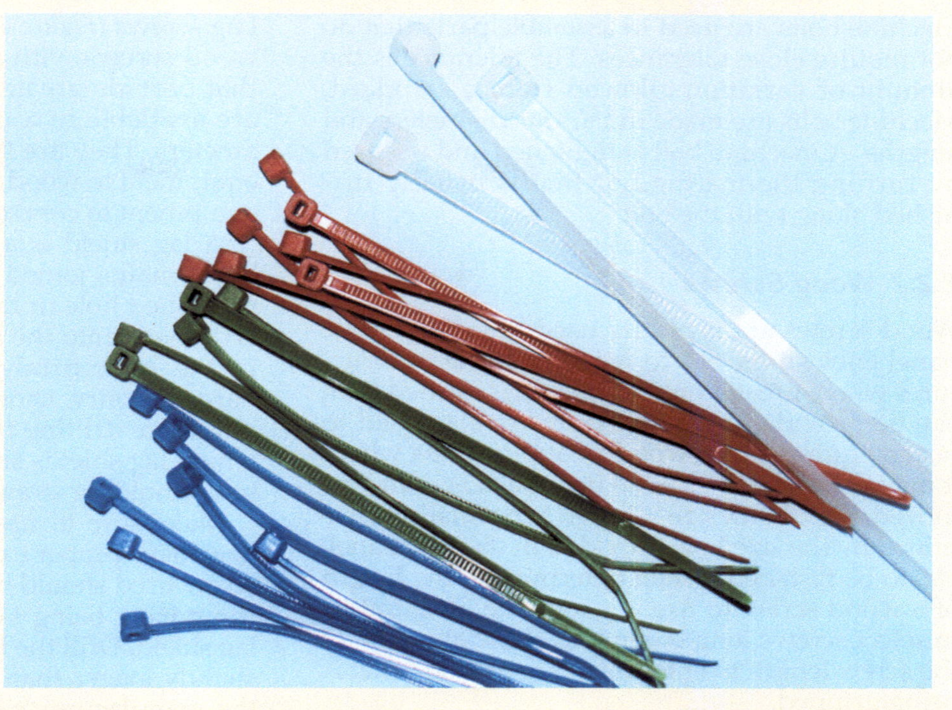

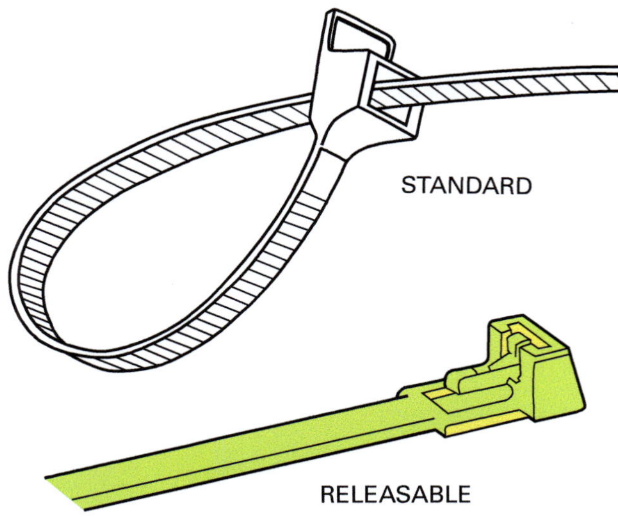

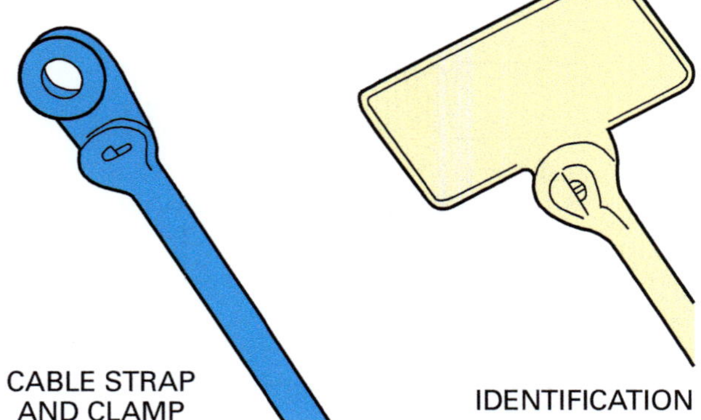

Figure 38 Tie wraps.

terial being fastened, always use a tool with the proper size and shape tip to fit the fastener.

Metric fasteners are sized differently than standard American (Imperial) fasteners. For example, a typical metric fastener size is M8 × 1. This means that the diameter of the threads is 8 mm and the thread pitch (space between the threads) is 1 mm. The length is indicated in centimeters. *Figure 39* compares the two types of fasteners.

Some of the more common types of fasteners are:

- Machine bolts
- Wood screws
- Lag screws
- Concrete/masonry screws
- Thread-forming and thread-cutting screws
- Drywall screws
- Drive screws

2.2.1 Machine Bolts

Machine bolts are used to assemble parts that do not require close tolerances. The tolerance is the amount of variation allowed from a standard. Machine bolts are made in various diameters and lengths. A machine bolt is tightened and released by turning the mating nut that is usually furnished along with the bolt.

2.2.2 Wood Screws

Wood screws are typically used to fasten boxes, panel enclosures, etc. to wood framing or structures where greater holding power is needed than can be provided by nails. They are also used to fasten equipment to wood in applications where it may occasionally need to be unfastened and removed. The shank size selected is normally determined by the size hole provided in the box, panel, etc. to be fastened. When determining the length of a wood screw to use, a good rule of thumb is to select screws long enough to allow about $^2/_3$ of the screw length to enter the piece of wood that is being gripped.

2.2.3 Lag Screws and Shields

Lag screws (*Figure 40*) or lag bolts are heavy-duty wood screws with square- or hex-shaped heads that provide greater holding power. Lag screws are available in a wide variety of sizes and diameters. They are typically used to fasten heavy equipment to wood, but can also be used to fasten equipment to concrete when a lag shield is used.

A lag shield is a tube that is split lengthwise but remains joined at one end. It is placed in a predrilled hole in the concrete. When a lag screw is screwed into the lag shield, the shield expands in the hole, firmly securing the lag screw. In hard masonry, short lag shields may be used to minimize drilling time. In soft or weak masonry, longer lag shields should be used to achieve maximum holding strength.

Make sure to use the proper length lag screw to achieve proper expansion. The length of the lag screw used should be equal to the thickness of the component being fastened plus the length of the lag shield. Drill the hole in the masonry to a depth slightly longer than the shield being used (follow the manufacturer's instructions). If the head of a

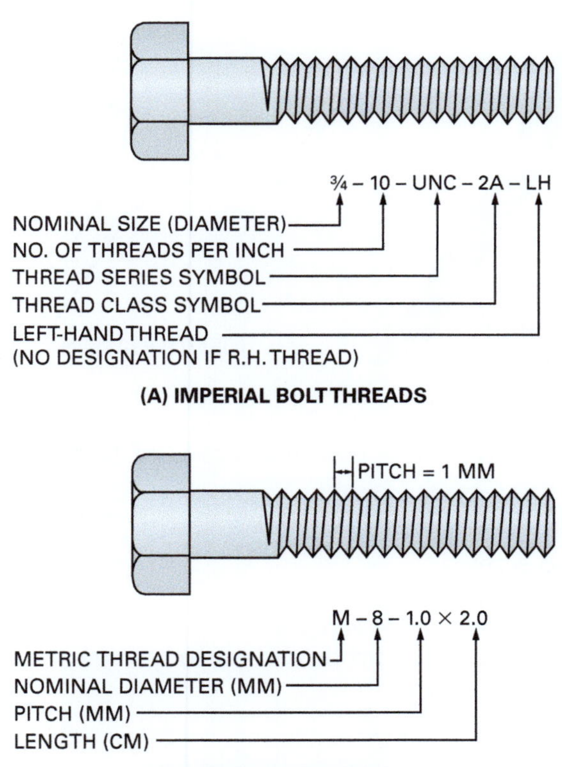

Figure 39 Imperial and metric bolt sizes.

Screws

In most applications, either threaded or nonthreaded fasteners such as nails could be used. However, threaded fasteners are sometimes preferred because they can usually be tightened and removed without damaging the surrounding material.

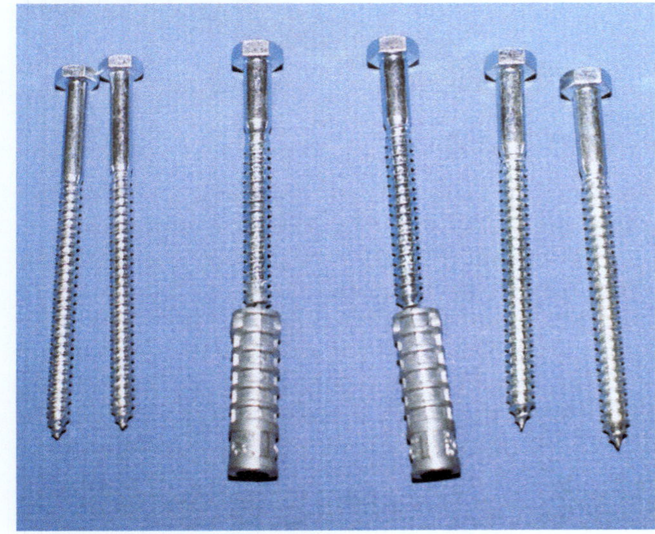

Figure 40 Lag screws and shields.

lag screw rests directly on wood when installed, a flat washer should be placed under the head to prevent the head from digging into the wood as the lag screw is tightened down. Be sure to take the thickness of any washers used into account when selecting the length of the screw.

2.2.4 Concrete/Masonry Screws

Concrete/masonry screws (*Figure 41*), commonly called self-threading anchors, are used to fasten a device or fixture to concrete, block, or brick. No anchor is needed. To provide a matched tolerance anchoring system, the screws are installed using specially designed carbide drill bits and installation tools made for use with the screws. These tools are typically used with a standard rotary drill hammer. The installation tool, along with an appropriate drive socket or bit, is used to drive the screws directly into predrilled holes that have a diameter and depth specified by the screw manufacturer. When being driven into the concrete, the widely spaced threads on the screws cut into the walls of the hole to provide a tight friction fit. Most types of concrete/masonry screws can be removed and reinstalled to allow for shimming and leveling of the fastened device.

> **WARNING!**
>
> Follow your company's rules for silica protection while drilling concrete.

2.2.5 Thread-Forming and Thread-Cutting Screws

Thread-forming screws (*Figure 42*), commonly called sheet metal screws, are made of hard metal. They form a thread as they are driven into the work. This thread-forming action eliminates the need to tap a hole before installing the screw. To achieve proper holding, it is important to make

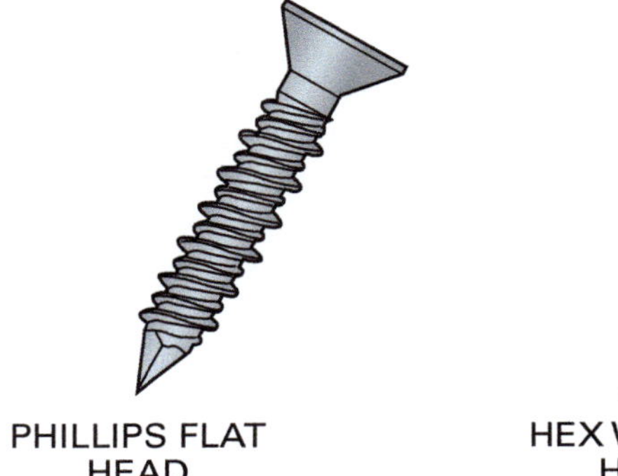

Figure 41 Concrete screws.

> ## Installing Wood Screws
>
> To maintain holding power, be careful not to drill your pilot hole too large. It's wise to drill a pilot hole deep enough to equal about two-thirds of the length of the threaded portion of the screw. Additionally, to lubricate screw threads, use soap, which makes the screw easier to drive.

sure to use the proper size bit when drilling pilot holes for thread-forming screws. The correct drill bit size used for a specific size screw is usually marked on the box containing the screws. Some types of thread-forming screws also drill their own holes, eliminating drilling, punching, and aligning parts. Thread-forming screws are primarily used to fasten light-gauge metal parts together. They are made in the same diameters and lengths as wood screws.

Hardened steel thread-cutting metal screws with blunt points and fine threads (*Figure 43*) are used to join heavy-gauge metals, metals of different gauges, and nonferrous metals. They are also used to fasten sheet metal to building structural members. These screws are made of hardened steel that is harder than the metal being tapped. They cut threads by removing and cutting a portion of the metal as they are driven into a pilot hole and through the material.

2.2.6 Drywall Screws

Drywall screws (*Figure 44*) are thin, self-drilling screws with bugle-shaped heads. Depending on the type of screw, it cuts through the wallboard

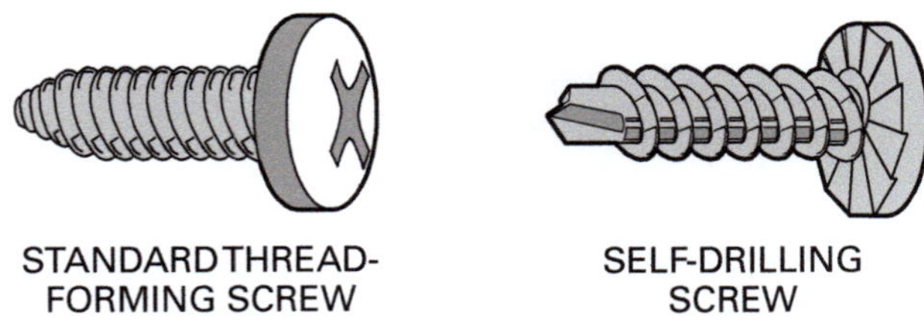

Figure 42 Thread-forming screws.

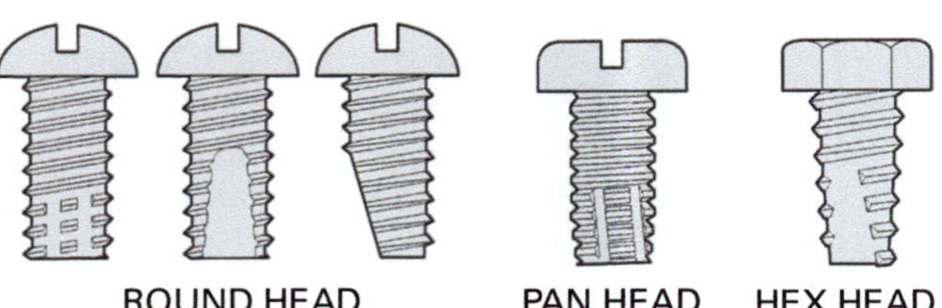

Figure 43 Thread-cutting screws.

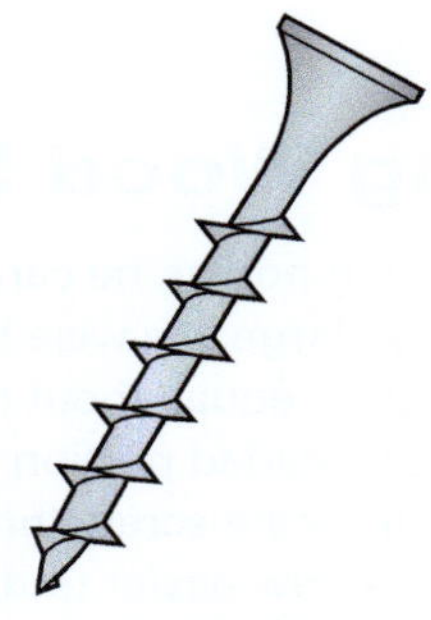

COARSE THREAD

FINE THREAD

HIGH-AND-LOW THREAD

Figure 44 Drywall screws.

and anchors itself into wood and/or metal studs, holding the wallboard tight to the stud. Coarse thread screws are normally used to fasten wallboard to wood studs. Fine thread and high-and-low thread types are generally used for fastening to metal studs. Some screws are made for use in either wood or metal. A Phillips or Robertson drive head allows the drywall screw to be countersunk without tearing the surface of the wallboard.

2.2.7 Drive Screws

Drive screws do not require that the hole be tapped. They are installed by hammering the screw into a drilled or punched hole of the proper size. Drive screws are mostly used to fasten parts that will not be exposed to much pressure. A typical use of drive screws is to attach permanent name plates on electric motors and other types of equipment. *Figure 45* shows a typical drive screw.

2.3.0 Hammer-Driven Pins and Studs

Hammer-driven pins or threaded studs (*Figure 46*) can be used to fasten wood or steel to concrete or block without the need to predrill holes. The pin or threaded stud is inserted into a hammer-driven tool designed for its use. The pin or stud is inserted in the tool point end out with the washer seated in the recess. The pin or stud is then positioned against the base material where it is to be fastened and the drive rod of the tool tapped lightly until the striker pin contacts the pin or stud. Following this, the tool's drive rod is struck using heavy blows with about a two-pound engineer's hammer. The force of the hammer blows is transmitted through the tool directly to the head of the fastener, causing it to be driven into the concrete or block. For best results, the drive pin or stud should be embedded a minimum of $\frac{1}{2}$" (13 mm) in hard concrete to $1\frac{1}{4}$" (32 mm) in softer concrete block.

2.4.0 Safety Requirements for Stud-Type Guns

Stud-type guns powered by powder, gas, rocket fuel, or spring power (*Figure 47*) can be used to drive a wide variety of specially designed pin and threaded stud-type fasteners into masonry and steel. Powder-actuated tools look and fire like a gun and use the force of a detonated gunpowder load (typically .22, .25, or .27 caliber) to drive the fastener into the material. The depth to which the pin or stud is driven is controlled by the density of the base material in which the pin or stud is being installed and by the power level or strength of the cased powder load.

Powder loads and their cases are designed for use with specific types and/or models of powder-actuated tools and are not interchangeable. Typically, powder loads are made in 12 increasing power or load levels used to achieve the proper penetration. The different power levels are identified by a color-code system and load case types. Note that different manufacturers may use different color codes to identify load strength. Power level 1 is the lowest power level while 12 is the highest. Higher number power levels are used when driving into hard materials or when a deeper penetration is needed. Powder loads are available as single-shot units for use with single-shot tools. They are also made in multi-shot strips or disks for semiautomatic tools.

OSHA Standard 29 CFR 1926.302(e) governs the use of powder-actuated tools and states that only those individuals who have been trained in the operation of the particular powder-actuated tool in use be allowed to operate it. Authorized instructors available from the various powder-actuated tool manufacturers generally provide such training and licensing. Trained operators must take precautions to protect both themselves and others in the area when using a powder-actuated driver tool:

- Always use the tool in accordance with the published tool operation instructions.
- Instructions should be kept with the tool. Never attempt to override the safety features of the tool.
- Never place your hand or other body parts over the front muzzle end of the tool.
- Use only fasteners, powder loads, and tool parts specifically made for use with the tool. Use of other materials can cause improper and unsafe functioning of the tool.
- Operators and bystanders must wear eye and hearing protection along with hard hats. Other personal safety gear, as required, must also be used.
- Always post warning signs that state Powder-Actuated Tool in Use within 50' (15 m) of the area where tools are used.
- Before using a tool, make sure it is unloaded and perform a proper function test. Check the functioning of the unloaded tool as described in the published tool operation instructions.
- Do not guess before fastening into any base material; always perform a center punch test.
- Always make a test firing into a suitable base material with the lowest power level recommended for the tool being used. If this does not set the fastener, try the next higher power level. Continue this procedure until the proper fastener penetration is obtained.
- Always point the tool away from operators or bystanders.

Think About It

Self-Drilling Screws

Can you name an electrical application for self-drilling screws?

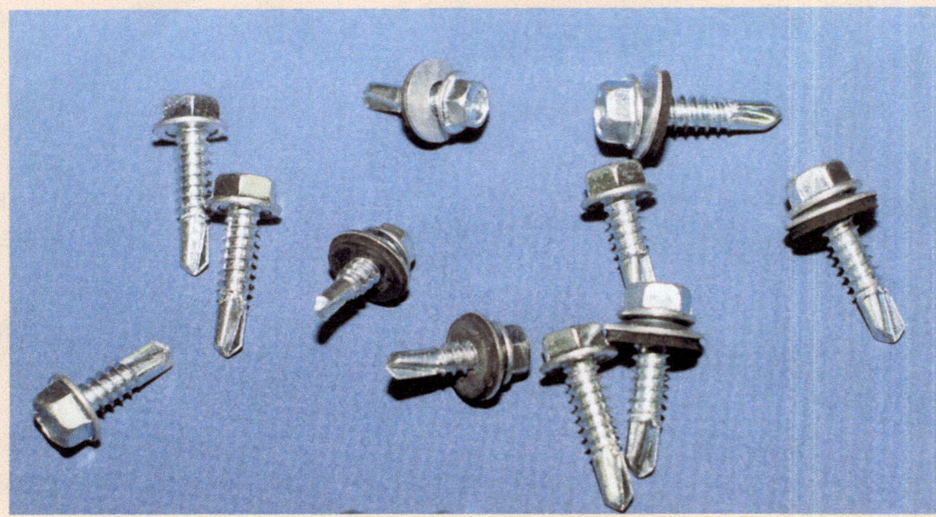

- Never use the tool in an explosive or flammable area.
- Never leave a loaded tool unattended. Do not load the tool until you are prepared to complete the fastening. Should you decide not to make a fastening after the tool has been loaded, always remove the powder load first, then the fastener. Always unload the tool before cleaning or servicing, when changing parts, prior to work breaks, and when storing the tool.

- Always hold the tool perpendicular to the work surface and use the spall (chip or fragment) guard or stop spall whenever possible.
- Always follow the required spacing, edge distance, and base material thickness requirements.
- Never fire through an existing hole or into a weld area.
- In the event of a misfire, always hold the tool depressed against the work surface for at least 30 seconds. If the tool still does not fire, follow the published tool instructions. Never carelessly discard or throw unfired powder loads into a trash receptacle.
- Always store the powder loads and unloaded tool under lock and key.

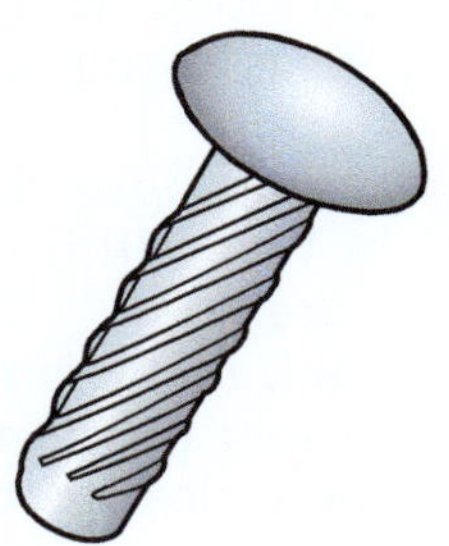

Figure 45 Drive screw.

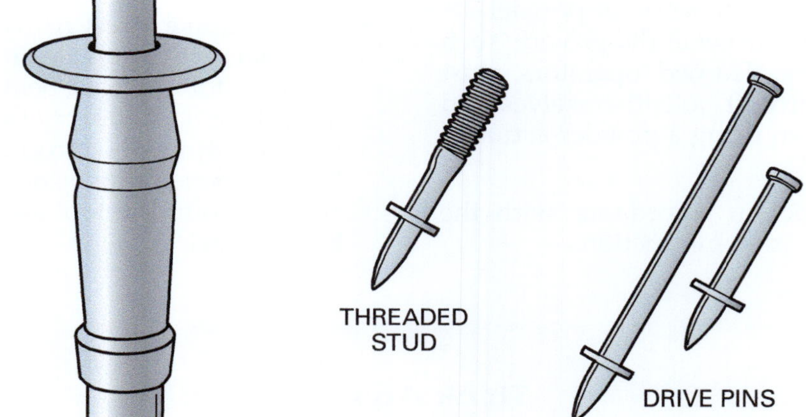

Figure 46 Hammer-driven pins and installation tool.

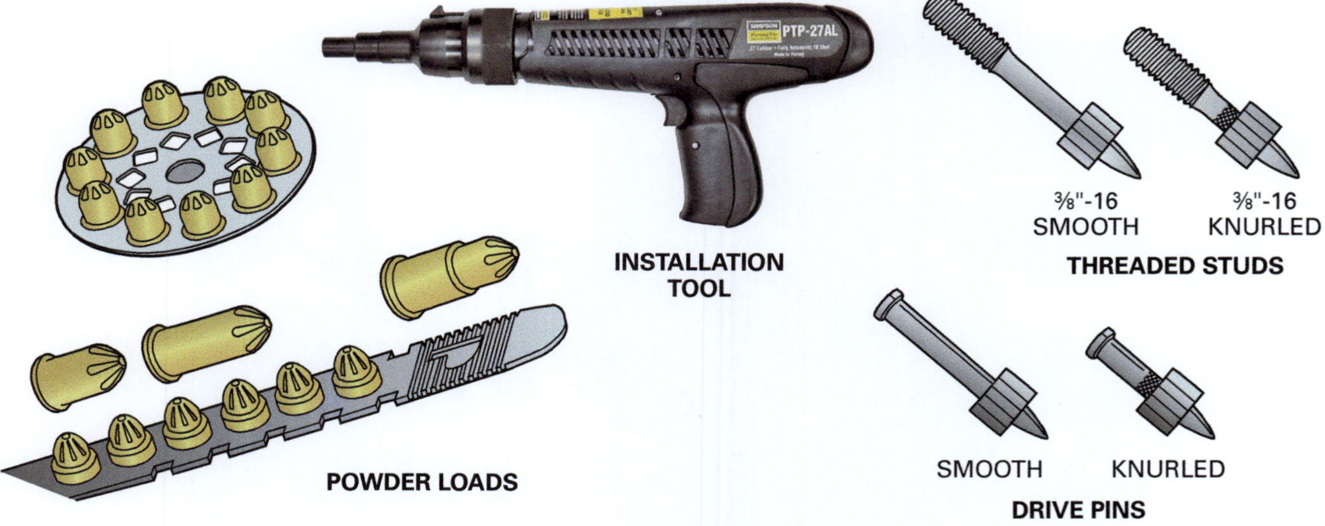

Figure 47 Powder-actuated installation tools and fasteners.

2.5.0 Masonry Anchors

Mechanical anchors are devices used to give fasteners a firm grip in a variety of materials, where the fasteners by themselves would otherwise have a tendency to pull out. Anchors can be classified in many ways by different manufacturers. In this module, anchors have been divided into four broad categories:

- One-step anchors
- Bolt anchors
- Screw anchors
- Self-drilling anchors

2.5.1 One-Step Anchors

One-step anchors (*Figure 48*) are designed so that they can be installed through the mounting holes in the component to be fastened. This is because the anchor and the drilled hole into which it is installed have the same size diameter. They are available in various diameters and lengths. Common types of one-step anchors include the following:

- *Wedge anchors* – Wedge anchors are heavy-duty anchors supplied with nuts and washers. The drill bit size used to drill the hole is the same diameter as the anchor. The depth of the hole is not critical as long as the minimum length recommended by the manufacturer is drilled. After the hole is blown clean of dust and other material, the anchor is inserted into the hole and driven with a hammer far enough so that at least six threads are below the top surface of the component. Then, the component is fastened by tightening the anchor nut to expand the anchor and tighten it in the hole.
- *Stud bolt anchors* – Stud bolt anchors are heavy-duty threaded anchors. Because this type of anchor is made to bottom in its mounting hole,

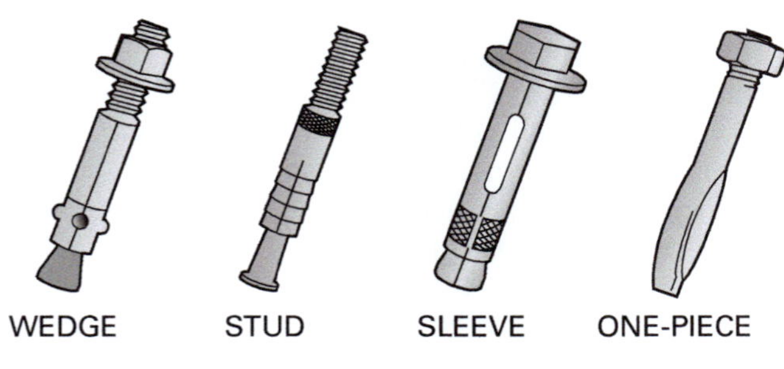

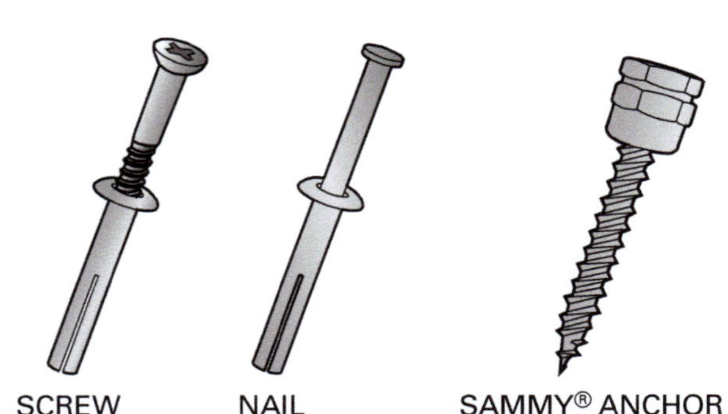

Figure 48 One-step anchors.

it is a good choice to use when jacking or leveling of the fastened component is needed. The depth of the hole drilled in the masonry must be as specified by the manufacturer in order to achieve proper expansion. After the hole is blown clean of dust and other material, the anchor is inserted in the hole with the expander plug end down. Following this, the anchor is driven into the hole with a hammer (or setting tool) to expand the anchor and tighten it in the hole. The anchor is fully set when it can no longer be driven into the hole. The component is fastened using the correct size and thread bolt for use with the anchor stud.

- *Sleeve anchors* – Sleeve anchors are multi-purpose anchors. The depth of the anchor hole is not critical as long as the minimum length recommended by the manufacturer is drilled. After the hole is blown clean of dust and other material, the anchor is inserted into the hole and tapped until flush with the component. Then, the anchor nut or screw is tightened to expand the anchor and tighten it in the hole.

- *One-piece anchors* – One-piece anchors are multi-purpose anchors. They work on the principle that as the anchor is driven into the hole, the spring force of the expansion mechanism is compressed and flexes to fit the size of the hole. Once set, it tries to regain its original shape. The depth of the hole drilled in the masonry must be at least deeper than the required embedment (follow the manufacturer's instructions). The proper depth is crucial. Overdrilling is as bad as underdrilling. After the hole is blown clean of dust and other material, the anchor is inserted through the component and driven with a hammer into the hole until the head is firmly seated against the component. It is important to make sure that the anchor is driven to the proper embedment depth. Note that manufacturers also make specially designed drivers and manual tools that are used instead of a hammer to drive one-piece anchors. These tools allow the anchors to be installed in confined spaces and help prevent damage to the component from stray hammer blows.

- *Hammer-set anchors* – Hammer-set anchors are made for use in concrete and masonry. There are two types: nail and screw. An advantage of the screw-type anchors is that they are removable. Both types have a diameter the same size as the anchoring hole. For both types, the anchor hole must be drilled to the diameter of the anchor and to a depth of at least $\frac{1}{4}$" (6 mm) deeper than that required for embedment (follow the manufacturer's instructions). After the hole is blown clean of dust and other material, the anchor is inserted into the hole through the mounting holes in the component to be fastened; then the screw or nail is driven into the anchor body to expand it. It is important to make sure that the head is seated firmly against the component and is at the proper embedment.

- *Threaded-rod anchors* – Threaded-rod anchors, such as the Sammy® anchor, are available for installation in concrete, steel, or wood. The anchor is designed to support a threaded rod, which is screwed into the head of the anchor after the anchor is installed. A special nut driver is available for installing the screws.

2.5.2 Bolt Anchors

Bolt anchors (*Figure 49*) are designed to be installed flush with the surface of the base material. They are used in conjunction with threaded machine bolts or screws. In some types, they can be used with threaded rod. Some commonly used types of bolt anchors include the following:

- *Drop-in anchors* – Drop-in anchors are typically used as heavy-duty anchors. There are two types of drop-in anchors. The first type is made for use in solid concrete and masonry, and has an internally threaded expansion anchor with a preassembled internal expander plug. The anchor hole must be drilled to the specific diameter and depth specified by the manufacturer. After the hole is blown clean of dust and other material, the anchor is inserted into the hole and tapped until it is flush with the surface. Following this, a setting tool supplied with the anchor is driven into the anchor to expand it. The component to be fastened is positioned in place and fastened by threading and tightening the correct size machine bolt or screw into the anchor.

- The second type, called a hollow-set drop-in anchor, is made for use in hollow concrete and masonry base materials. Hollow-set drop-in anchors have a slotted, tapered expansion sleeve and a serrated expansion cone. They come in various lengths compatible with the outer wall thickness of most hollow base materials. They can also be used in solid concrete and masonry. The anchor hole must be drilled to the specific diameter specified by the manufacturer. When installed in hollow base materials, the hole is drilled into the cell or void. After the hole is blown clean of dust and other material, the anchor is inserted into the hole and tapped until it is flush with the surface. Following this, the component to be fastened is positioned in place; then the proper size machine bolt or screw is threaded into the anchor and tightened to expand the anchor in the hole.

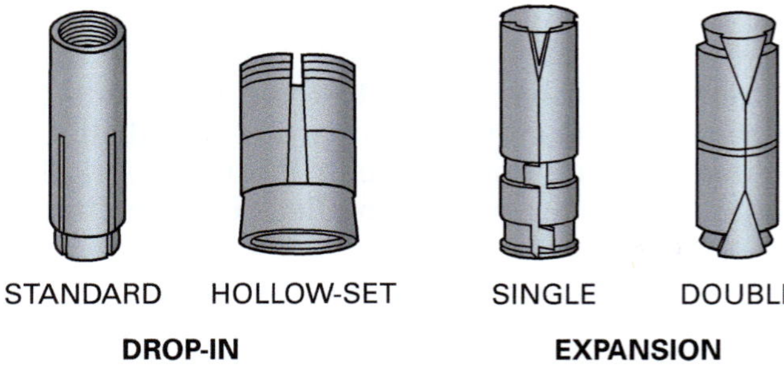

Figure 49 Bolt anchors.

- *Single- and double-expansion anchors* – Single- and double-expansion anchors are both made for use in concrete and other masonry. The double-expansion anchor is used mainly when fastening into concrete or masonry of questionable strength. For both types, the anchor hole must be drilled to the specific diameter and depth specified by the manufacturer. After the hole is blown clean of dust and other material, the anchor is inserted into the hole, threaded cone end first. It is then tapped until it is flush with the surface. Following this, the component to be fastened is positioned in place; then the proper size machine bolt or screw is threaded into the anchor and tightened to expand the anchor in the hole.

2.5.3 Screw Anchors

Screw anchors are lighter-duty anchors made to be installed flush with the surface of the base material. They are used in conjunction with sheet metal, wood, or lag screws depending on the anchor type. Fiber and plastic anchors are common types of screw anchors (*Figure 50*). The lag shield anchor used with lag screws was described earlier in this module.

Fiber and plastic anchors are typically used in concrete and masonry. Plastic anchors are also commonly used in wallboard and similar base materials. The installation of all types is simple. The anchor hole must be drilled to the diameter specified by the manufacturer. The minimum depth of the hole must equal the anchor length. After the hole is blown clean of dust and other material, the anchor is inserted into the hole and tapped until it is flush with the surface. Following this, the component to be fastened is positioned in place; then the proper type and size screw is driven through the component mounting hole and into the anchor to expand the anchor in the hole.

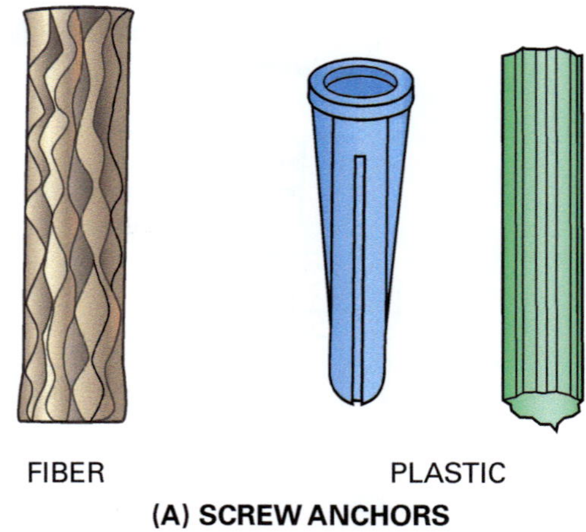

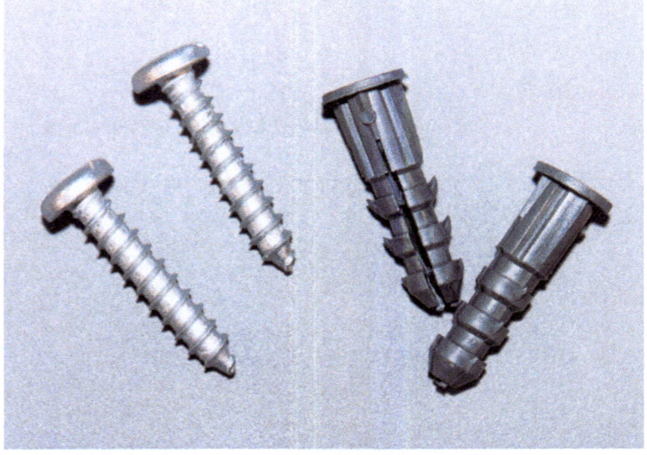

Figure 50 Screw anchors and screws.

2.5.4 Self-Drilling Anchors

Some anchors made for use in masonry are self-drilling anchors. *Figure 51* is typical of those in common use. This fastener has a cutting sleeve that is first used as a drill bit and later becomes the expandable fastener itself. A rotary hammer is used to drill the hole in the concrete using the anchor sleeve as the drill bit. After the hole is drilled, the anchor is pulled out and the hole cleaned. This is followed by inserting the anchor's expander plug into the cutting end of the sleeve. The anchor sleeve and expander plug are driven back into the hole with the rotary hammer until they are flush with the surface of the concrete. As the fastener is hammered down, it hits the bottom, where the tapered expander causes the fastener to expand and lock into the hole. The anchor is then snapped off at the shear point with a quick lateral movement of the hammer. The component to be fastened can then be attached to the anchor using the proper size bolt.

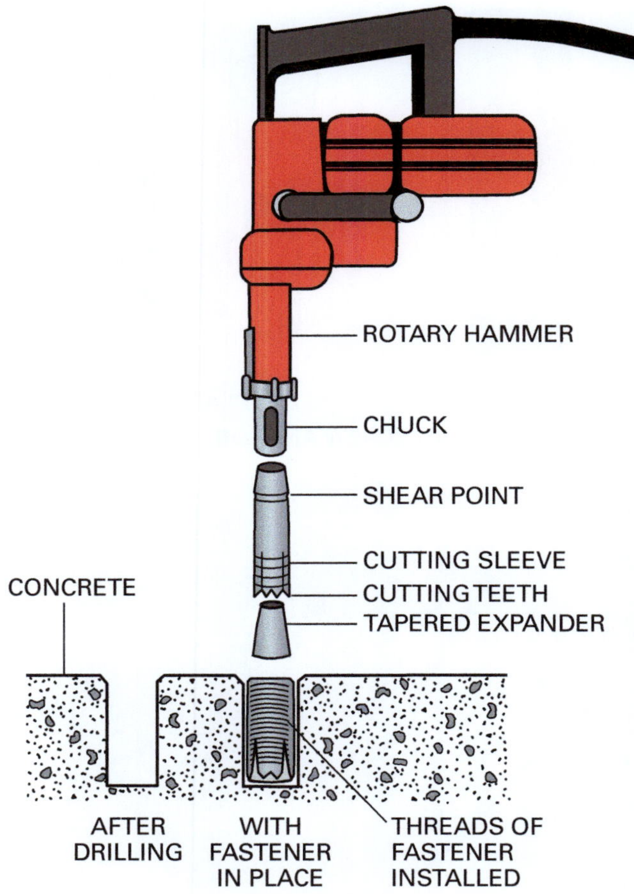

Figure 51 Self-drilling anchor.

2.5.5 Drilling Anchor Holes in Hardened Concrete or Masonry

When selecting masonry anchors, regardless of the type, always take into consideration and follow the manufacturer's recommendations pertaining to hole diameter and depth, minimum embedment in concrete, maximum thickness of material to be fastened, and the pullout and shear load capacities.

When installing anchors and/or anchor bolts in hardened concrete, make sure the area where the equipment or component is to be fastened is smooth so that it will have solid footing. Uneven footing might cause the equipment to twist, warp, not tighten properly, or vibrate when in operation. Before starting, carefully inspect the rotary hammer or hammer drill and the drill bit(s) to ensure they are in good operating condition. Be sure to use the type of carbide-tipped masonry or percussion drill bits recommended by the drill/hammer or anchor manufacturer because these bits are made to take the higher impact of the masonry materials. Also, it is recommended that the drill or hammer tool depth gauge be set to the depth of the hole needed. The trick to using masonry drill bits is not to force them into the material by pushing down hard on the drill. Use a little pressure and let the drill do the work. For large holes, start with a smaller bit, then change to a larger bit.

The methods for installing the different types of anchors in hardened concrete or masonry have been briefly described. Always install the selected anchors according to the manufacturer's directions. Here is an example of a typical procedure used to install many types of expansion anchors in hardened concrete or masonry (refer to *Figure 52* as you study the procedure):

Step 1 Drill the anchor bolt hole the same size as the anchor bolt. The hole must be deep enough for six threads of the bolt to be below the surface of the concrete. Clean out the hole using a squeeze bulb.

Step 2 Drive the anchor bolt into the hole using a hammer. Protect the threads of the bolt with a nut that does not allow any threads to be exposed.

Step 3 Put a washer and nut on the bolt, and tighten the nut with a wrench until the anchor is secure in the concrete.

2.6.0 Hollow-Wall Anchors

Hollow-wall anchors are used in hollow materials such as concrete plank, block, structural steel, wallboard, and plaster. Some types can also be used in solid materials. Toggle bolts, sleeve-type wall anchors, wallboard anchors, and metal drive-in anchors are common anchors used when fastening to hollow materials.

When installing anchors in hollow walls or ceilings, regardless of the type, always follow the manufacturer's recommendations pertaining to use, hole diameter, wall thickness, grip range (thickness of the anchoring material), and the pullout and shear load capacities.

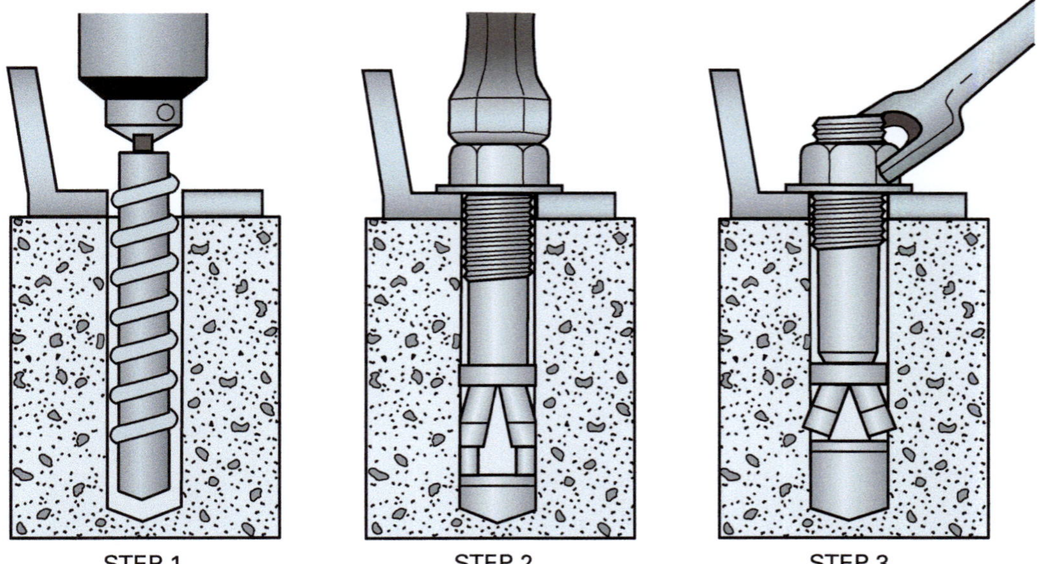

Figure 52 Installing an anchor bolt in hardened concrete.

2.6.1 Toggle Bolts

Toggle bolts (*Figure 53*) are used to fasten equipment, hangers, supports, and similar items into hollow surfaces such as walls and ceilings. They consist of a slotted bolt or screw and spring-loaded wings. When the bolt is inserted through the item to be fastened, then through a predrilled hole in the wall or ceiling, the wings spring apart and provide a firm hold on the inside of the hollow wall or ceiling as the bolt is tightened. Note that the hole drilled in the wall or ceiling should be just large enough for the compressed wing-head to pass through. Once the toggle bolt is installed, be careful not to completely unscrew the bolt because the wings will fall off, making the fastener useless. Screw-actuated plastic toggle bolts are also made. These are similar to metal toggle bolts, but they come with a pointed screw and do not require as large a hole. Unlike the metal version, the plastic wings remain in place if the screw is removed.

Toggle bolts are used to fasten a part to hollow block, wallboard, plaster, panel, or tile. The following general procedure can be used to install toggle bolts:

Step 1 Select the proper size drill bit or punch and toggle bolt for the job.

Step 2 Check the toggle bolt for damaged or dirty threads or a malfunctioning wing mechanism.

Step 3 Drill a hole completely through the surface to which the part is to be fastened.

Step 4 Insert the toggle bolt through the opening in the item to be fastened.

Step 5 Screw the toggle wing onto the end of the toggle bolt, ensuring that the flat side of the toggle wing is facing the bolt head.

Step 6 Fold the wings completely back and push them through the drilled hole until the wings spring open.

Step 7 Pull back on the item to be fastened in order to hold the wings firmly against the inside surface to which the item is being attached.

Step 8 Tighten the toggle bolt with a screwdriver until it is snug.

> **WARNING!**
>
> Follow all safety precautions when using an electric drill.

Safety

Be sure to wear safety goggles whenever you tackle any fastening project, regardless of how small the job may seem. Remember, you can never replace lost eyesight.

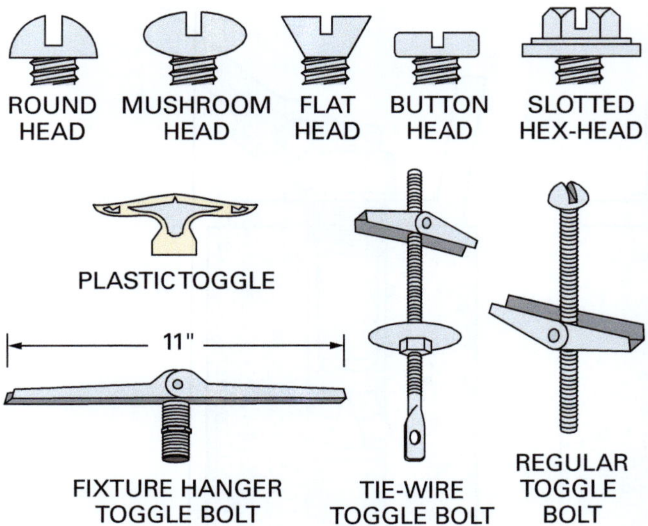

Figure 53 Toggle bolts.

2.6.2 Sleeve-Type Wall Anchors

Sleeve-type wall anchors (*Figure 54*) are suitable for use in concrete, block, plywood, wallboard, hollow tile, and similar materials. The two types made are standard and drive. The standard type is commonly used in walls and ceilings and is installed by drilling a mounting hole to the required diameter. The anchor is inserted into the hole and tapped until the gripper prongs embed in the base material. Following this, the anchor's screw is tightened to draw the anchor tight against the inside of the wall or ceiling. Note that the drive-type anchor is hammered into the material without the need for drilling a mounting hole. After the anchor is installed, the anchor screw is removed, the component being fastened is positioned in place, then the screw is reinstalled through the mounting hole in the component and into the anchor. The screw is tightened into the anchor to secure the component.

2.6.3 Wallboard Anchors

Wallboard anchors (*Figure 54*) are self-drilling medium- and light-duty anchors used for fastening in wallboard. The anchor is driven into the wall with a Phillips head manual or cordless screwdriver until the head of the anchor is flush with

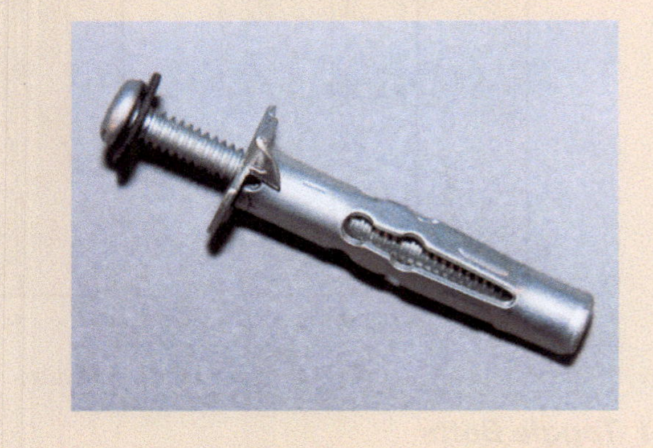

the wall or ceiling surface. Following this, the component being fastened is positioned over the anchor, then secured with the proper size sheet metal screw driven into the anchor.

2.6.4 Metal Drive-In Anchors

Metal drive-in anchors (*Figure 54*) are used to fasten light to medium loads to wallboard. They have two pointed legs that stay together when the anchor is hammered into a wall and spread out against the inside of the wall when a sheet metal screw is driven in.

2.7.0 Epoxy Anchoring Systems

Epoxy resin compounds can be used to anchor threaded rods, dowels, and similar fasteners in solid concrete, hollow wall, and brick. For one manufacturer's product, a two-part epoxy is packaged in a two-chamber cartridge that keeps the resin and hardener ingredients separated until use. This cartridge is placed into a special tool similar to a caulking gun. When the gun handle is pumped, the epoxy resin and hardener components are mixed within the gun; then the epoxy is ejected from the gun nozzle.

To use the epoxy to install an anchor in solid concrete (*Figure 55*), a hole of the proper size is drilled in the concrete and cleaned using a nylon (not metal) brush. Following this, a small amount of epoxy is dispensed from the gun to make sure that the resin and hardener have mixed properly. This is indicated by the epoxy being of a uniform

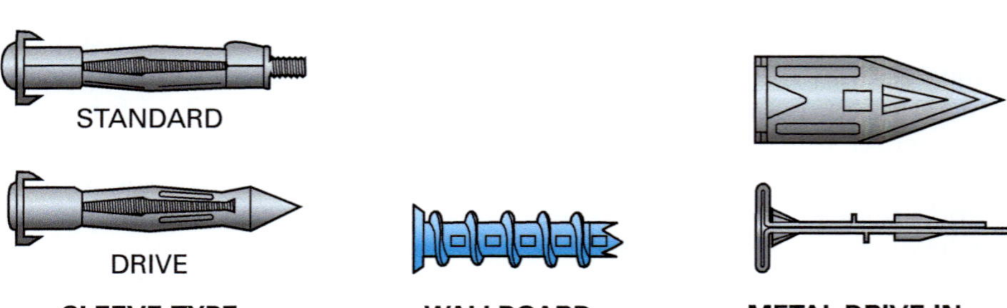

Figure 54 Sleeve-type, wallboard, and metal drive-in anchors.

Case History

Installation Requirements

In a college dormitory, battery-powered emergency lights were anchored to sheetrock hallway ceilings with sheetrock screws, with no additional support. These fixtures weigh 8-10 pounds each and might easily have fallen out of the ceiling, causing severe injury. When the situation was discovered, the contractor had to remove and replace dozens of fixtures.

The Bottom Line: Incorrect anchoring methods can be both costly and dangerous.

Think About It

Ceiling Installations

In the dormitory problem discussed earlier, which of the following fasteners could have been used to safely secure the emergency lights?

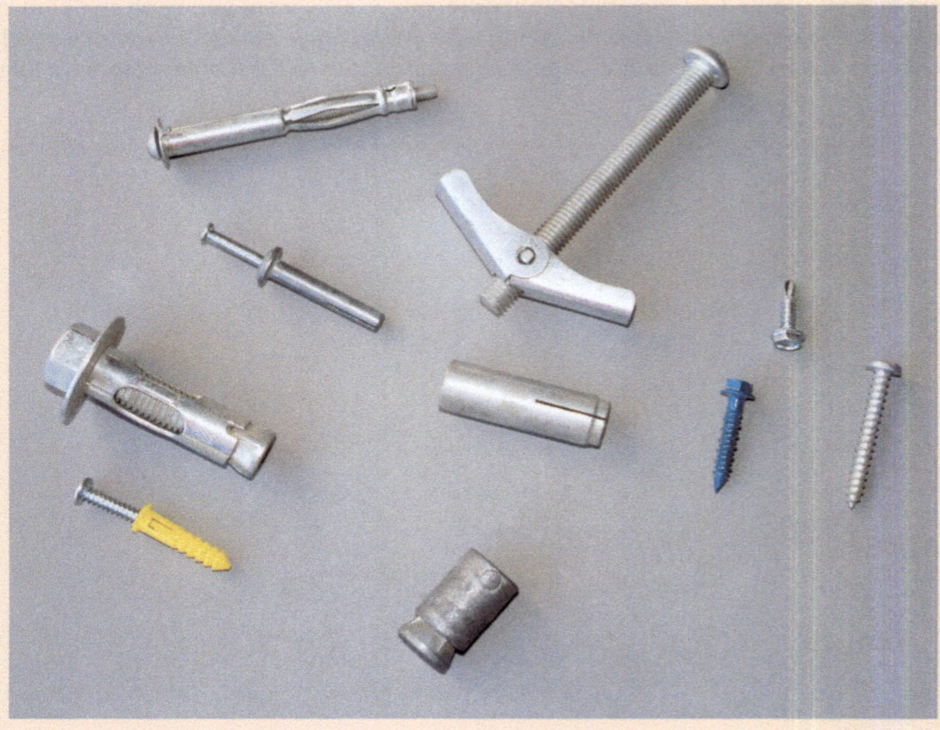

color. The gun nozzle is then placed into the hole, and the epoxy is injected into the hole until half the depth of the hole is filled. Following this, the selected fastener is pushed into the hole with a slow twisting motion to make sure that the epoxy fills all voids and crevices, then is set to the required plumb (or level) position. After the recommended cure time for the epoxy has elapsed, the fastener nut can be tightened to secure the component or fixture in place.

The procedure for installing a fastener in a hollow wall or brick using epoxy is basically the same as that described for solid concrete. The difference is that the epoxy is first injected into an anchor screen to fill the screen, then the anchor screen is installed into the drilled hole. Use of the anchor screen is necessary to hold the epoxy intact in the hole until the anchor is inserted into the epoxy.

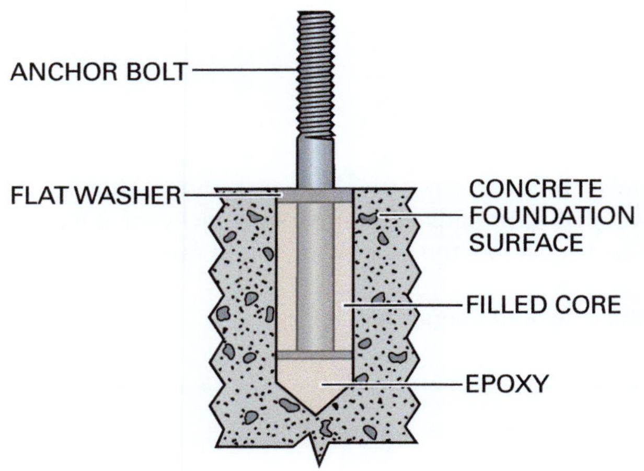

Figure 55 Fastener anchored in epoxy.

Use the Proper Tool for the Application

To avoid damaging fasteners, use the correct tool for the job. For example, don't use pliers to install bolts, and don't use a screwdriver that is too large or too small.

Using Epoxy

Once mixed, epoxy has a limited working time. Therefore, mix exactly what you need and work quickly. After the working time is up, epoxy requires a specific curing time. Always give epoxy its recommended curing time. Because epoxy is so strong and sets so quickly, you'll be tempted to stress the bond before it's fully cured.

NCCER – *Maritime Electrical*

Additional Resources

*Concrete Fastening Systems.***www.confast.com**.

National Electrical Code®Handbook, Latest Edition. **Quincy, MA: National Fire Protection Association.**

2.0.0 Section Review

1. The best tie wrap for outdoor use is _____.

   a. white
   b. green
   c. orange
   d. black

2. Which of the following is true regarding screw gauges and shank?

   a. The lower the gauge number, the larger the diameter of the shank.
   b. The lower the gauge number, the longer the length of the shank.
   c. The higher the gauge number, the larger the diameter of the shank.
   d. The higher the gauge number, the longer the length of the shank.

3. When installing hammer-driven pins and studs, _____.

   a. the predrilled hole should be slightly smaller than the pin or stud
   b. the predrilled hole should be the same size as the pin or stud
   c. the predrilled hole should be slightly larger than the pin or stud
   d. a predrilled hole is not required

4. When using a powder-actuated tool, _____.

   a. post warning signs within 50' (15 m) of use
   b. power level 12 is the lowest
   c. color codes are often used to indicate the manufacturer
   d. all loads are interchangeable among makes and models

5. A type of fastener that uses the anchor itself as a drill bit is a _____.

   a. wedge anchor
   b. stud anchor
   c. one-piece anchor
   d. self-drilling anchor

6. A type of fastener that does not require predrilling is a _____.

   a. toggle bolt
   b. sleeve-type anchor
   c. drive-type anchor
   d. stud anchor

7. Epoxy anchoring systems are commonly used to anchor fasteners in _____.

   a. wood
   b. wallboard
   c. plaster
   d. brick

SECTION THREE

3.0.0 WIREWAYS AND OTHER SPECIALTY RACEWAYS

Objective

Select and install wireways and other specialty raceways.

   a. Identify types of wireways and their components.
   b. Install wireway supports.
   c. Identify and install specialty raceways.

Trade Term

Trough: A long, narrow box used to house electrical connections that could be exposed to the environment.

A wireway is a sheet metal *trough* provided with a hinged or screw-on removable cover. Like other types of raceways, wireways are used for housing electric wires and cables. Wireways are available in various lengths to allow runs of different lengths without cutting the wireway ducts.

Metal wireways are covered in *NEC Article 376*. As listed in *NEC Section 376.22(A)*, the sum of the cross-sectional areas of all contained conductors and cables at any cross section of a wireway shall not exceed 20% of the interior cross-sectional area of the wireway. The derating factors in *NEC Table 310.15(B)(3)(a)* shall be applied to wireways only where the number of current-carrying conductors exceeds 30, including neutral conductors classified as current-carrying under the provisions of *NEC Section 310.15(B)(5)*. Conductors for signaling or controller conductors between a motor and its starter used only for starting duty shall not be considered current-carrying conductors.

It is also noted in *NEC Section 376.56(A)* that conductors, together with splices and taps, must not fill the wireway to more than 75% of its cross-sectional area. No conductor larger than that for which the wireway is designed shall be installed in any wireway. Be sure to check *NEC Article 378* for the requirements of nonmetallic wireways.

NEC Section 376.23(A) requires that the dimensions of *NEC Table 312.6(A)* be applied where insulated conductors are deflected in a wireway. *NEC Section 376.23(B)* requires that the provisions of *NEC Section 314.28* apply where metal wireways are used as pull boxes.

An auxiliary gutter is a wireway that is intended to add to wiring space at switchboards, meters, and other distribution locations. Auxiliary gutters are dealt with in *NEC Article 366*. Even though the component parts of wireways and auxiliary gutters are identical, you should be familiar with the differences in their use. Auxiliary gutters are used as parts of complete assemblies of apparatus such as switchboards, distribution centers, and control equipment. However, an auxiliary gutter may only contain conductors or busbars, even though it looks like a surface metal raceway that may contain devices and equipment. Unlike auxiliary gutters, wireways represent a type of wiring because they are used to carry conductors between points located considerable distances apart.

The allowable ampacities for insulated conductors in wireways and gutters are given in *NEC Tables 310.15(B)(16) and 310.15(B)(18)*. It should be noted that these tables are used for raceways in general. These *NEC®* tables and the notes are often used to determine if the correct materials are on hand for an installation. They are also used to determine if it is possible to add conductors in an existing wireway or gutter.

In many situations, it is necessary to make extensions from the wireways to wall receptacles and control devices. In these cases, *NEC Section 376.70* specifies that these extensions be made using any wiring method presented in *NEC Chapter 3* that includes a means for equipment grounding. Finally, as required in *NEC Section 376.120*, wireways must be marked in such a way that their manufacturer's name or trademark will be visible.

As you can see in *Figure 56*, a wide range of fittings is required for connecting wireways to one another and to fixtures such as switchboards, power panels, and conduit.

3.1.0 Types of Wireways and Their Components

Rectangular duct-type wireways come as either hinged-cover or screw-cover troughs. They are available in a variety of lengths to avoid cut-

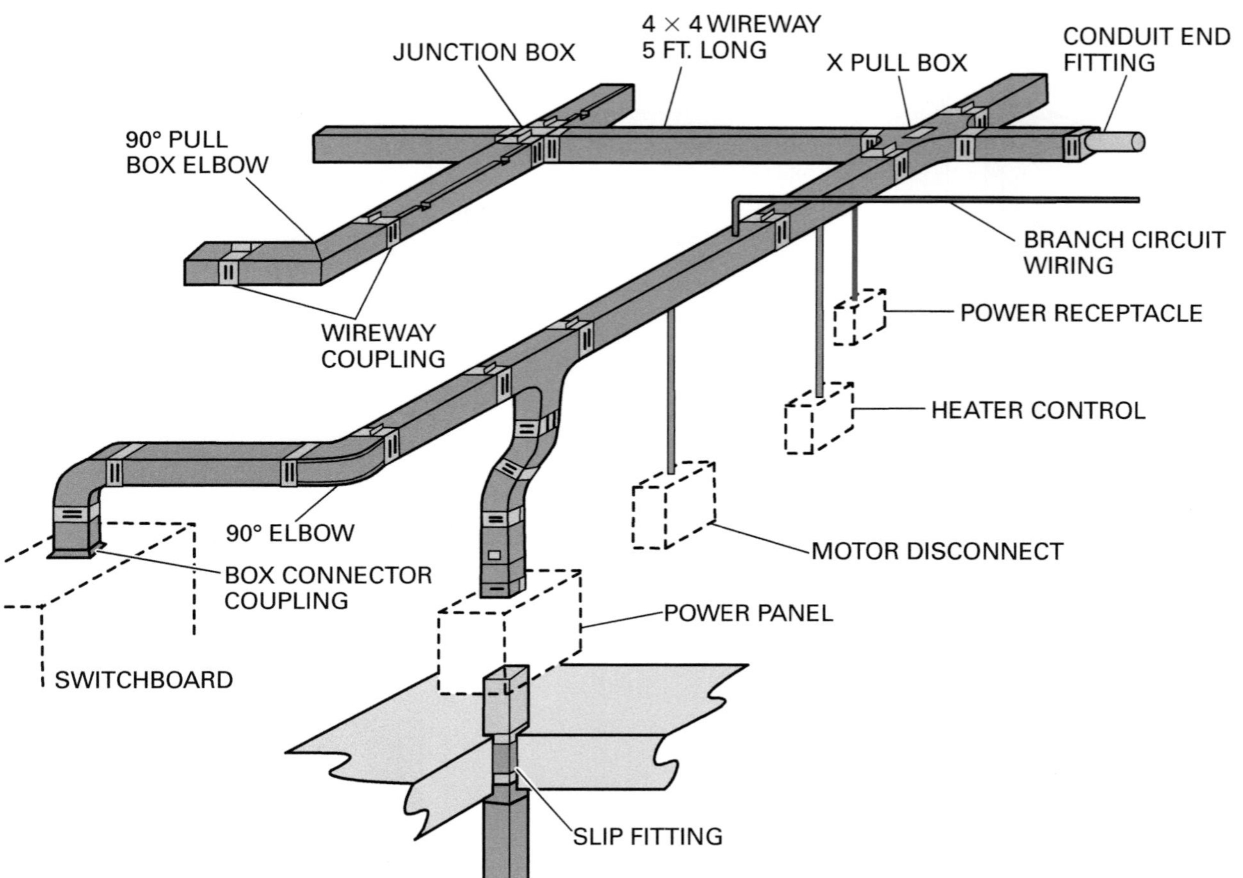

Figure 56 Wireway system layout.

ting. Raintight troughs are used in environments where moisture is not permitted within the raceway. However, a raintight trough should not be confused with the raintight lay-in wireway, which has a hinged cover. *Figure 57* shows a raintight trough with a removable side cover.

Wireway troughs are exposed when first installed. Whenever possible, they are mounted on the ceilings or walls, although they may sometimes be suspended from the ceiling. Note that in *Figure 58*, the trough has knockouts similar to those found on junction boxes. After the wireway system has been installed, branch circuits are brought from the distribution panels using conduit. The conduit is joined to the wireway at the most convenient knockout possible.

Wireway components such as trough crosses, 90° internal elbows, and tee connectors serve the same function as fittings on other types of raceways. The fittings are attached to the duct using slip-on connectors. All attachments are made with nuts and bolts or screws. When assembling wireways, always place the head of the bolt on the inside and the nut on the outside so that the

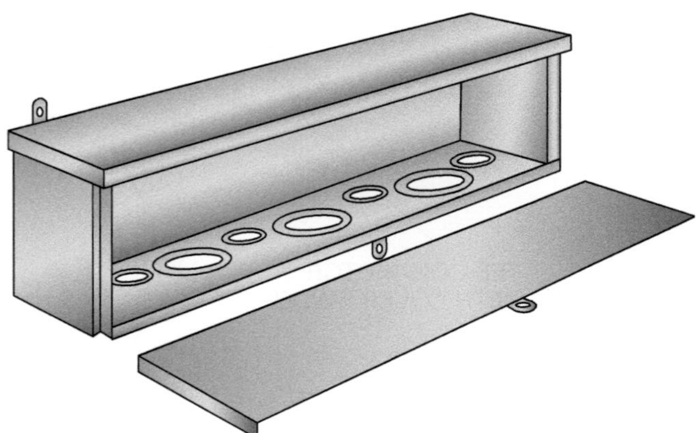

Figure 57 Raintight trough.

conductors will not be resting against a sharp edge. It is usually best to assemble sections of the wireway system on the floor, and then raise the sections into position. An exploded view of a section of wireway is shown in *Figure 59*. Both the wireway fittings and the duct come with screw-on, hinged, or snap-on covers to permit conductors to be laid in or pulled through.

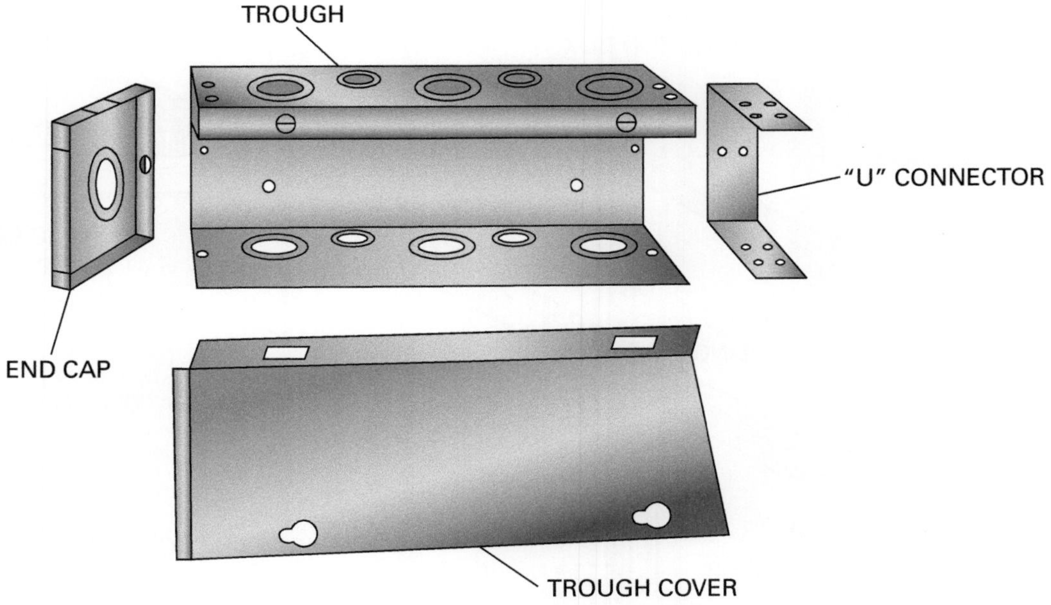

Figure 58 Trough.

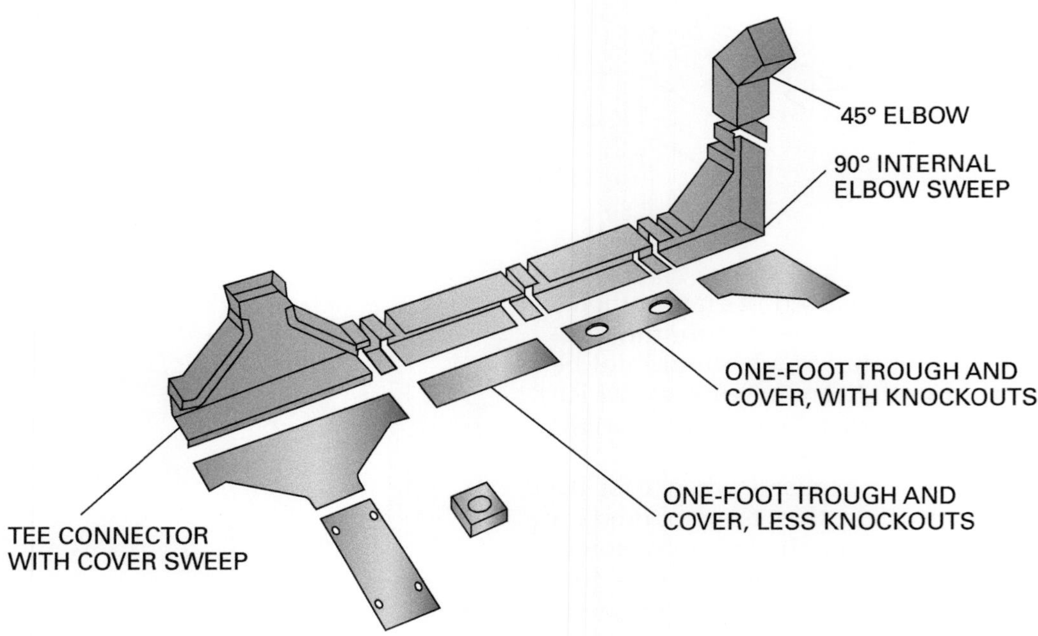

Figure 59 Wireway sections.

The *NEC*® specifies that wireways may be used only for exposed work. Therefore, they cannot be used in underfloor installations. If they are used for outdoor work, they must be of an approved raintight construction. It is important to note that wireways must not be installed where they are subject to severe physical damage, corrosive vapors, or hazardous locations.

Wireway troughs must be installed so that they are supported at distances not exceeding 5' (1.5 m). When specially approved supports are used, the distance between supports must not exceed 10' (3 m).

3.1.1 Wireway Fittings

Many different types of fittings are available for wireways, especially for use in exposed, dry locations. The following sections explain fittings commonly used in the electrical craft.

NCCER – *Maritime Electrical*

Wireway or Trough?

A raintight lay-in wireway has a hinged cover, as shown here. A raintight trough simply has a removable cover.

3.1.2 Connectors

Connectors (*Figure 60*) are used to join wireway sections and fittings. Connectors are slipped inside the end of a wireway section and are held in place by small bolts and nuts. Alignment slots allow the connector to be moved until it is flush with the inside surface of the wireway. After the connector is in position, it can be bolted to the wireway. This helps to ensure a strong rigid connection. Connectors have a friction hinge that helps hold the wireway cover open when needed.

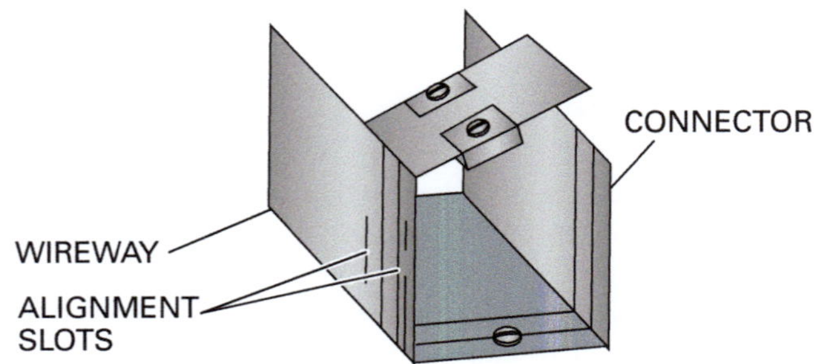

Figure 60 Connector.

3.1.3 End Plates

End plates, or closing plates (*Figure 61*), are used to seal the ends of wireways. They are inserted into the end of the wireway and fastened by screws and bolts. End plates contain knockouts so that conduit or cable may be extended from the wireway.

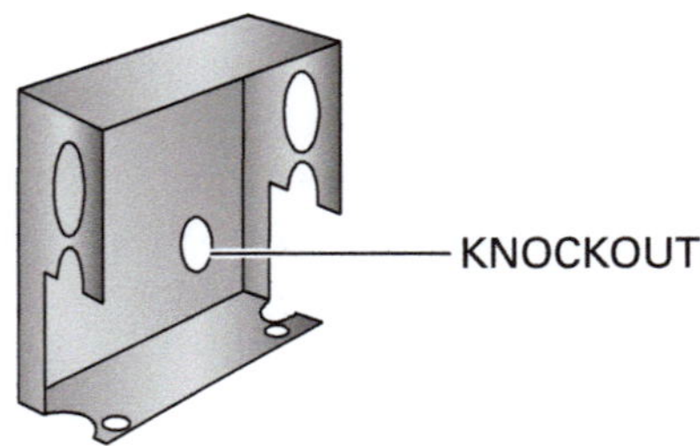

Figure 61 End plate.

3.1.4 Tees

Tee fittings (*Figure 62*) are used when a tee connection is needed in a wireway system. A tee connection is used where circuit conductors may branch in different directions. The tee fitting's covers and sides can be removed for access to splices and taps. Tee fittings are attached to other wireway sections using standard connectors.

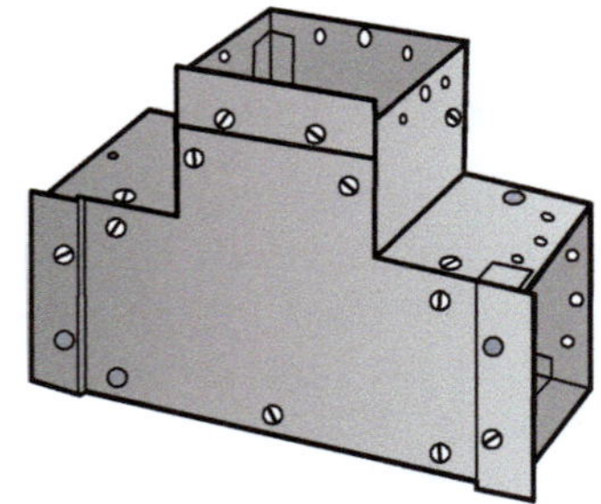

Figure 62 Tee.

3.1.5 Crosses

Crosses (*Figure 63*) have four openings and are attached to other wireway sections with standard connectors. The cover is held in place by screws and can be easily removed for laying in wires or for making connections.

3.1.6 Elbows

Elbows are used to make a bend in the wireway. They are available in angles of 22 $\frac{1}{2}$°, 45°, or 90°, and are either internal or external. They are attached to wireway sections with standard connectors. Covers and sides can be removed for wire installation. The inside corners of elbows are rounded to prevent damage to conductor insulation. An inside elbow is shown in *Figure 64*.

3.1.7 Telescopic Fittings

Telescopic or slip fittings may be used between lengths of wireway. Slip fittings are attached to standard lengths by setscrews and usually adjust from $\frac{1}{2}$" to 11 $\frac{1}{2}$" (12.7 mm to 292.1 mm). Slip fittings have a removable cover for installing wires and are similar in appearance to a nipple.

3.2.0 Wireway Supports

Horizontal wireway runs must be securely supported at each end and at intervals of no more than 5' (1.5 m) or for individual lengths greater than 5' (1.5 m) at each end or joint, unless listed for other support intervals. In no case shall the support distance be greater than 10' (3 m), in accordance with *NEC Section 376.30(A)*. If possible, wireways can be mounted directly to a surface. Otherwise, wireways are supported by hangers or brackets.

3.2.1 Suspended Hangers

In many cases, the wireway is supported from a ceiling, beam, or other structural member. In such installations, a suspended hanger (*Figure 65*) may be used to support the wireway.

The wireway is attached to or laid in the hanger. The hanger is suspended by a threaded rod. One end of the rod is attached to the hanger with hex nuts. The other end of the rod is attached to a beam clamp or anchor.

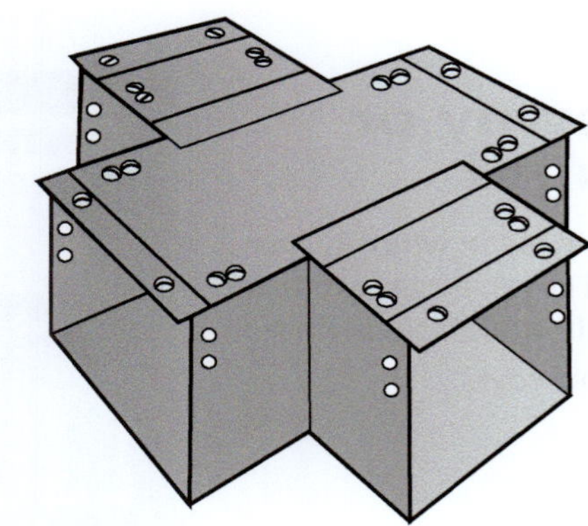

Figure 63 Cross.

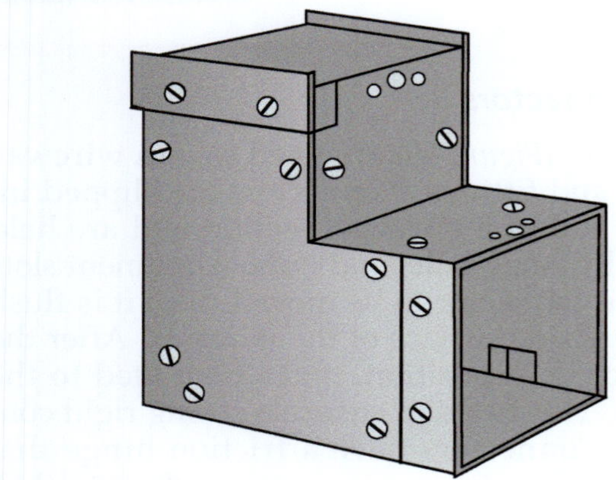

Figure 64 90° inside elbow.

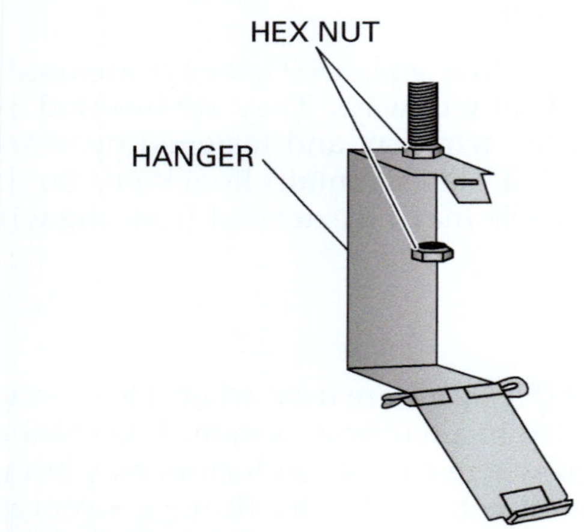

Figure 65 Suspended hanger.

3.2.2 Gusset Brackets

Another type of support used to mount wireways is a gusset bracket (*Figure 66*). This is an L-type bracket that is mounted to a wall. The wireway rests on the bracket and is attached by screws or bolts.

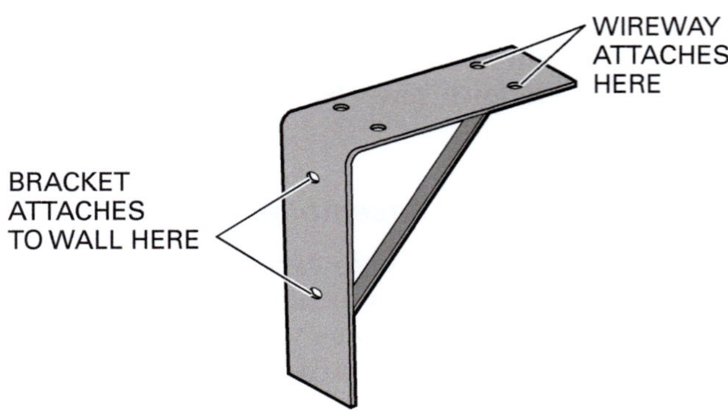

Figure 66 Gusset bracket.

3.2.3 Standard Hangers

Standard hangers (*Figure 67*) are made in two pieces. The two pieces are combined in different ways for different installation requirements. The wireway is attached to the hanger by bolts and nuts.

3.2.4 Wireway Hangers

When a larger wireway must be suspended, a wireway hanger may be used. A wireway hanger is made by suspending a piece of strut from a ceiling, beam, or other structural member. The strut is suspended by threaded rods attached to beam clamps or other ceiling anchors, as shown in *Figure 68*.

3.3.0 Specialty Raceways

Specialty raceways include enclosures such as surface metal and nonmetallic raceways, underfloor raceways, and underground ducts.

3.3.1 Surface Metal and Nonmetallic Raceways

Surface metal raceways consist of a wide variety of special raceways designed primarily to carry power and communications wiring to locations on the surface of ceilings or walls of building interiors.

Installation specifications of both surface metal raceways and surface nonmetallic raceways are listed in detail in *NEC Articles 386 and 388*, respectively. All these raceways must be installed in dry, interior locations. The number of conductors, their amperage, and the allowable cross-sectional area of the conductors, as well as regulations for combination raceways, are specified in *NEC Tables 310.15(B)(16) and 310.15(B)(18)* and *NEC Articles 386 and 388*.

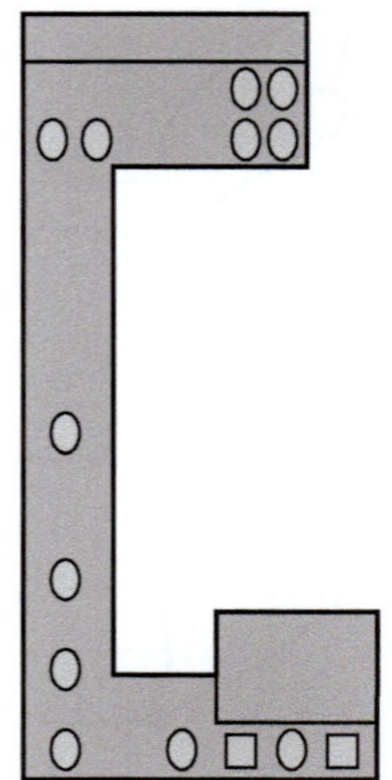

Figure 67 Standard hanger.

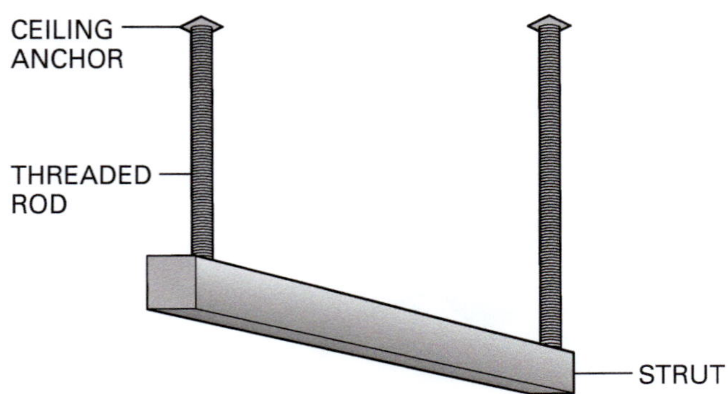

Figure 68 Wireway hanger.

One use of surface metal raceways is to protect conductors that run to non-accessible outlets.

Surface metal and nonmetallic raceways are divided into subgroups based on the specific purpose for which they are intended. There are three small surface raceways that are primarily used for extending power circuits from one point to another. In addition, there are six larger surface raceways that have a much wider range of applications. Typical cross sections of the first three smaller raceways are shown in *Figure 69*.

Additional surface metal raceway designs are referred to as pancake raceways, because their flat cross sections resemble pancakes. Their primary use is to extend power, lighting, telephone, or signal wire across a floor to locations away from the walls of a room without embedding them under the floor. Pancake raceways are shown in *Figure 70*.

There are also surface metal raceways available that house two or three different conductor raceways. These are referred to as twin-duct or triple-duct. These raceways permit different circuits, such as power and signal, to be placed within the same raceway.

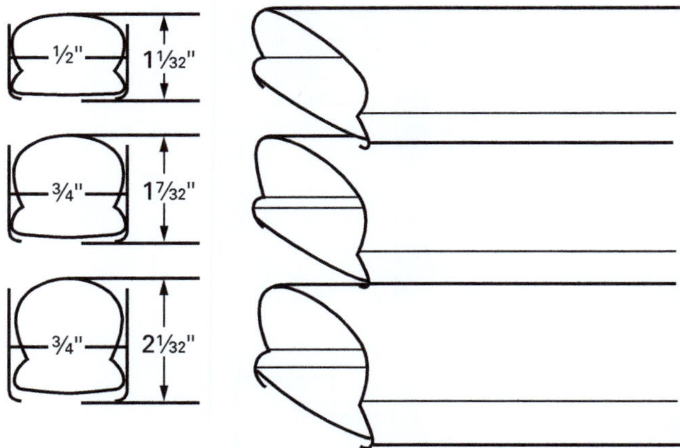

Figure 69 Smaller surface raceways.

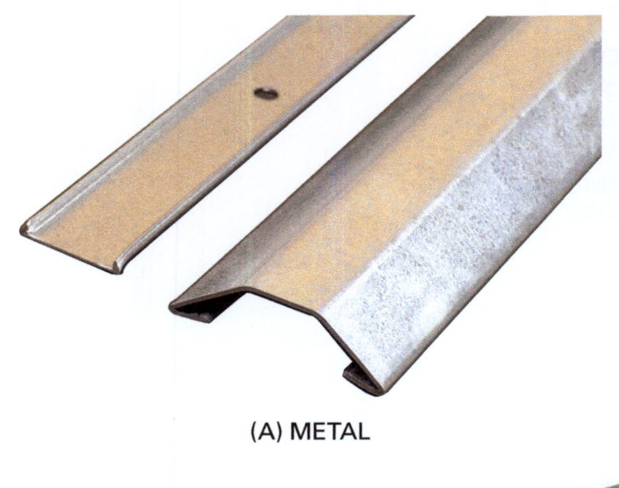

(A) METAL

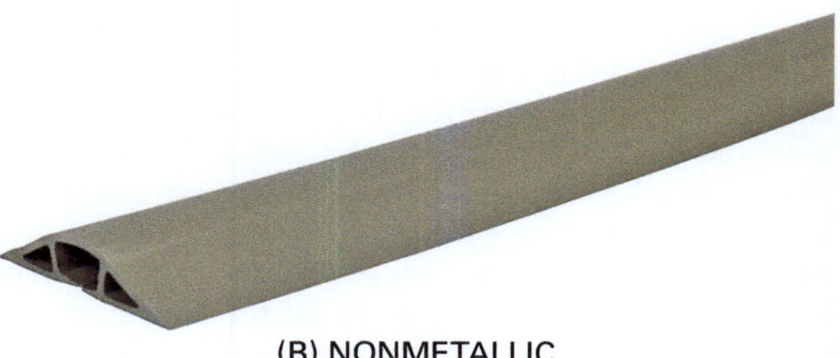

(B) NONMETALLIC

Figure 70 Pancake raceways.

Nonmetallic raceways come in a variety of styles. The perimeter raceways shown in *Figure 71* are available in sizes ranging from $\frac{3}{4}$" to more than 7" wide. Many of these raceways contain barriers that allow them to carry both low voltage and power wiring.

The number and types of conductors permitted to be installed and the capacity of a particular surface raceway must be calculated and matched with *NEC®* requirements, as discussed previously. *NEC Tables 310.15(B)(16) through 310.15(B)(18)* are used for surface raceways in the same manner in which they are used for wireways. For surface raceway installations with more than three conductors in each raceway, particular reference must be made to *NEC Table 310.15(B)(3)(a)*.

3.3.2 Multi-Outlet Assemblies

Manufacturers offer a wide variety of multi-outlet surface raceways. Their function is to hold receptacles and other devices within the raceway. When surface raceways are used in this manner, the assembly is referred to as a multi-outlet assembly. Multi-outlet assemblies (*Figure 72*) are covered in *NEC Article 380*. Multi-outlet systems are either wired in the field or come pre-wired from the factory.

3.3.3 Pole Systems

In many situations, power and other electric circuits must be carried from overhead wiring systems to devices that are not located near existing wall outlets or control circuits. This type of wiring is typically used in open office spaces where cubicles are provided by temporary dividers. Poles are used to accomplish this. The poles usually come in lengths suitable for different ceiling heights. *Figure 73* shows a typical pole base.

3.3.4 Underfloor Systems

Underfloor raceway systems were developed to provide a practical means of bringing conductors for lighting, power, and signaling to cabinets and consoles. Underfloor raceways are available in 10' (3 m) lengths and widths of 4" and 8" (100 mm and 200 mm). The sections are made with inserts spaced every 24" (600 mm). The inserts can be removed for outlet installation. These are explained in *NEC Article 390*.

> **NOTE**
>
> Inserts must be installed so that they are flush with the finished grade of the floor.

Junction boxes are used to join sections of underfloor raceways. Conduit is also used with under-floor raceways by using a raceway-to-conduit connector (conduit adapter). A typical underfloor raceway duct with fittings is shown in *Figure 74*. This wiring method makes it possible to place a desk or table in any location where it will always be over, or very near to, a duct line. The wiring method for lighting and power between cabinets

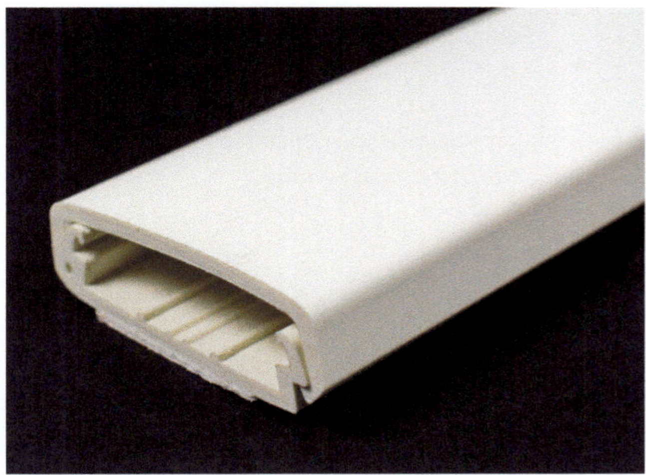

(A) SINGLE CHANNEL

(B) DUAL CHANNEL

Figure 71 Examples of surface raceway.

Figure 72 Multi-outlet assembly.

and the raceway junction boxes may be conduit, underfloor raceway, wall elbows, and cabinet connectors. *NEC Article 390* covers the installation of underfloor raceways.

3.3.5 Cellular Metal Floor Raceways

A cellular metal floor raceway is a type of floor construction designed for use in steel-frame buildings. In these buildings, the members supporting the floor between the beams consist of sheet steel rolled into shapes. These shapes are combined to form cells, or closed passageways, which extend across the building. The cells are of various shapes and sizes, depending upon the structural strength required. The cells of this type of floor construction form the raceways, as shown in *Figure 75*.

Connections to the cells are made using headers that extend across the cells. A header connects only to those cells to be used as raceways for

Figure 73 Power pole.

conductors. A junction box or access fitting is necessary at each joint where a header connects to a cell. Two or three separate headers, connecting to different sets of cells, may be used for different systems. For example, light and power, signaling systems, and public telephones would each have a separate header. A special elbow fitting is used to extend the headers up to the distribution equipment on a wall or column. *NEC Article 374* covers the installation of cellular metal floor raceways.

Surface Raceways

Surface raceways with multiple channels are commonly used in computer networking applications to provide conductors for AC power to the computers, as well as telephone and other low-voltage wiring.

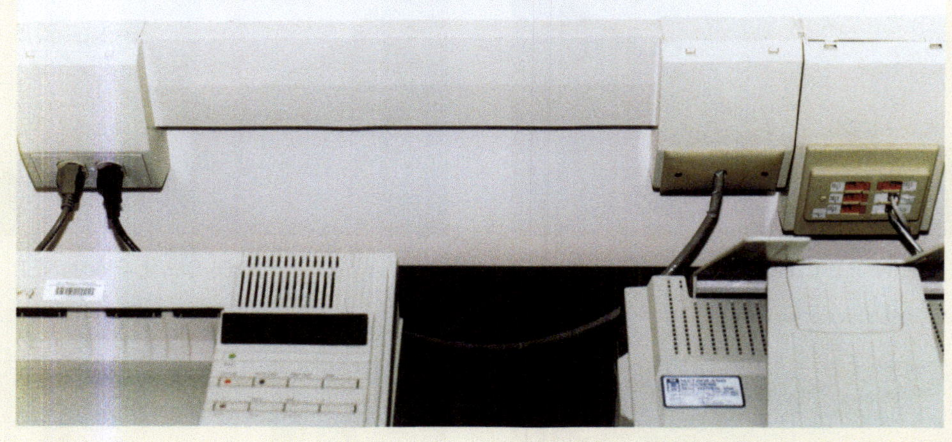

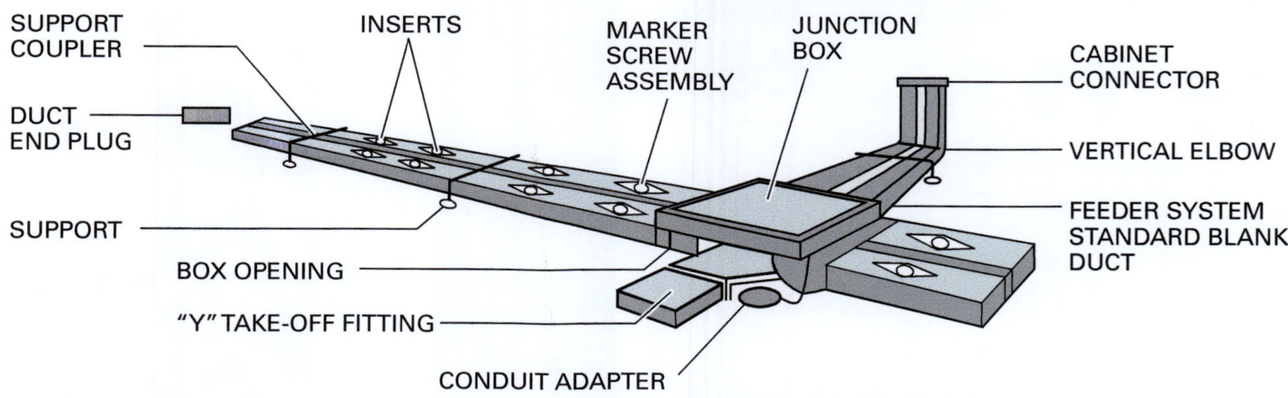

Figure 74 Underfloor raceway duct.

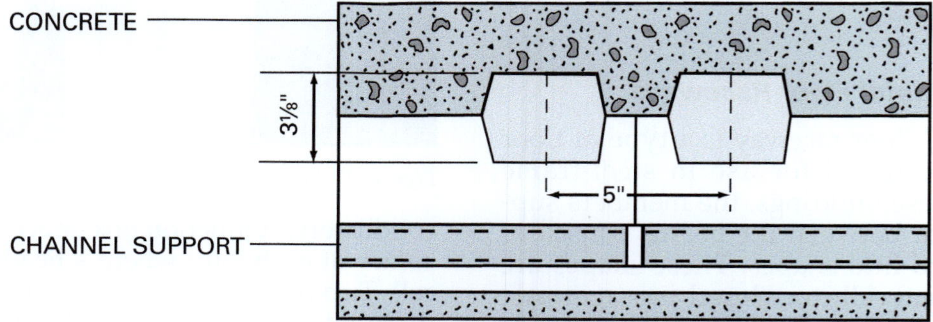

Figure 75 Cross section of a cellular floor.

3.3.6 Cellular Concrete Floor Raceways

The term *precast cellular concrete floor* refers to a type of floor used in steel-frame, concrete-frame, and wall-bearing construction. In this type of system, the floor members are precast with hollow voids that form smooth, round cells. The cells form raceways, which can be adapted, using fittings, for use as underfloor raceways. A precast cellular concrete floor is fire-resistant and requires no further fireproofing. The precast reinforced concrete floor members form the structural floor and are supported by beams or bearing walls. Connections to the cells are made with headers that are secured to the precast concrete floor. *NEC Article 372* covers the installation of cellular concrete floor raceways.

3.3.7 Duct Systems

In the common vocabulary of the electrical trade, a duct is a single enclosed raceway, or runway, through which conductors or cables can be led. Basically, ducting is a system of ducts. However, underground duct systems include manholes, transformer vaults, and risers.

There are several reasons for running power lines underground rather than overhead. In some situations, an overhead high-voltage line would be dangerous, or the space may not be adequate. For aesthetic reasons, architectural plans may require buried lines throughout a subdivision or a planned community. Tunnels may already exist, or be planned, for carrying steam or water lines. In any of these situations, underground installations are appropriate. Underground cables may be buried directly in the ground or run through tunnels or raceways, including conduit and recognized ducts.

In underground construction, a duct system provides a safe passageway for power lines, communication cables, or both. A duct consists of conduit or an approved duct system (such as HDPE) placed in a trench and covered with earth or concrete. The minimum depth at which the duct will be placed is determined using *NEC Table 300.5*. Encasing the duct in concrete or other materials provides mechanical strength and helps dissipate heat. *Figure 76* shows a duct bank in place and ready for backfill. In this case, it will be covered in concrete.

Manholes are set at intervals in an underground duct run. *Figure 77* shows a manhole with pull strings installed and tied off in preparation for the conductor installation. Manholes provide access through throats (sometimes called

Figure 76 Duct bank.

Figure 77 Manhole.

chimneys). At ground level, or street surface level, a manhole cover closes off the manhole area tightly. A duct line may consist of a single conduit or several, each carrying a cable length from one manhole to the next.

Manholes provide room for conductor installation and maintenance. Workers enter a manhole from above. In a two-way manhole, cables enter and leave in only two directions. There are also three-way and four-way manholes. Often manholes are located at the intersection of two streets so that they can be used for cables leaving in four directions. Manholes are usually constructed of brick or concrete. Their design must provide room for drainage and for workers to move around inside them. A similar opening known as a handhole is sometimes provided for splicing on lateral two-way duct lines.

Transformer vaults house power transformers, voltage regulators, network protectors, meters, and circuit breakers. A cable may end at a transformer vault. Other cables end at a customer's substation or terminate as risers that connect with overhead lines.

Underground duct lines can be made of fiber, vitrified tile, rigid metal or nonmetallic conduit, or poured concrete. The inside diameter of the ducting for a specific job is determined by the size of the cable that will be drawn into the duct. Inside diameters of 2" to 6" (MD 53 to MD 155) are available for most types of ducting.

Rigid nonmetallic conduit may be made of PVC (polyvinyl chloride), PE (polyethylene), or styrene. Because this type of conduit is available in longer lengths, fewer couplings are needed than with other types of ducting. PVC is popular because it is easy to install, requires less labor than other types of conduit, and is low in cost.

Monolithic concrete duct is poured at the job site. Multiple duct lines can be formed using rubber tubing cores on spacers. The cores may be removed after the concrete has set. A die containing steel tubes, known as a boat, can also be used to form ducts. It is pulled slowly through the trench on a track as concrete is poured from the top. Poured concrete ducting made by either method is relatively expensive, but offers the advantage of creating a very clean duct interior with no residue that can decay. The rubber core method is especially useful for curving or turning part of a duct system.

One of the most popular duct types is the cable-in-duct. This type of duct comes from the manufacturer with cables already installed. The duct comes in a reel and can be laid in the trench with ease. The installed cables can be withdrawn in the future, if necessary. This type of duct, because of the form in which it comes, reduces the need for fittings and couplings. It is most frequently used for street lighting systems.

Additional Resources

National Electrical Code®Handbook, Latest Edition. Quincy, MA: National Fire Protection Association.

3.0.0 Section Review

1. Which of the following is true regarding wireways?

   a. Leave knockouts open in raintight troughs to allow for drainage.
   b. Raintight lay-in wireways have a hinged cover.
   c. Raintight lay-in wireways are the same as raintight troughs.
   d. Troughs do not require knockouts because the conductors are fully enclosed.

2. The maximum distance between wireway supports is ______.

   a. 5' (1.5 m)
   b. 10' (3 m)
   c. 15' (4.5 m)
   d. 20' (6 m)

3. Which of the following is true regarding surface metal raceways?

   a. They are typically used to carry service-entrance cable.
   b. They can be used indoors or outdoors.
   c. They are used in dry, interior locations only.
   d. They are typically installed on floors.

SECTION FOUR

4.0.0 CABLE TRAYS

Objective

Select and install cable trays.
  a. Identify cable tray types and fittings.
  b. Install cable tray supports.

Trade Term

Cable trays: Rigid structures used to support electrical conductors.

Cable trays function as a support for conductors and tubing (*NEC Article 392*). A cable tray has the advantage of easy access to conductors, and thus lends itself to installations where the addition or removal of conductors is a common practice.

4.1.0 Cable Tray Types and Fittings

Cable trays are fabricated from aluminum, steel, and fiberglass. Cable trays are available in two basic forms: ladder and trough. Ladder tray, as the name implies, consists of two parallel channels connected by rungs. Trough consists of two parallel channels (side rails) having a corrugated, ventilated bottom, or a corrugated, solid bottom. There is also a special center-rail cable tray available for use in light-duty applications such as telephone and sound wiring. Cable trays are available in a variety of lengths, widths, and load depths to suit different applications.

Cable trays may be used in most electrical installations. Cable trays may be used in air handling ceiling space, but only to support the wiring methods permitted in such spaces by *NEC Section 300.22(C)(1)*. Also, cable trays may be used in Class 1, Division 2 locations according to *NEC Section 501.10(B)*. Cable trays may also be used above a suspended ceiling that is not used as an air handling space. Some manufacturers offer an aluminum cable tray that is coated with PVC for installation in caustic environments. A typical cable tray system with fittings is shown in *Figure 78*.

Wire and cable installation in cable trays is defined by the *NEC*. Read *NEC Article 392* to become familiar with the requirements and restrictions made by the *NEC* for safe installation of wire and cable in a cable tray.

Metallic cable trays that support electrical conductors must be grounded as required by *NEC Article 250*. Where steel and aluminum cable tray systems are used as an equipment grounding conductor, all of the provisions of *NEC Section 392.60* must be complied with.

> **WARNING!**
>
> Do not stand on, climb in, or walk on a cable tray.

Cable tray fittings are part of the cable tray system and provide a means of changing the direction or dimension of the different trays. Some of the uses of horizontal and vertical tees, horizontal and vertical bends, horizontal crosses, reducers, barrier strips, covers, and box connectors are shown in *Figure 78*.

4.2.0 Installing Cable Tray Supports

Cable trays are usually supported in one of five ways: direct rod suspension, trapeze mounting, center hung, wall mounting, and pipe rack mounting.

4.2.1 Direct Rod Suspension

The direct rod suspension method of supporting cable tray uses threaded rods and hanger clamps. One end of the threaded rod is connected to the ceiling or other overhead structure. The other end

---

Think About It

Cable Trays and Wireways

What is the difference between a wireway and a cable tray? What kinds of conductors would you expect to find in a cable tray as compared to a wireway?

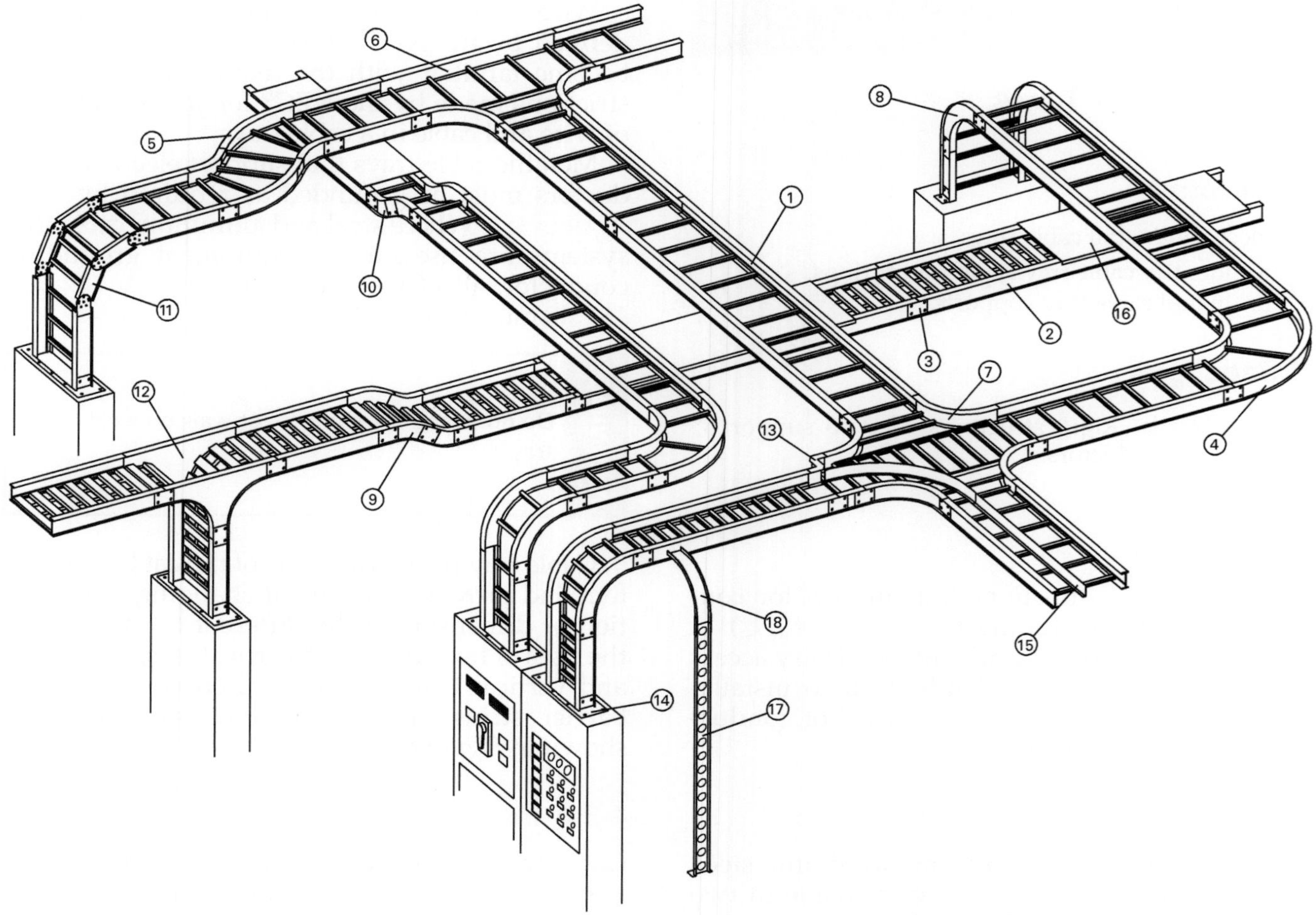

Legend

1. LADDER CABLE TRAY
2. VENTILATED TROUGH CABLE TRAY
3. STRAIGHT SPLICE PLATE
4. 90° HORIZONTAL BEND, LADDER CABLE TRAY
5. 45° HORIZONTAL BEND, LADDER CABLE TRAY
6. HORIZONTAL TEE, LADDER CABLE TRAY
7. HORIZONTAL CROSS, LADDER CABLE TRAY
8. 90° VERTICAL OUTSIDE BEND, LADDER CABLE TRAY
9. 45° VERTICAL OUTSIDE BEND, VENTILATED CABLE TRAY
10. 30° VERTICAL INSIDE BEND, LADDER CABLE TRAY
11. VERTICAL BEND SEGMENT (VBS)
12. VERTICAL TEE DOWN, VENTILATED TROUGH CABLE TRAY
13. LEFT HAND REDUCER, LADDER CABLE TRAY
14. FRAME-TYPE BOX CONNECTOR
15. BARRIER STRIP STRAIGHT SECTION
16. SOLID FLANGED TRAY COVER
17. VENTILATED CHANNEL STRAIGHT SECTION
18. CHANNEL CABLE TRAY, 90° VERTICAL OUTSIDE BEND

Figure 78 Cable tray system.

is connected to hanger clamps that are attached to the cable tray side rails. A direct rod suspension assembly is shown in *Figure 79*.

4.2.2 Trapeze Mounting and Center Hung Support

Trapeze mounting of cable tray is similar to direct rod suspension mounting. The difference is in the method of attaching the cable tray to the threaded rods. A structural member, usually a steel chan-nel or strut, is connected to the vertical supports to provide an appearance similar to a swing or trapeze. The cable tray is mounted to the structural member. Often, the underside of the channel or strut is used to support conduit. A trapeze mounting assembly is shown in *Figure 80*.

A method that is similar to trapeze mounting is a center hung tray support. In this case, only one rod is used and it is centered between the cable tray side rails.

4.2.3 Wall Mounting

Wall mounting is accomplished by supporting the cable tray with structural members attached to the wall (*Figure 81*). This method of support is often used in tunnels and other underground or sheltered installations where large numbers of conductors interconnect equipment that is separated by long distances.

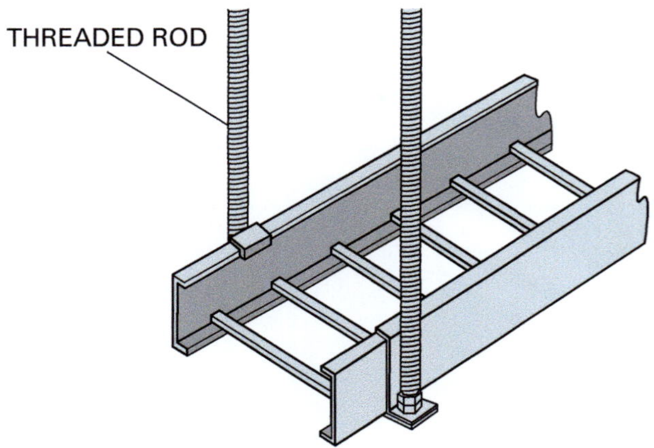

Figure 79 Direct rod suspension.

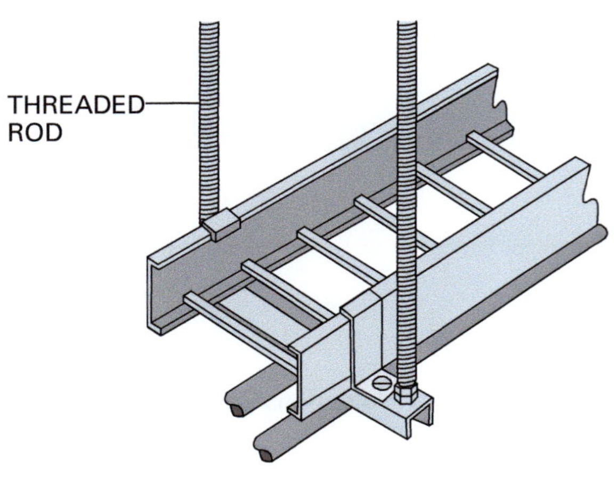

Figure 80 Trapeze mounting and center hung support.

4.2.4 Pipe Rack Mounting

Pipe racks are structural frames used to support piping that interconnects equipment in outdoor industrial facilities. Usually, some space on the rack is reserved for conduit and cable tray. Pipe rack mounting of cable tray is often used when power distribution and electrical wiring is routed over a large area.

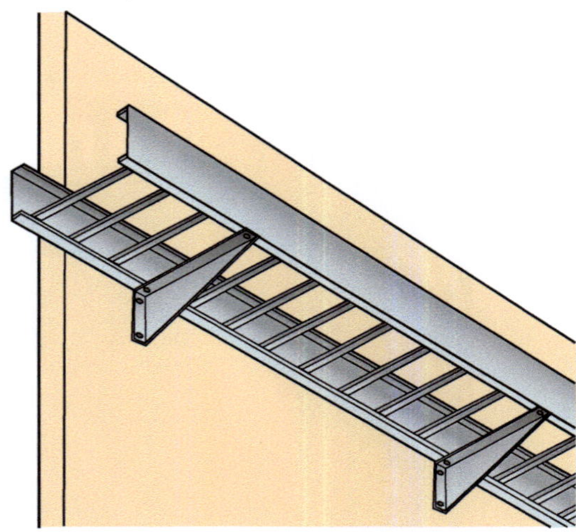

Figure 81 Wall mounting.

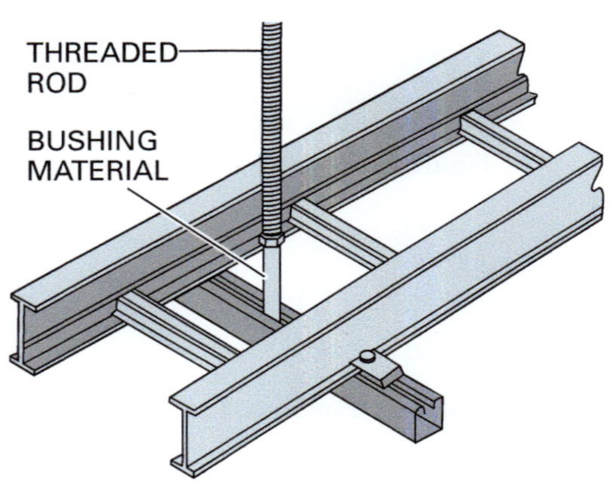

Cable Tray Systems

Cable tray systems must be continuous and grounded. One of the advantages of using a cable tray system is that it makes it easy to expand or modify the wiring system following installation. Unlike conduit systems, wires can be added or changed by simply laying them into (or lifting them out of) the tray.

Figure Credit: Jim Mitchem

Additional Resources

National Electrical Code®Handbook, Latest Edition. **Quincy, MA: National Fire Protection Association.**

4.0.0 Section Review

1. Special light-duty center-rail cable tray is most often used for _____.

   a. industrial power requirements
   b. telephone or sound wiring
   c. outdoor power distribution
   d. large conductors

2. The type of cable tray support most likely to be used when power distribution and electrical wiring is routed over a large outdoor area is _____.

   a. trapeze mounting
   b. wall mounting
   c. pipe rack mounting
   d. direct rod suspension

5.0.0 HANDLING AND STORING RACEWAYS

Objective

Handle and store raceways.
  a. Handle raceways.
  b. Store raceways.

Proper and safe methods of storing conduit, wireways, raceways, and cable trays may sound like a simple task, but improper storage techniques can result in wasted time and damage to the raceways, as well as personal injury. Storing raceways correctly will help avoid costly damage, save time in identifying stored raceways, and reduce the chance of personal injury.

5.1.0 Handling Raceways

Raceway is made to strict specifications. It can be easily damaged by careless handling. From the time raceway is delivered to a job site until the installation is complete, use proper and safe handling techniques. These are a few basic guidelines for handling raceway that will help avoid damaging or contaminating it:

- Never drag raceway off a delivery truck or off other lengths of raceway.
- Never drag raceway on the ground or floor. Dragging raceway can cause damage to the ends.
- Keep the thread protection caps on when handling or transporting conduit raceway.
- Keep raceway away from any material that might contaminate it during handling.

- Flag the ends of long lengths of raceway when transporting it to the job site.
- Never drop or throw raceway when handling it.
- Never hit raceway against other objects when transporting it.
- Always use two people when carrying long pieces of raceway. Make sure that you both stay on the same side and that the load is balanced. Each person should be about one-quarter of the length of the raceway from the end. Lift and put down the raceway at the same time.

5.2.0 Storing Raceways

Pipe racks are commonly used for storing conduit. The racks provide support to prevent bending, sagging, distorting, scratching, or marring of conduit surfaces. Most racks have compartments where different types and sizes of conduit can be separated for ease of identification and selection. The storage compartments in racks are usually elevated to help avoid damage that might occur at floor level. Conduit that is stored at floor level is easily damaged by people and other materials or equipment in the area.

The ends of stored conduit should be sealed to help prevent contamination and damage. Conduit ends can be capped, taped, or plugged.

Always inspect raceway before storing it to make sure that it is clean and not damaged. It is discouraging to get raceway for a job and find that it is dirty or damaged. Also, make sure that the raceway is stored securely so that when someone comes to get it for a job, it will not fall in any way that could cause injury.

To prevent contamination and corrosion of stored raceway, it should be covered with a tarpaulin or other suitable covering. It should also be separated from noncompatible materials such as hazardous chemicals.

Wireways, surface metal raceways, and cable trays should always be stored off the ground on boards in an area where people will not step on it and equipment will not run over it. Stepping on or running over raceway bends the metal and makes it unusable.

Putting It All Together

Think about the effort that goes into the design of a large industrial installation. If you were to design a large complex, such as the one shown here, where would you start and why?

5.0.0 Section Review

1. When moving long pieces of raceway, it is best to _____.

   a. drag it
   b. use two people
   c. remove the caps
   d. use a forklift

2. When storing raceway, it is best to _____.

   a. leave it uncovered
   b. store it at ground level
   c. separate different types
   d. use it as a temporary walkway

SUMMARY

This module discussed the various types of raceways, boxes, and fittings, including their uses and procedures for installation. The primary purpose of raceways is to house electric wire used for power distribution, communication, or electronic signal transmission. Raceways provide protection to the wiring and even a means of identifying one type of wire from another when run adjacent to each other. This process requires proper planning to allow for current needs, future expansion, and a neat and orderly appearance.

1. The lightest duty and most widely used non-flexible metal conduit is ______.

   a. electrical metallic tubing
   b. rigid metal conduit
   c. aluminum conduit
   d. plastic-coated RMC

2. Which of the following is often referred to as thinwall conduit?

   a. IMC
   b. EMT
   c. RMC
   d. Galvanized rigid steel conduit

3. *NEC Article 358* covers ______.

   a. RMC
   b. IMC
   c. EMT
   d. ENT

4. In order to resist corrosion in wet environments, EMT is ______.

   a. galvanized
   b. rubber-coated
   c. enamel-coated
   d. dipped in aluminum

5. RMC is made of ______.

   a. cast iron
   b. steel or aluminum
   c. copper or aluminum
   d. PVC

6. A type of conduit that should be used where corrosion is a factor is ______.

   a. PVC
   b. black enamel steel conduit
   c. IMC
   d. RMC

7. Which of the following is true with regard to PVC?

   a. It can be used as an equipment grounding conductor.
   b. It increases the voltage drop of the conductors.
   c. It requires the use of expansion joints to avoid damage due to temperature changes.
   d. It must be threaded.

8. A type of conduit approved for direct burial is ______.

   a. EMT
   b. HDPE
   c. IMC
   d. RMC

9. A type of conduit used to connect machines that vibrate during operation is ______.

   a. aluminum
   b. black enamel steel conduit
   c. flexible metal conduit
   d. RMC

10. Which of the following is an acceptable combination of conduit bends between pull points?

    a. four 90-degree bends
    b. three 90-degree bends and three 45-degree bends
    c. six 45-degree bends and two 90-degree bends
    d. eight 30-degree bends and two 90-degree bends

11. The fitting used to protect conductors from the sharp edges of conduit where it enters a box is called a ______.

    a. bushing
    b. locknut
    c. coupling
    d. nipple

12. Ungrounded conductors entering a raceway must be protected by bushings when they are sized No. ______.

    a. 10 AWG and larger
    b. 8 AWG and larger
    c. 6 AWG and larger
    d. 4 AWG and larger

13. To avoid having to make an offset bend, use a fitting known as a(n) ______.

    a. nipple
    b. conduit body
    c. bushing
    d. hub

14. A Type LB conduit body has a cover on ______.

    a. the left
    b. the right
    c. the back
    d. both sides

15. A type of conduit body used to provide a pull point on a straight conduit run is ______.

    a. Type C
    b. Type L
    c. Type T
    d. Type X

16. A type of conduit body used at four intersecting conduits is ______.

    a. Type LR
    b. Type LL
    c. Type T
    d. Type X

17. A Type L conduit body that has a cover on both sides is a(n) ______.

    a. Type LL
    b. Type LRL
    c. LB
    d. LX

18. Where metal conduit is to be installed in a wet location, a clamp back strap can be used to maintain the minimum required distance from the wall surface, which is ______.

    a. $\frac{1}{4}$ inch
    b. $\frac{1}{2}$ inch
    c. $\frac{3}{4}$ inch
    d. 1 inch

19. The preferred box depth for use in concrete construction is ______.

    a. 3 inches
    b. 4 inches
    c. 5 inches
    d. 6 inches

20. Which of the following is true when installing boxes in a metal stud environment?

    a. Boxes are always mounted flush with the studs.
    b. You must know the thickness of the finished surface before installing the boxes.
    c. PVC is the most common type of conduit used in metal stud environments.
    d. MC cable is not used in metal stud environments.

21. Which of the following regulations applies to drilling of wood joists and beams?

    a. An engineered beam cannot be drilled unless approved by the manufacturer.
    b. They can only be drilled in the center third.
    c. The hole must be at least 1" from an edge.
    d. The hole diameter must not exceed one-half of the depth of the girder or joist.

22. Hammer-set anchors are designed for use in ______.

    a. wood
    b. metal studs
    c. concrete
    d. structural steel

23. *NEC Section 376.56(A)* limits wireway fill to no more than _____.

   a. 40% of the cross-sectional area of the wireway
   b. 60% of the cross-sectional area of the wireway
   c. 75% of the cross-sectional area of the wireway
   d. 90% of the cross-sectional area of the wireway

24. The cross-sectional areas of all conductors at any cross section of a wireway shall not exceed _____.

   a. 20% of the interior cross-section of the wireway
   b. 35% of the interior cross-section of the wireway
   c. 40% of the interior cross-section of the wireway
   d. 75% of the interior cross-section of the wireway

25. Raceways designed to extend conductors across a floor without embedding it in the floor are called _____.

   a. cellular raceways
   b. raceway ducts
   c. cellular ductways
   d. pancake raceways

Trade Terms Quiz

Fill in the blank with the correct term that you learned from your study of this module.

1. A(n) __________ area is one that can be reached for service or repair.

2. When something is in a(n) __________, it is not permanently closed in by the structure or finish of a building.

3. When materials meet a regulatory agency's requirements, the material is then said to be __________.

4. __________ is a regulatory agency that evaluates and approves electrical components and equipment.

5. A(n) __________ is used to make a continuous grounding path between equipment and ground.

6. __________ are rigid structures, either suspended or mounted, that are used to support electrical conductors.

7. Similar to pipe, __________ is a round raceway that houses conductors.

8. A(n) __________ is a bend made in a piece of conduit to alter its course.

9. __________ are enclosed channels that are used to house wires and cables.

10. __________ are steel troughs designed to carry electrical wire and cable.

11. A(n) __________ is the connection of two or more conductors.

12. An intermediate point on a main circuit where another wire is connected to supply electrical current to another circuit is called a(n) __________.

13. Electrical connectors that could be exposed to the environment are housed in a long, narrow box, or a __________.

Trade Terms

Accessible
Approved
Bonding wire
Cable trays
Conduit

Exposed location
Kick
Raceways
Splice
Tap

Trough
Underwriters Laboratories, Inc. (UL)
Wireways

1. Most electrical equipment that has a metal frame must be __________.

   a. When installing EMT in a wet location, what type of fittings must be used?

   b. Setscrew

   c. Compression

   d. Raintight

   e. Steel

2. Conductors, along with splices and taps, must not fill a wireway to more than __________ of its cross-sectional area.

   a. 20%

   b. 30%

   c. 75%

   d. 90%

3. Which of the following covers the installation requirements for cable tray?

   a. *NEC Article 333*

   b. *NEC Article 338*

   c. *NEC Article 392*

   d. *NEC Article 394*

4. True or False? IMC conduit has the same internal diameter as RMC.

5. All of the following can be used as an equipment grounding conductor, *except* __________.

   a. rigid metal conduit

   b. rigid nonmetallic conduit

   c. electrical metallic tubing

   d. intermediate metal conduit

6. Flexible metal conduit can be connected to rigid metal conduit using a(n) __________ coupling.

7. Which of the following covers grounding provisions for metal boxes?

   a. *NEC Section 500.8(A)*

   b. *NEC Section 250.30(A)*

   c. *NEC Section 250.53(C)*

   d. *NEC Section 314.40(D)*

8. What type of rigid nonmetallic conduit may be installed where exposed to physical damage?

   a. Type DB Schedule 80

   b. Type DB Schedule 40

   c. Type EB

   d. Type 1

9. RMC is commonly used in __________ locations.

10. What type of conduit body is used to provide a junction point for three intersecting conduits?

    a. Type LL

    b. Type LR

    c. Type C

    d. Type T

11. When installing metal to conduit in a wet area, a(n) __________ air space must be provided between it and the supporting surface.

12. Conductors installed in PVC are subject to a(n) __________.

    a. increase in operating temperature

    b. decrease in operating temperature

    c. increase in ampacity

    d. reduced voltage drop

13. Flexible metal conduit must be supported within __________ of each end.

14. When installing a cable system near a metal corrugated sheet roofing deck, a spacing of __________ must be maintained from any point of the roof system.

Leonard "Skip" Layne
Rust Constructors Inc.

How did you choose a career in the electrical field?

I think the electrical field chose me. My father was a contractor for several years before closing shop and accepting a job as an electrical superintendent with the Rust Engineering Company. That happened when I was nine years old. After being moved around the country for the next several years and working as an apprentice on Dad's projects during my college summers, I couldn't think of anything that I would rather do.

Tell us about your apprenticeship experience.

I've never attended a formal apprenticeship school. There are probably several in our group who might say that they suspected this. My electrical education came from field work exposure and several electrical and engineering courses and seminars I've attended over the years.

I'm happy to say that I'm still learning and I've learned a great deal while working on the NCCER Electrical Committee and from my association with the other subject matter experts.

What positions have you held in the industry?

I started as a field apprentice on a tire plant in Madison, Tennessee, in 1959. I've held field positions as an apprentice, journeyman, field engineer, start-up manager, and superintendent. I spent a number of years estimating work, and I established the material control department for another major open-shop contractor several years ago. I managed the project controls group on a nuclear project for another open-shop contractor. I even spent a few years as vice president with an underground utility/treatment plant contractor.

What would you say is the primary factor in achieving success?

Keep learning. Work hard. I've had to work sixteen-hour days on the job site and in the office in order to meet the schedule and incorporate changes. Do what is asked of you and do it well.

What does your current job involve?

My job title says that I'm the Construction Engineering Manager for Rust, but the lack of a definitive job title means that I do whatever the company needs me to do at the time. I qualify the company's electrical licenses in seventeen states where we work.

Recently, Rust volunteered my services to the Gulf Coast Workforce Initiative, a business roundtable initiative to train 20,000 new construction workers for the Gulf Coast area devastated by hurricanes Katrina and Rita.

Do you have any advice for someone just entering the trade?

Get all of the classroom learning you can. Go through all four levels of the Electrical program while working in the field. Ask questions and try to get assigned to as many new and different tasks as you can. All of our larger ABC contractors have excellent supervisory training programs and you need to get into those after your craft training. Be adaptable and keep learning.

Al Hamilton
Willmar Electric Service

How did you become an electrician?

I was in college and met Ed, an electrician who told me about wiring buildings. I went to work for him part time at first and liked it which led to my becoming his apprentice. He was tough taskmaster who cared about his apprentices and he was an excellent communicator and teacher.

How did you get your training?

My training was 100% on the job. Everything was learned "hands on."

I was very fortunate to work for Ed and other experienced electricians who were true craftsmen.

What factor or factors have contributed to your success?

Work ethic, relationships, and reading. A good work ethic has allowed me to overcome mistakes and keep working to learn our craft. I was able to learn about electrical work, business, and life through relationships. I have always looked for successful people to listen and learn from. I discovered that many successful people are happy to share their ideas and that proved to be the key to personal growth. I was told long ago that if I could force myself to read just 10 pages a day that I could read a book a month because the average book is 300 pages. I have done this for many years and I read books on many subjects including the Bible, business, history, biographies, and money and, of course, the National Electrical Code. Through reading you can educate yourself and become a more knowledgeable and interesting person who others will look to for advice.

What does your current job entail?

I work for a leading electrical company, Willmar Electric Service Corp. and my responsibilities are business development and estimating. This includes maintaining relationships with our existing customers and finding new customers to work for. Our estimating team uses estimating software to bid jobs. We do competitive bidding to win jobs and we also estimate for design/build projects and projects that are negotiated with customers.

Any advice for apprentices just beginning their careers?

Work for an organization that shares your values and will recognize and reward your efforts. Don't ever give up! Who you become will depend largely on the people you meet and the books that you read. Find honest, ethical people who have demonstrated success and get to know them. If you have not been reading, start today—make yourself do it. Get involved in helping others by becoming a leader in your company, helping out at church, and passing on what you have learned.

Trade Terms Introduced in This Module

Accessible: Able to be reached, as for service or repair.

Approved: Meeting the requirements of an appropriate regulatory agency.

Bonding wire: A wire used to make a continuous grounding path between equipment and ground.

Cable trays: Rigid structures used to support electrical conductors.

Conduit: A round raceway, similar to pipe, that houses conductors.

Exposed location: Not permanently closed in by the structure or finish of a building; able to be installed or removed without damage to the structure.

Kick: A bend in a piece of conduit, usually less than 45°, made to change the direction of the conduit.

Raceways: Enclosed channels designed expressly for holding wires, cables, or busbars, with additional functions as permitted in the *NEC*®.

Splice: Connection of two or more conductors.

Tap: Intermediate point on a main circuit where another wire is connected to supply electrical current to another circuit.

Trough: A long, narrow box used to house electrical connections that could be exposed to the environment.

Underwriters Laboratories, Inc. (UL): An agency that evaluates and approves electrical components and equipment.

Wireways: Steel troughs designed to carry electrical wire and cable.

Additional Resources

This module presents thorough resources for task training. The following resource material is suggested for further study.

Benfield Conduit Bending Manual, 2nd Edition. Overland Park, KS: EC&M Books.
*Concrete Fastening Systems.***www.confast.com**.
National Electrical Code®Handbook, Latest Edition. Quincy, MA: National Fire Protection Association.

Figure Credits

Section Review Answer Key

Answer	Section Reference	Objective
Section One		
1. c	1.1.0	1a
2. b	1.2.0	1b
3. d	1.3.5	1c
4. c	1.4.0	1d
5. d	1.5.4	1e
6. a	1.6.1	1f
Section Two		
1. d	2.1.0	2a
2. a	2.2.0	2b
3. d	2.3.0	2c
4. a	2.4.0	2d
5. d	2.5.4	2e
6. c	2.6.2	2f
7. d	2.7.0	2g
Section Three		
1. b	3.1.0	3a
2. b	3.2.0	3b
3. c	3.3.1	3c
Section Four		
1. b	4.1.0	4a
2. c	4.2.4	4b
Section Five		
1. b	5.1.0	5a
2. c	5.2.0	5b

NCCER CURRICULA — USER UPDATE

NCCER makes every effort to keep its textbooks up-to-date and free of technical errors. We appreciate your help in this process. If you find an error, a typographical mistake, or an inaccuracy in NCCER's curricula, please fill out this form (or a photocopy), or complete the online form at **www.nccer.org/olf**. Be sure to include the exact module ID number, page number, a detailed description, and your recommended correction. Your input will be brought to the attention of the Authoring Team. Thank you for your assistance.

Instructors – If you have an idea for improving this textbook, or have found that additional materials were necessary to teach this module effectively, please let us know so that we may present your suggestions to the Authoring Team.

NCCER Product Development and Revision

13614 Progress Blvd., Alachua, FL 32615

Email: curriculum@nccer.org
Online: www.nccer.org/olf

❏ Trainee Guide   ❏ Lesson Plans   ❏ Exam   ❏ PowerPoints   Other ___________________

Craft / Level: _______________________________________ Copyright Date: _______________

Module ID Number / Title: ____

Section Number(s): ____

Description: ____

____

____

____

Recommended Correction: ____

____

____

____

Your Name: ____

Address: ____

Email: ____ Phone: _______________________

Device Boxes

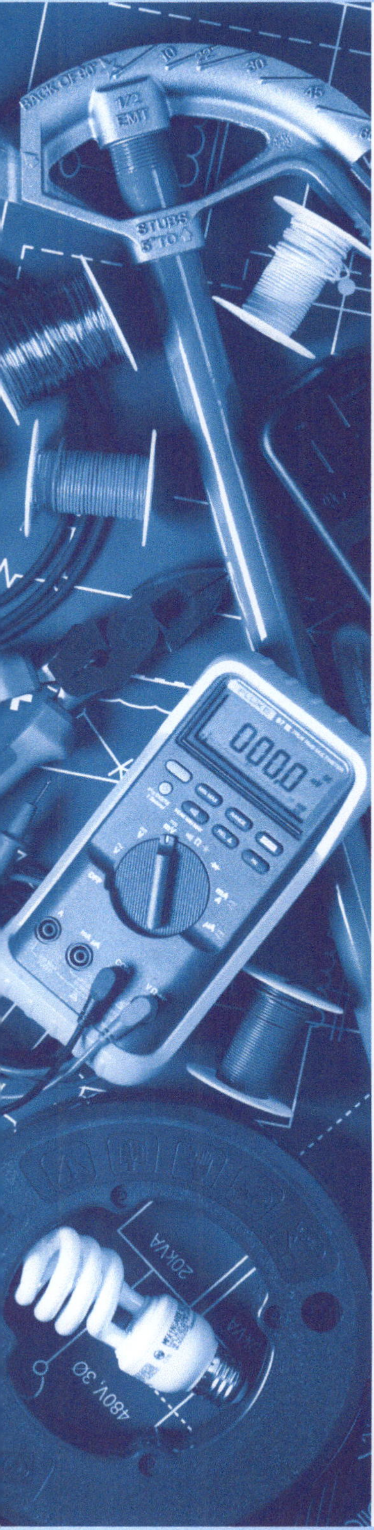

OVERVIEW

Electricians work with device boxes almost every day on every project, making a thorough understanding of the types of boxes available and their applications essential. This module describes the various types of boxes and explains how to calculate the *NEC*® fill requirements for outlet and junction boxes under 100 cubic inches (1,650 cubic centimeters).

Module 26106-17

Trainees with successful module completions may be eligible for credentialing through the NCCER Registry. To learn more, go to **www.nccer.org** or contact us at 1.888.622.3720. Our website has information on the latest product releases and training, as well as online versions of our *Cornerstone* magazine and Pearson's product catalog.

Your feedback is welcome. You may email your comments to **curriculum@nccer.org**, send general comments and inquiries to **info@nccer.org**, or fill in the User Update form at the back of this module.

This information is general in nature and intended for training purposes only. Actual performance of activities described in this manual requires compliance with all applicable operating, service, maintenance, and safety procedures under the direction of qualified personnel. References in this manual to patented or proprietary devices do not constitute a recommendation of their use.

DEVICE BOXES

Objectives

When you have completed this module, you will be able to do the following:

1. Size and install outlet boxes.
   a. Identify boxes and their applications.
   b. Size outlet boxes.
   c. Install outlet boxes.
2. Size and install pull and junction boxes.
   a. Size pull and junction boxes.
   b. Install pull and junction boxes.

Performance Tasks

Under the supervision of the instructor, you should be able to do the following:

1. Identify the appropriate box type and size for a given application.
2. Select the minimum size pull or junction box for the following applications:
   - Conduit entering and exiting for a straight pull.
   - Conduit entering and exiting at an angle.

Trade Terms

Connector
Explosion-proof
Handy box
Junction box
Outlet box

Pull box
Raintight
Watertight
Weatherproof

Industry Recognized Credentials

If you are training through an NCCER-accredited sponsor, you may be eligible for credentials from NCCER's Registry. The ID number for this module is 26106-17. Note that this module may have been used in other NCCER curricula and may apply to other level completions. Contact NCCER's Registry at 888.622.3720 or go to **www.nccer.org** for more information.

Note

NFPA 70®, *National Electrical Code®* and *NEC®* are registered trademarks of the National Fire Protection Association, Quincy, MA.

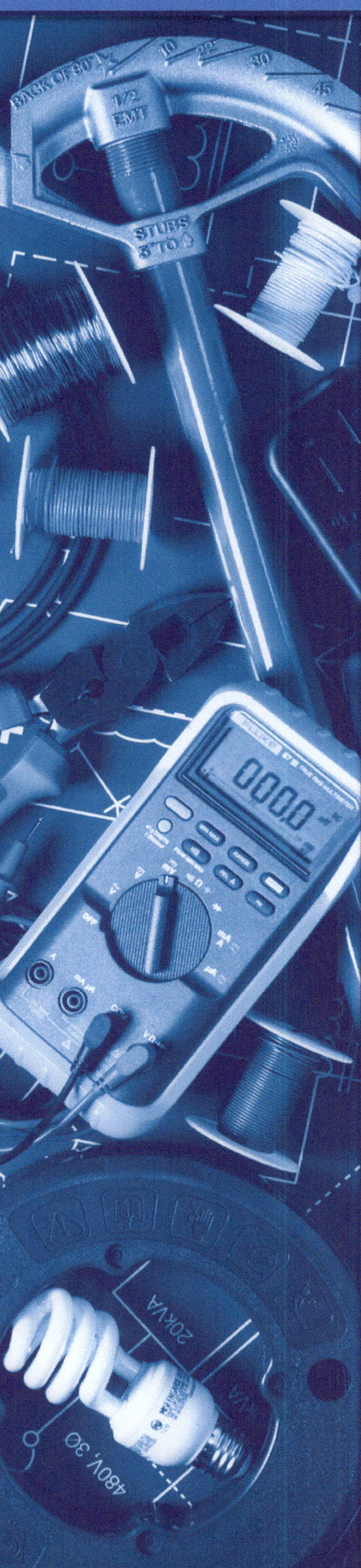

Contents

Figures

1.0.0 OUTLET BOXES

Objective

Size and install outlet boxes.

a. Identify boxes and their applications.
b. Size outlet boxes.
c. Install outlet boxes.

Performance Task

1. Identify the appropriate box type and size for a given application.

Trade Terms

Connector: Device used to physically connect conduit or cable to an outlet box, cabinet, or other enclosure.

Explosion-proof: Designed and constructed to withstand an internal explosion without creating an external explosion or fire.

Handy box: Single-gang outlet box used for surface mounting to enclose receptacles or wall switches on concrete or concrete block construction of industrial and commercial buildings; nongangable; also made for recessed mounting; also known as a utility box.

Outlet box: A metallic or nonmetallic box installed in an electrical wiring system from which current is taken to supply some apparatus or device.

Raintight: Constructed or protected so that exposure to a beating rain will not result in the entrance of water under specified test conditions.

Watertight: Constructed so that moisture will not enter the enclosure under specified test conditions.

Weatherproof: Constructed or protected so that exposure to the weather will not interfere with successful operation.

On every job, boxes are required. All of these must be sized, installed, and supported to meet current *NEC*® requirements. Because the *NEC*® limits the number of conductors, fittings, and devices allowed in each outlet or switch box according to its size, you must install boxes that are large enough to accommodate the number of conductors that must be spliced in the box or fed through it. Therefore, a knowledge of the various types of boxes and the volume of each is essential.

1.1.0 Identifying Boxes and Their Applications

Besides being able to calculate the required box sizes, you must also know how to select the proper type of box for any given application. For example, metallic boxes used in concrete deck pours are different from those used as device boxes in residential or commercial buildings. Boxes used for the support of lighting fixtures or for securing devices in outdoor installations will be different from the two types just mentioned. Boxes for use in certain hazardous locations will further differ in construction; many must be rated as being explosion-proof.

You must also know what fittings are available for terminating the various wiring methods in these boxes.

Electrical drawings rarely indicate the exact type of outlet box to be used in a given area, with the possible exception of boxes used in hazardous locations. Electricians who lack practical on-the-job experience may not always choose the best box for a given application. The use of improper boxes and other ill-adapted materials will cause excessive time to be taken on the job. For example, outlet boxes for use with only Type AC cable should contain built-in clamps; many times boxes will be ordered with knockouts only. This latter case requires extra connector usage. Each connection may require only a few additional seconds, but when these are added up over the period of a large project, much additional time is wasted. Because the cost of labor is an expensive item, any excess labor required will more than offset any savings gained from the use of inadequate materials.

Labor is often wasted due to obstructions from debris that enter raceways during the general construction work. Most of these obstructions can be avoided by plugging all raceway openings with capped bushings or similar means of protection (*Figure 1*). Care should also be taken to thoroughly tighten all fittings, couplings, and so on. All openings in boxes must be closed as specified by *NEC Section 314.17(A)*. Knockout closures are used for this purpose (*Figure 1*).

Outlet boxes normally fall into three categories:

- Pressed steel boxes with knockouts of various sizes for raceway or cable entrances
- Cast iron, aluminum, or brass boxes with threaded hubs of various sizes and locations for raceway entrances
- Nonmetallic boxes

Pressed steel boxes also fall into two categories:

- Boxes with conduit, electric metallic tubing, and cable
- Boxes designed for use with specific types of surface metal raceways

Outlet boxes vary in size and shape depending upon their use, the size of the raceway, the number of conductors entering the box, the type of building construction, the atmospheric conditions of the area, and special requirements.

Outlet box covers are usually required to adapt the box to the particular use it is to serve. For example, a 4" (100 mm) square box is adapted to one-gang or two-gang switches or receptacles by the use of either one-gang or two-gang flush device covers. A one-gang cast hub box can be adapted to provide a vapor-proof switch or a vapor-proof receptacle cover. Special outlet box hangers are available to facilitate their installation, particularly in frame building construction.

> **NOTE**
>
> Metric conversions vary depending on rounding and whether the dimension is matched to a standard metric size for a particular box, conductor, raceway, or other product. All metric conversions in this module are taken from the *NEC®*.

The types of enclosures used as outlet and device boxes for the support of fixtures, or for securing devices such as switches, receptacles, or other equipment on the same yoke or strap, are available in various sizes and shapes. These enclosures may be used in the one-gang, two-gang, three-gang, or four-gang types. Ceiling outlet boxes are available in various shapes. Device boxes are those that are usually installed to support receptacles and switches.

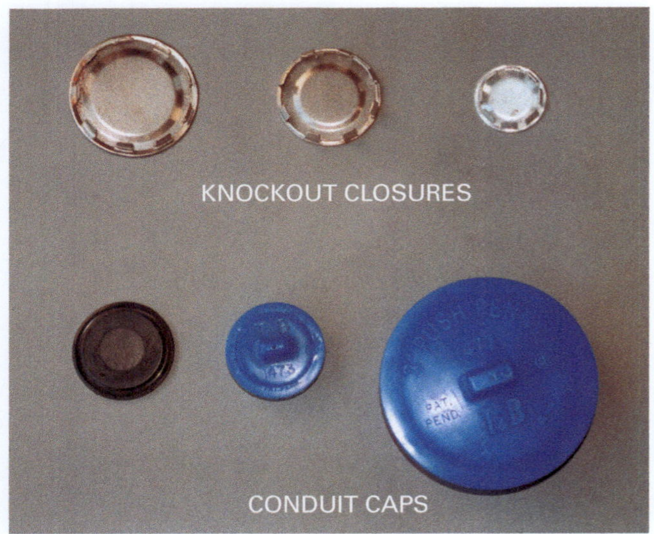

Figure 1 Conduit caps and knockout closures.

Boxes installed for the support of luminaires (lighting fixtures) are required to be listed for the purpose. Most device boxes are typically not designed or listed for use to support luminaires. The use of device boxes to support luminaires is addressed in *NEC Section 314.27(A)*.

A floor box that is listed specifically for installation in a floor is required where receptacles or junction boxes are installed in a floor. Listed floor boxes are provided with covers and gaskets to exclude surface water and cleaning compounds.

A box used at fan outlets is not permitted to be used as the sole support for ceiling (paddle) fans, unless it is listed for the application as the sole means of support. Where a ceiling fan does not exceed 70 lbs (32 kg) in weight, it is permitted to be supported by outlet boxes listed and identified for such use. Boxes designed to support more than 35 lbs (16 kilograms) must be marked with the maximum weight to be supported. See *NEC Section 314.27(C)*. These boxes must be rigidly supported from a structural member of the building. A paddle fan box and its related accessories are shown in *Figure 2*.

NEC Section 314.27(D) states that boxes used for support of utilization equipment other than ceiling-suspended (paddle) fans shall meet the requirements of *NEC Section 314.27(A)* for support of a luminaire that is the same size and weight.

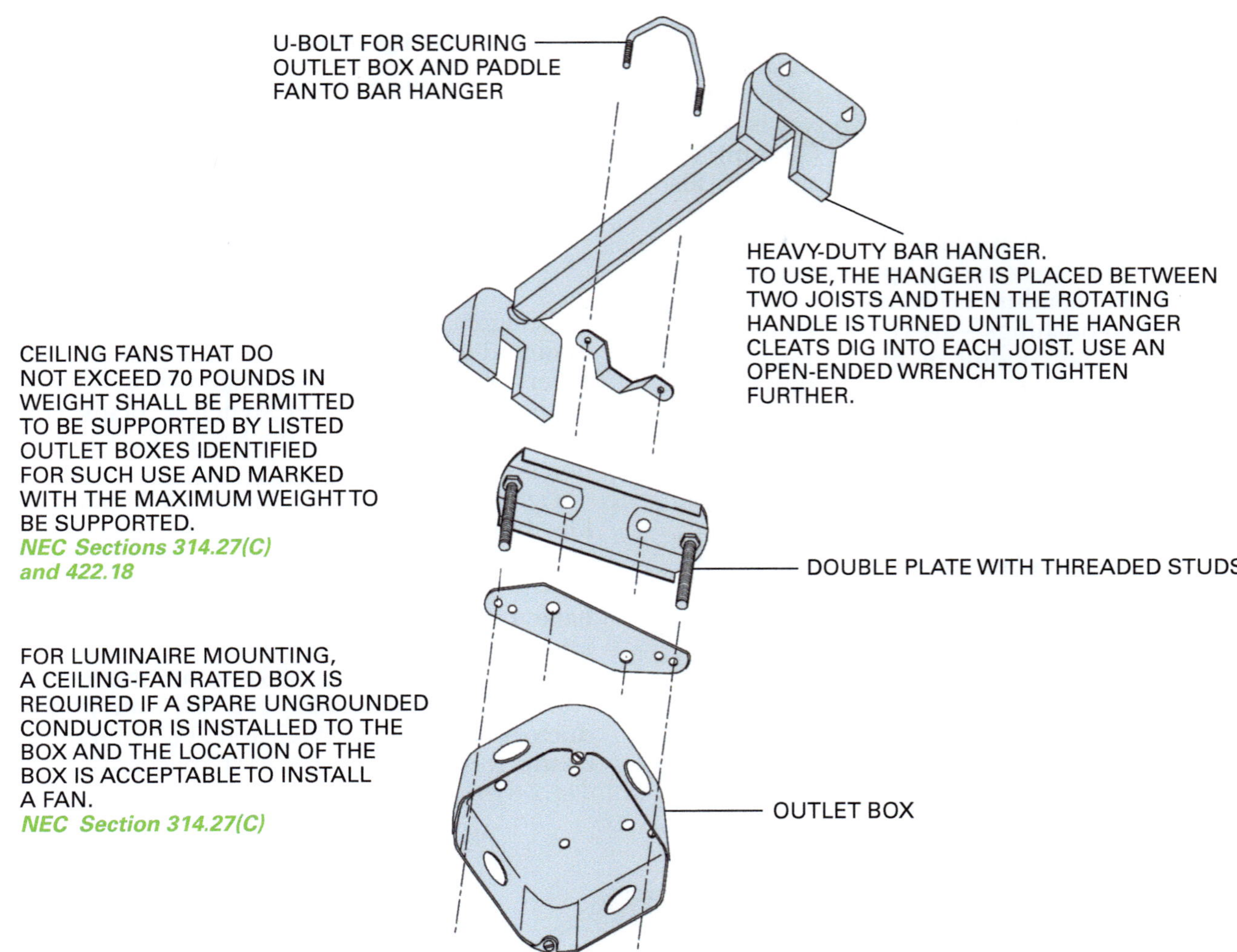

Figure 2 Typical fan box for installation in an existing ceiling.

Special Considerations

Before you can effectively plan and route conductors to their termination points and then select and install the correct types and sizes of boxes, you will need to have the following information:

- Length and number of conductors
- Conductor ampacity
- Allowances for voltage drops
- Environment

Outlet Boxes in Poured Concrete

When installing outlet boxes in poured concrete structures, stuff the inside of the boxes with newspaper and cover the openings with duct tape or use capped bushings to keep concrete out of the raceway system.

1.1.1 Octagon and Round Boxes

Octagon boxes are available with knockouts for use with either conduit (using locknuts and bushings) or cable box connectors. They are also available with both Type AC and NM cable clamps. The standard width of octagon boxes is 4" (100 mm), with depths available in 1¼" (32 mm), 1½" (38 mm), or 2⅛" (54 mm). Extension rings are also available for increasing the depth.

Round boxes are available in the same dimensions, but *NEC Section 314.2* prevents using such boxes where conduit or connectors—requiring locknuts and bushings—are connected to the side of the box. The conduit or box connector must terminate in the top of such boxes. Round boxes with cable clamps, however, are permitted for use with Type AC and NM cables that may terminate in either the side or top of the box. Nonmetallic round boxes are permitted only with open wiring on insulators, concealed knob-and-tube wiring, Type NM cable, and nonmetallic raceways.

Figure 3 shows typical metallic octagon boxes and an octagon extension ring. *Figure 3 (A)* shows a box with concentric knockouts for conduit or box connectors, while *Figure 3 (B)* shows a box that utilizes cable clamps. *Figure 3 (C)* shows the extension ring. *Figure 4* shows a nonmetallic round box and fixture ring with a bar hanger for mounting between studs or joists.

Octagon and round boxes are used for wall-mounted lighting fixtures (luminaires). However, covers are available for octagon boxes that will support receptacles and switches. Blank covers are also available when the box is used as a junction box.

Figure 5 shows a cross section of a round shallow box used for supporting lighting fixtures that have integral wire termination space. *NEC Section 314.24(A)* requires that this and all boxes have a minimum depth of ½" (12.7 mm). Boxes intended to enclose flush devices must not have a depth of less than ¹⁵⁄₁₆" (23.8 mm) for conductors No. 14 AWG and smaller, 1³⁄₁₆" (30.2 mm) for conductors No. 12 or 10 AWG, and 2¹⁄₁₆" (52.4 mm) for conductors No. 8, 6, or 4 AWG per *NEC Section 314.24(B)*.

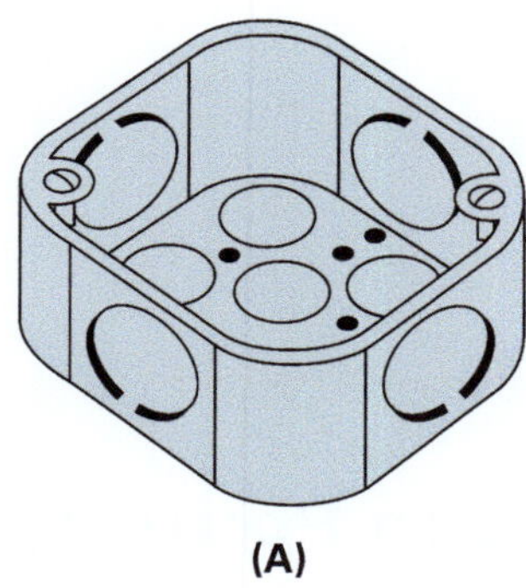

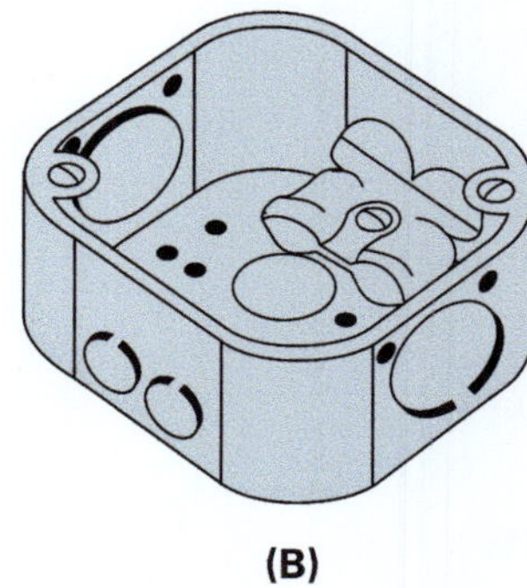

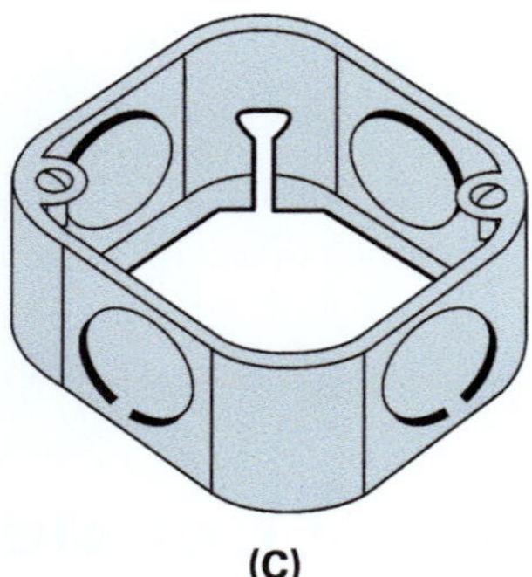

Figure 3 Typical octagon boxes and octagon extension ring.

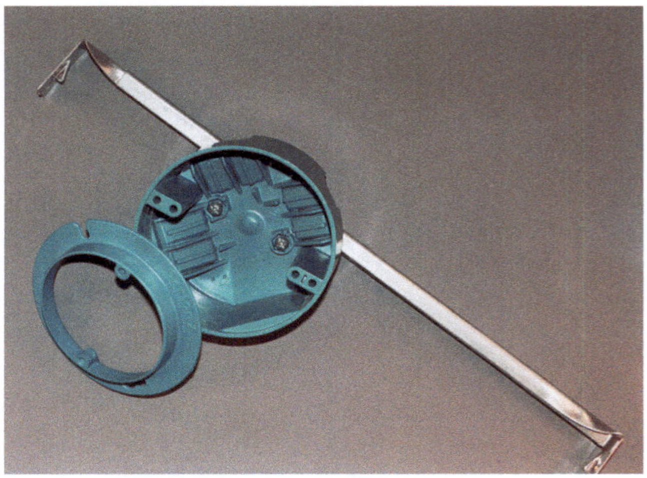

Figure 4 Nonmetallic round box and fixture ring with bar hanger.

1.1.2 *Square Boxes*

Square boxes are typically available in 4" (100 mm) and $4^{11}/_{16}$" (120 mm) square sizes. Both are available in depths of $1^{1}/_{4}$" (32 mm), $1^{1}/_{2}$" (38 mm), and $2^{1}/_{8}$" (54 mm). Extension rings are also available to further increase the depth. These boxes are available with or without mounting brackets for fastening to structural members. Boxes designed for use with cable may have either Type AC or NM clamps for securing the cable at the box entrance points. Square boxes may be used with a single or two-gang device ring (e.g., plaster ring or tile ring) for mounting receptacles or switches. A ring with a round opening is also available for mounting lighting fixtures. Blank covers are available when the boxes are used as junction boxes.

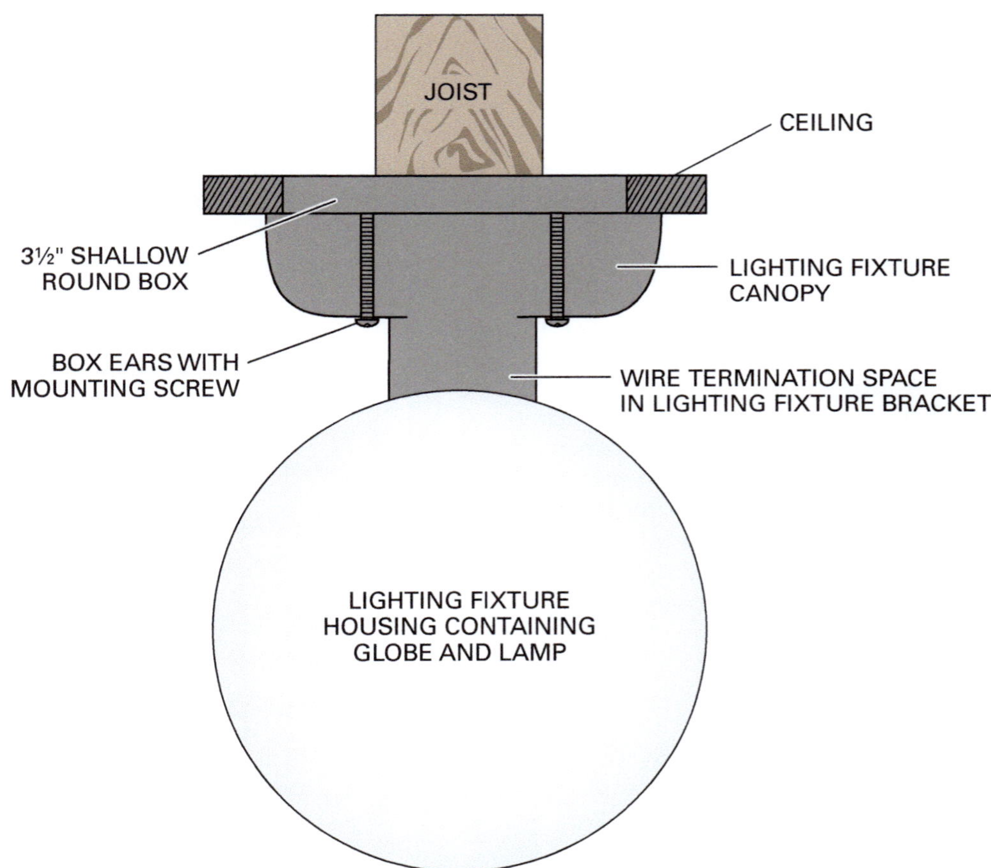

Figure 5 Shallow round box used for mounting a lighting fixture.

Figure 6 shows a square box with a rectangular extension ring that is used to bring the box to the finished surface. Square boxes can be used as junction boxes, and when the number of conductors warrants more capacity than is available in other types of boxes.

1.1.3 Device Boxes

Device boxes house switches and receptacles. They are designed for flush mounting mainly in residential and some commercial applications. Device boxes are available with or without cable clamps and brackets for mounting to wooden structural members. This type of box is also available with plaster ears for installation in finished wall partitions. Three types of device boxes are shown in *Figure 7*. As you can see, some boxes include integral nails for direct installation into wall studs. Boxes used in metal stud environments are installed using self-tapping sheet metal screws, while those installed in concrete may require the use of special powder-actuated fasteners.

Figure 6 Square box with extension ring.

A special single-gang box with the mounting ears on the inside of the box is called a **handy box** (also known as a utility box). Such boxes are available in depths of $1\frac{1}{2}$" (38 mm), $1\frac{7}{8}$" (48 mm), and $2\frac{1}{8}$" (54 mm). Care must be exercised when using these boxes because their limited volume restricts the number of conductors permitted in the box.

1.1.4 Masonry Boxes

Special boxes known as masonry or concrete boxes are used in flat-slab construction jobs. These boxes consist of a sleeve with external ears and a plate that is attached after the sleeve is nailed to the deck.

Masonry boxes are manufactured in different heights. Care should be taken to use boxes of sufficient height to allow the knockouts to come well above the reinforcing rods. This eliminates

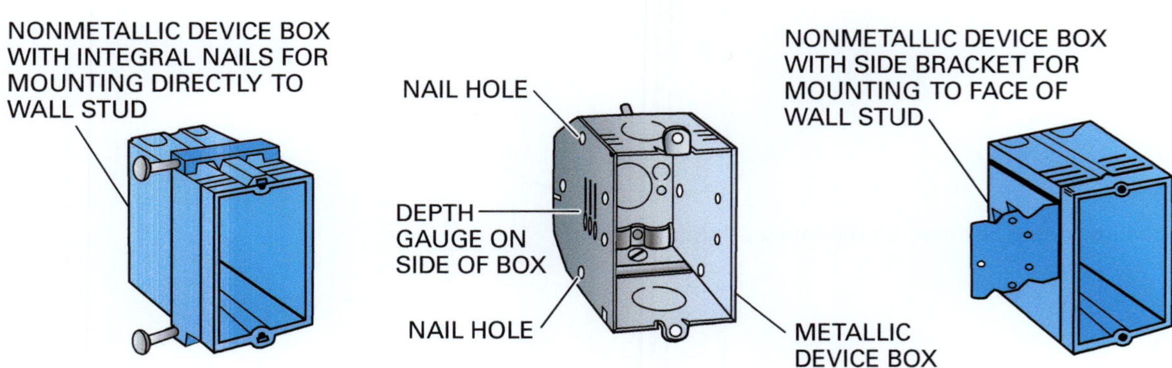

Figure 7 Typical device boxes.

the need for offsets in the conduit where it enters a box. *Figure 8* shows a practical application of a masonry box.

1.1.5 Boxes for Wet and Damp Locations

In damp or wet locations, boxes and fittings must be placed or equipped to prevent moisture or water from entering and accumulating within the box or fitting. It is recommended that approved boxes of nonconductive material be used with nonmetallic sheathed cable or approved nonmetallic conduit when the cable or conduit is used in locations where there is likely to be occasional moisture present. Boxes installed in wet locations must be listed for such use as stated in *NEC Section 314.15*.

A wet location is any location subject to saturation with water or other liquids, such as locations exposed to weather or water, washrooms, garages, and interiors that might be hosed down. Underground installations or those in concrete slabs or masonry in direct contact with the earth must be considered wet locations. **Raintight** or **watertight** equipment (including fittings) may satisfy the requirements for **weatherproof** equipment. Boxes with threaded conduit hubs and gasketed covers will normally prevent water from entering the box except for condensation within the box.

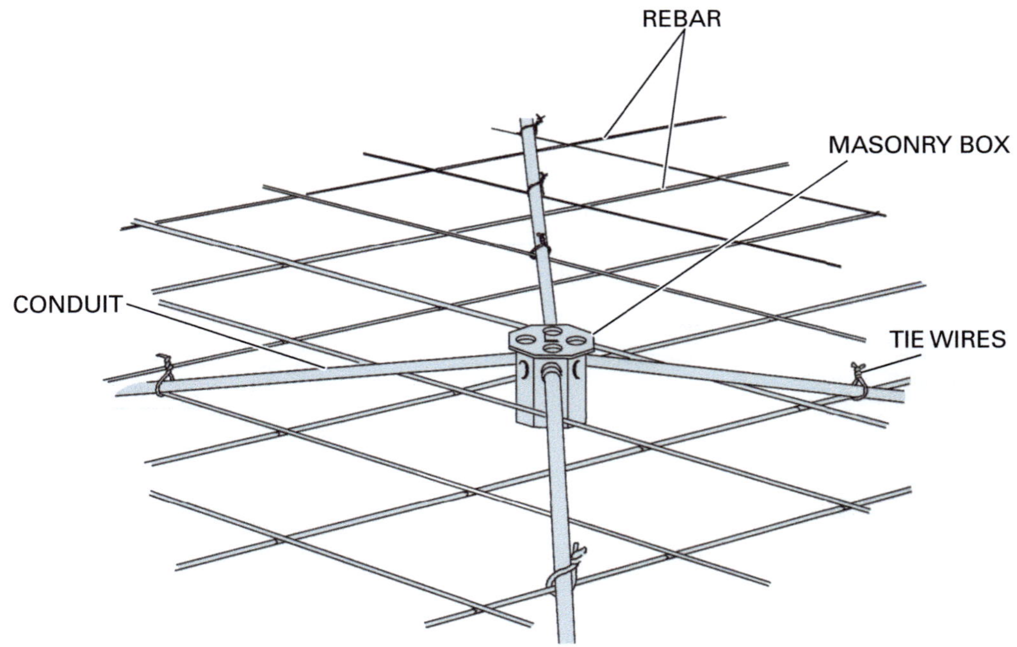

Figure 8 Masonry box.

A damp location is a location subject to some degree of moisture. Such locations include partially protected outdoor locations—such as under canopies, marquees, and roofed open porches. It also includes interior locations subject to moderate degrees of moisture—such as some basements, some barns, and cold storage warehouses.

Weatherproof covers for outdoor receptacles must be chosen with care. All 15A and 20A 125–250V receptacles in wet locations require a cover that is weatherproof whether or not the receptacle is in use. Other receptacles may or may not require these covers, depending on the rating. *NEC Sections 406.9(A)* and *(B)* cover installation of receptacles in damp or wet locations.

1.2.0 Sizing Outlet Boxes

In general, the maximum number of conductors permitted in standard outlet boxes is listed in *NEC Table 314.16(A)*. These figures apply where no fittings or devices such as fixture studs, cable clamps, switches, or receptacles are contained in the box and where no grounding conductors are part of the wiring within the box. Obviously, in all modern residential wiring systems there will be one or more of these items contained in the outlet box. Therefore, where one or more of the above-mentioned items are present, the number of conductors is reduced by one less than that shown in the table for each type of fitting and by two for

Outdoor Boxes

Outdoor wiring must be able to resist the entry of water. Outdoor boxes are either drip tight, which means sealed against falling water from above, or watertight, which means sealed against water from any direction. Drip-tight boxes simply have lids that deflect rain; they are not waterproof. Watertight boxes are sealed with gaskets to prevent the entry of water from any angle.

each device strap. For example, a deduction of two conductors must be made for each strap containing a device such as a switch or duplex receptacle; a further deduction of one conductor shall be made for one or more grounding conductors entering the box. For example, a 3" × 2" × 3½" (75 mm × 50 mm × 90 mm) box is listed in the table as containing a maximum number of eight No. 12 wires. If the box contains a cable clamp and a duplex receptacle, three wires will have to be deducted from the total of eight—providing for only five No. 12 wires. If a ground wire is used, only four No. 12 wires may be used, which might be the case when a three-wire cable with ground is used to feed a three-way wall switch. Also, each looped conductor over 12" (300 mm) used in the box counts as two conductors.

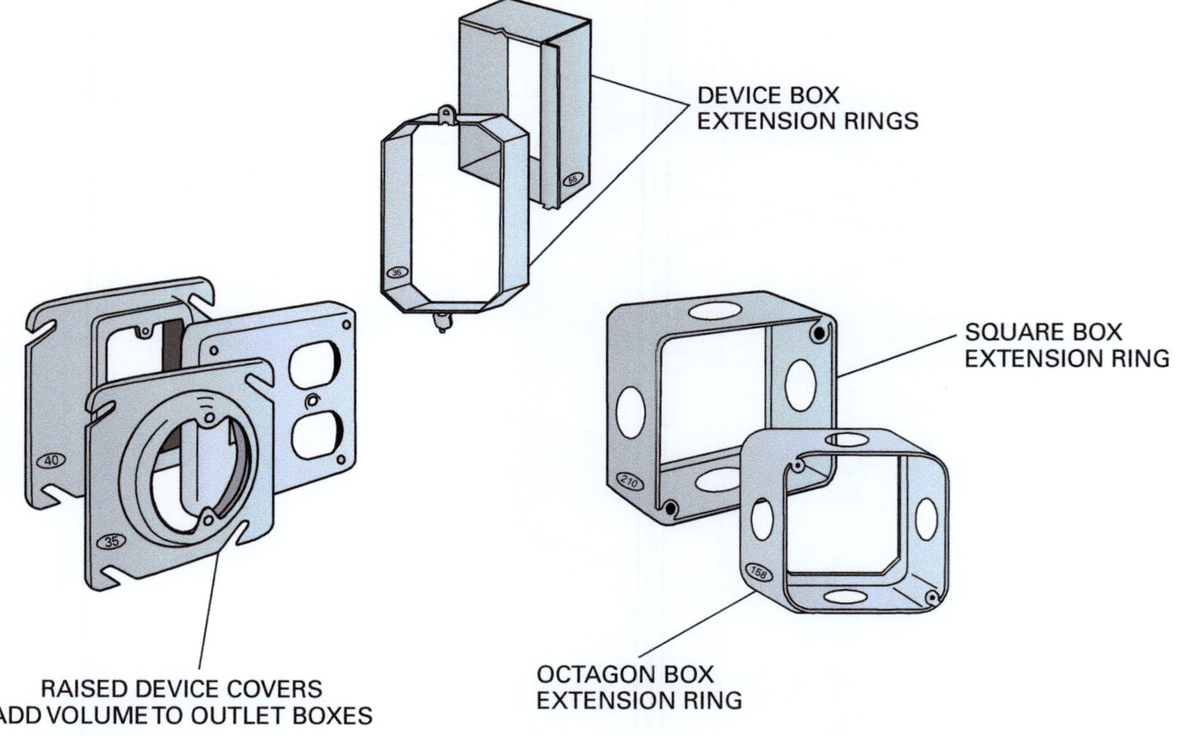

Figure 9 Devices or components that add to outlet box capacity.

A pictorial definition of stipulated conditions as they apply to *NEC Section 314.16(A)* is shown in *Figure 9* through *Figure 11*. *Figure 9* illustrates an assortment of raised covers and outlet box extensions. These components, when combined with the appropriate outlet boxes, serve to increase the usable space. Each type is marked with its capacity, which may be added to the figures in *NEC Table 314.16(A)* to calculate the increased number of conductors allowed.

Figure 10 shows typical wiring configurations and devices that must be counted as conductors when calculating the total capacity of outlet boxes. A wire passing through the box without a splice or tap is counted as one conductor. Therefore, a cable containing two wires that passes in and out of an outlet box without a splice or tap is counted as two conductors. However, a wire that enters a box and is either spliced or connected to a terminal, and then exits again,

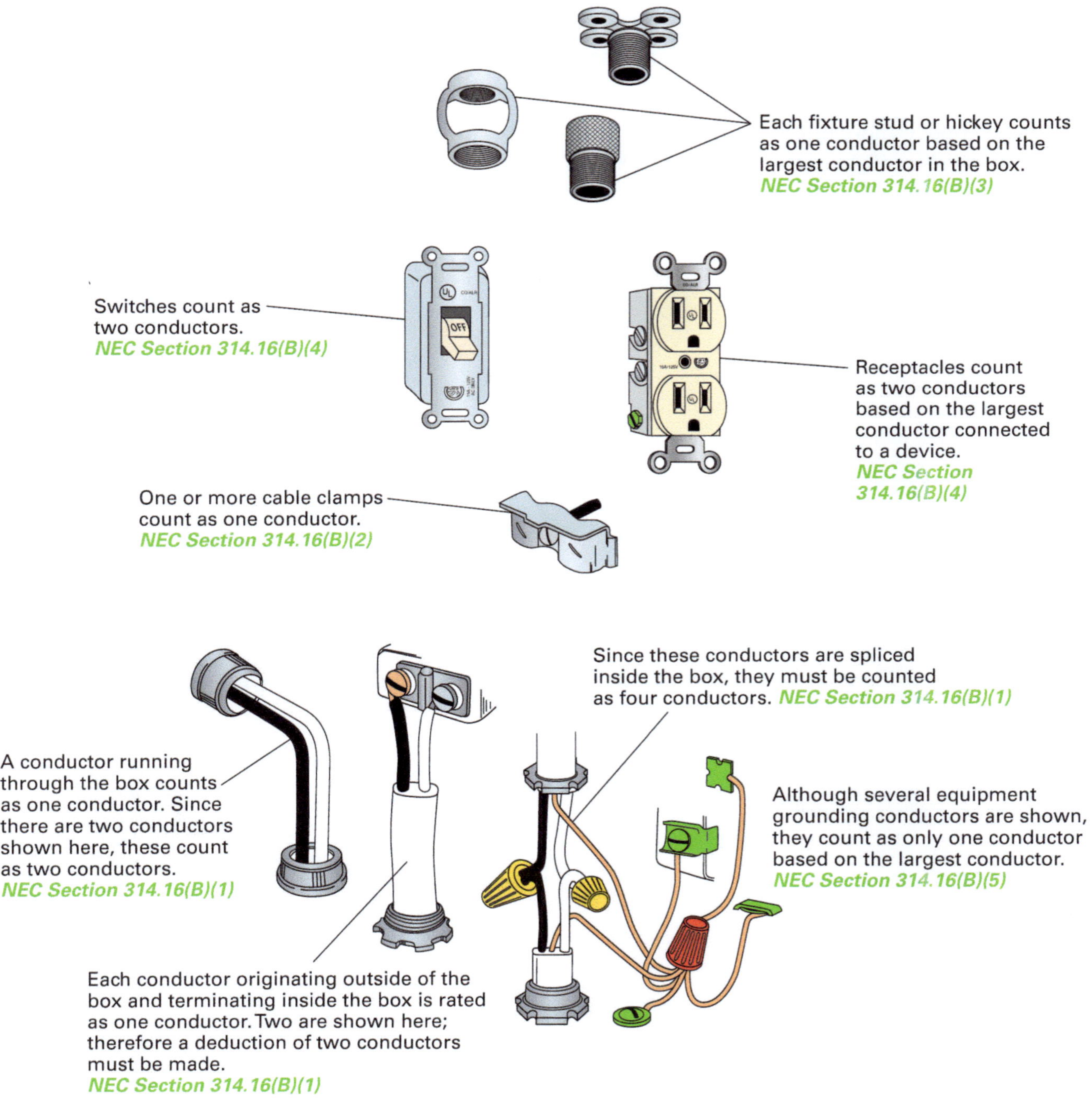

Figure 10 Devices and components that require deductions in outlet box capacity.

is counted as two conductors. In the case of two cables that each have two wires, the total conductors counted will be four. Wires that enter and terminate in the same box are counted as individual conductors and in this case, the total count would be two conductors. Remember, when one or more grounding wires enter the box and are joined, a deduction of only one is required, regardless of their number.

Further components that require deduction adjustments from those specified in *NEC Table 314.16(A)* include fixture studs and hickeys [*NEC Section 314.16(B)(3)*]. One conductor must be deducted from the total for each type of fitting used. Two conductors must be deducted for each strap-mounted device, such as duplex receptacles and wall switches; a deduction of one conductor is made when one or more internally mounted cable clamps are used [*NEC Section 314.16(B)(2)* and

(4)]. When mixed conductor sizes are installed on a yoke, the deduction is based on the largest wire size in the box per *NEC Table 314.16(B)*.

Figure 11 shows components that may be used in outlet boxes without affecting the total number of conductors. Such items include grounding clips and screws, wire nuts, and cable connectors when the latter are inserted through knockout holes in the outlet box and secured with locknuts. Prewired fixture wires are not counted against the total number of allowable conductors in an outlet box; neither are conductors originating and terminating in the box, such as pigtails.

To better understand how outlet boxes are sized, use the following as an example: Two No. 12 AWG conductors are installed in trade size ½" EMT (MD 12) and terminate into a metallic outlet box containing one duplex receptacle. What size outlet box will meet *NEC*® requirements?

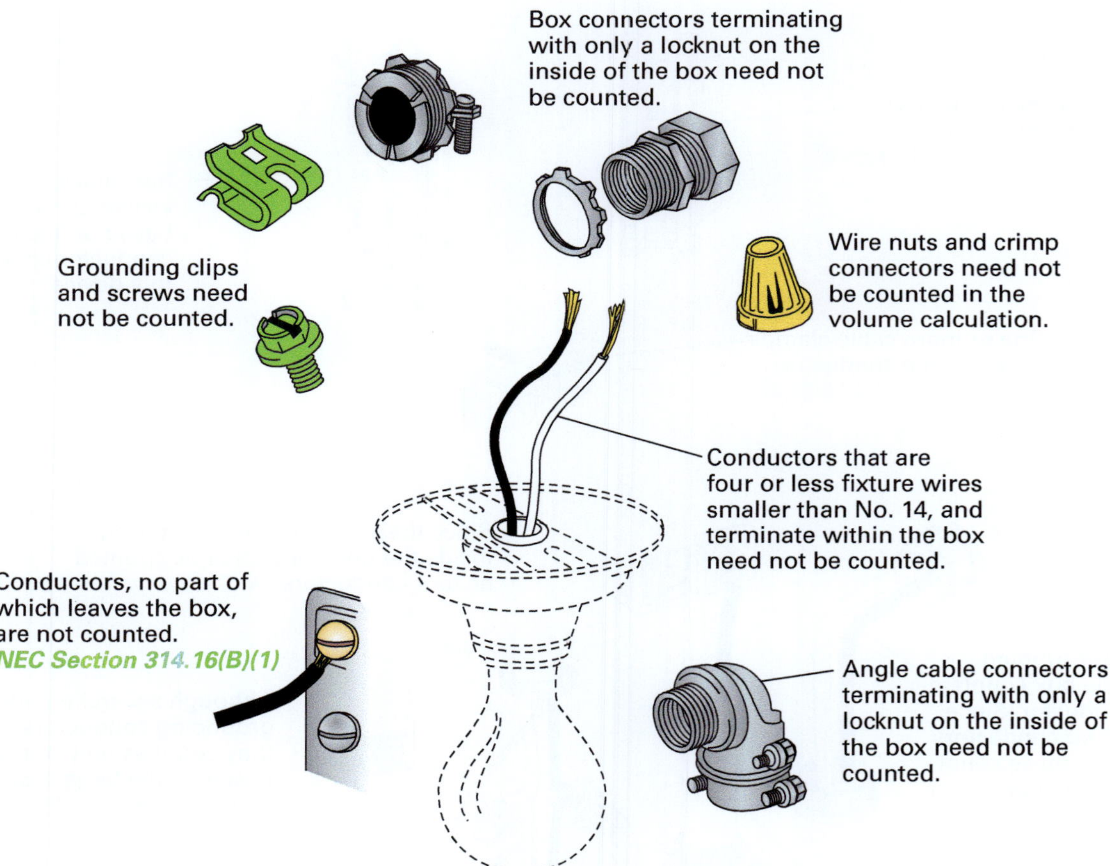

Figure 11 Items that may be disregarded when calculating outlet box capacity.

The first step is to count the total number of conductors and equivalents that will be used in the box (*NEC Section 314.16*). The following steps describe how to determine the size of the outlet box:

Step 1 Calculate the total number of conductors and their equivalents:

$$\begin{array}{r} \text{One receptacle} = 2 \\ +\ \text{Two \#12 conductors} = 2 \\ \hline \text{Total \#12 conductors} = 4 \end{array}$$

Step 2 Determine the amount of space required for each conductor. *NEC Table 314.16(B)* gives the box volume required for each conductor:

US Measure:
No. 12 AWG = 2.25 in³

Metric:
No. 12 AWG = 36.9 cm³

Step 3 Calculate the outlet box space required by multiplying the volume required for each conductor by the number of conductors found in Step 1 above.

US Measure:
4 × 2.25 in³ = 9.00 in³

Metric:
4 × 36.9 cm³ = 147.6 cm³

Step 4 Once you have determined the required box capacity, refer to *NEC Table 314.16(A)* and note that a 3" x 2" x 2" (75 mm x 50 mm x 50 mm) box comes closest to our requirements. This box size is rated for 10.0 cubic inches (164 cubic centimeters).

For another example, if four No. 12 conductors enter the box, two additional No. 12 conductors must be added to our previous count for a total of six conductors.

US Measure:
6 × 2.25 in³ = 13.5 in³

Metric:
6 × 36.9 cm³ = 221 cm³

Again, refer to *NEC Table 314.16(A)* and note that a 3" × 2" × 2¾" (75 mm × 50 mm × 70 mm) device box with a rated capacity of 14.0 cubic inches (230 cubic centimeters) is the closest device box that meets *NEC* requirements. Of course, any box with a larger capacity is permitted.

1.3.0 Installing Outlet Boxes

In addition to box fill, there are a number of other considerations when installing boxes. These include additional *NEC*® requirements and making the actual wiring connections.

Some of the general *NEC*® requirements are as follows:

- The box selected must be listed for the given application (for example, a box used in a wet location must be listed for use in that location).
- As discussed previously, the box must have sufficient volume, as listed in *NEC Table 314.16(A)*, and must allow sufficient free space for conductors, as listed in *NEC Table 314.16(B)*.
- Conductors entering boxes and fittings must be protected from abrasion.
- Boxes must be installed and supported properly, and the finished installation must be accessible for later repair or maintenance.

You must also consider the type of box cover or canopy to be used, as well as the type of box. Refer to *NEC Sections 314.15 through 314.30* for additional details.

1.3.1 *NEC*® Requirements for Receptacle Locations

NEC Section 210.52 states the minimum requirements for the location of receptacles in dwelling units. It specifies that in each kitchen, family room, dining room, living room, parlor, library, den, sunroom, bedroom, recreation room, or similar room or area, receptacle outlets shall be installed so that no point along the floor line in any wall space is more than 6' (1.8 m), measured horizontally, from an outlet in that space, including any wall space 2' (600 mm) or more in width and the wall space occupied by fixed panels in walls, but excluding sliding panels (*Figure 12*). When spaced in this manner, a 6' (1.8 m) extension cord will reach a receptacle from any point along the wall line. Receptacle outlets shall, insofar as practicable, be spaced equal distances apart. Receptacle outlets in floors shall not be counted as part of the required number of receptacle outlets unless located within 18" (450 mm) of the wall.

The *NEC*® defines wall space as a wall that is unbroken along the floor line by doorways, fireplaces, or similar openings. Each wall space that is 2' (600 mm) or more in width must be treated individually and separately from other wall spaces within the room. This minimizes the use of cords across doorways, fireplaces, and similar openings.

At least one receptacle is required in each laundry area, within 3' (900 mm) of bathroom basins, on the outside of the building at the front and back (GFCI protected), on any balcony, deck, or porch accessible from the inside of the house, in each basement, in each attached and detached garage or vehicle bay, in each hallway 10' (3 m) or more in length, and at an accessible location for servicing any HVAC equipment.

Although no actual *NEC*® requirements exist for mounting heights and positioning of receptacles, other than the prohibition against mounting receptacles face up on countertops and similar work surfaces, there are certain *NEC*® requirements regarding receptacle placement. For example, *NEC Section 210.52(C)(5)* states that receptacles shall be not more than 20" above a kitchen countertop. Also, where allowed, receptacles may not be located more than 12" (300 mm) below the countertop surface. See *NEC Section 210.52(C)(5) Exception*. In addition to these *NEC*® guidelines, certain installation methods have become standard in the electrical industry. *Figure 13* shows common mounting heights of duplex receptacles used on conventional residential and small commercial installations. However, these dimensions are frequently varied to suit the building structure. For example, ceramic tile might be placed above a kitchen or bathroom countertop. If the dimensions in *Figure 13* put the receptacle part of the way out of the tile, the mounting height should be adjusted to place the receptacle either completely in the tile or completely out of the tile, as shown in *Figure 14*.

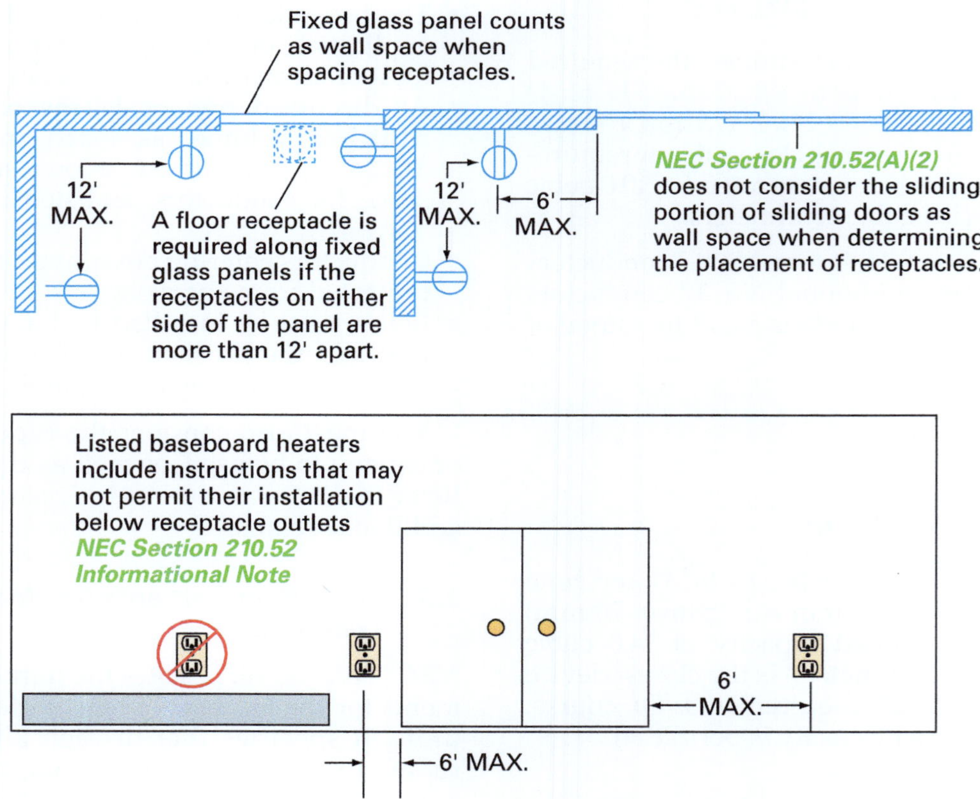

Figure 12 NEC ® requirements for receptacle locations in dwelling units.

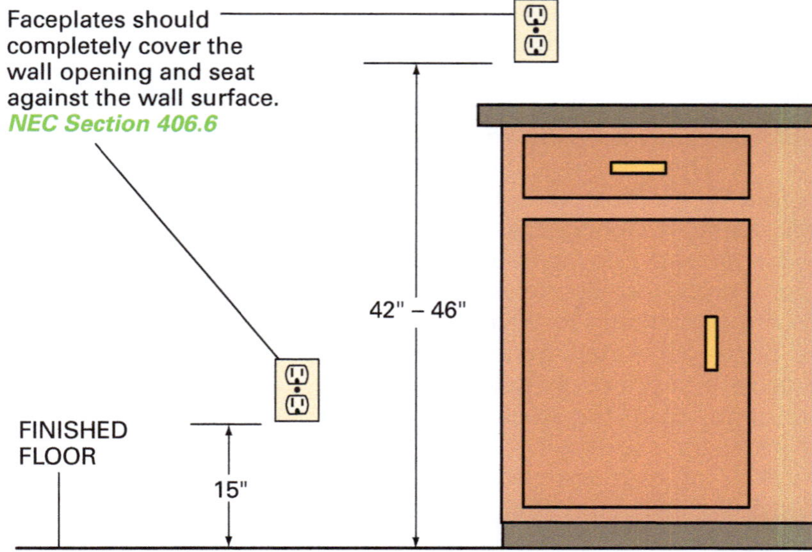

Figure 13 Mounting heights of duplex receptacles.

Figure 14 Adjusting mounting heights.

Refer again to *Figure 13* and note that the mounting heights are given to the bottom of the outlet box. Many dimensions on electrical drawings are given to the center of the outlet box or receptacle. However, during the actual installation, workers installing the outlet boxes can mount them more accurately (and in less time) by lining up the bottom of the box with a chalk mark rather than trying to eyeball this mark to the center of the box.

A decade or so ago, most electricians mounted receptacle outlets 12" (300 mm) from the finished floor to the center of the outlet box. However, a survey of more than 500 homeowners shows that they prefer a mounting height of 15" (380 mm) from the finished floor to the bottom of the outlet box. It is easier to plug and unplug the cord assemblies at this height. However, always check

the working drawings, written specifications, details of construction, and local codes for measurements that may affect the mounting height of a particular receptacle outlet. For example, persons using wheelchairs may require more specific receptacle height locations to fit their individual needs.

NEC Section 314.20 requires all outlet boxes installed in walls or ceilings of concrete, tile, or other noncombustible material, such as plaster or drywall, to be installed in such a manner that the front edge of the box or fitting is not set back from the finished surface by more than $\frac{1}{4}$" (6 mm). Where walls and ceilings are constructed of wood or other combustible materials, outlet boxes and fittings must be flush with the finished surface of the wall.

Wall surfaces such as drywall or plaster that contain wide gaps or are broken, jagged, or otherwise damaged, must be repaired so there will be no gaps or open spaces greater than ⅛" (3 mm) between the outlet box and the wall material (*NEC Section 314.21*). These repairs should be made prior to installing the faceplate. Such repairs are best made using a noncombustible caulking or spackling compound. See *Figure 15*.

Many industrial/commercial outlet boxes are surface-mounted and connect directly to the electrical raceway. These boxes are often mounted to the building steel or to a support channel or strut that is welded or clamped to the steel. The installation of outlet boxes or raceways should be coordinated so as not to interfere with piping, ductwork, fireproofing, and other items that may or may not already be in place.

1.3.2 Making Connections

Before any installation begins, study the electrical floor plan and consult with the builder or architect to ensure that no last-minute changes have been made in the electrical system. Install all boxes in accordance with the electrical drawings. The spacing should be as even as possible. The box center is the midpoint on the vertical dimension of the box. Make sure that you check the door swing direction so that the switches are not installed behind a door. Measure the height of the switch boxes from the floor so that they will be at the proper height when installed. After the boxes are installed, the wires must be spliced.

Insulated spring connectors, commonly called wire nuts or Wirenuts®, are solderless connectors made in various color-coded sizes that allow for splicing the hundreds of different solid or stranded wire combinations typically encountered in branch-circuit and fixture splicing applications (*Figure 16*). Several varieties of wire nuts are available, but the following are the ones used most often:

- Those for use on wiring systems 300V and under
- Those for use on wiring systems 600V and under (1,000V in lighting fixtures and signs)

Some types of wire nuts have thin wings on each side of the connector to facilitate their installation. Wire nuts are normally made in sizes to accommodate conductors as small as No. 22 AWG up to as large as No. 10 AWG, with practically any combination of those sizes in between.

Figure 15 Gaps or openings around outlet boxes must be repaired.

Figure 16 Wire nuts.

Always check the listing requirements on the package prior to use. Some wire nuts are only listed for specific applications, such as copper to copper only. The maximum temperature rating is 105°C (221°F).

The general procedure for splicing wires with wire nuts is as follows:

Step 1 Select the proper size wire nut to accommodate the wires being spliced. Wire nut packages contain charts that list the allowable combinations of wires by size. Refer to the label on the wire nut box or container for this information.

Step 2 Select the appropriate tool (*Figure 17*), and then strip the insulation from the ends of the wires to be spliced. The length of insulation stripped off is typically about ½" (13 mm) (*Figure 18*), but may vary depending on the wire size and the wire nut being used. Follow the manufacturer's directions given on the wire nut package.

Step 3 Stick the ends of the wires into the wire nut and turn clockwise until tight. The wire nut draws the conductors and insulation into the body of the connector. See *Figure 19*.

Step 4 After making the connections within a box, tuck the wires neatly into the back of the box. That way, when the painters and plasterers come along to finish the walls, the wires will not be covered in drywall compound or paint.

> **NOTE**
>
> Some manufacturers of wire nuts require that the wires be pre-twisted before screwing on the nut. Also, some manufacturers recommend using a nut driver to tighten the wire nut. Always follow the manufacturer's instructions.

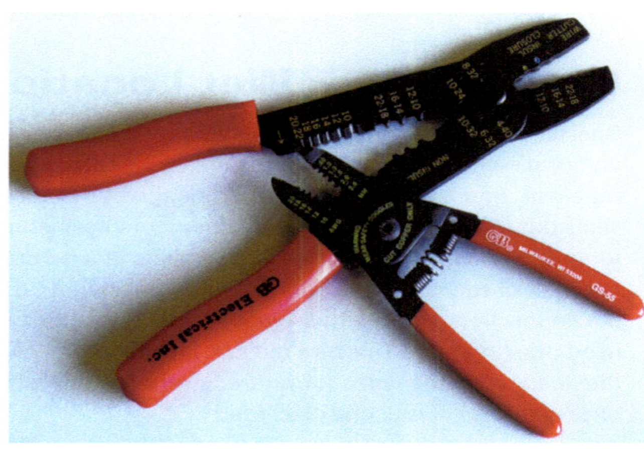

Figure 17 Stripping tools.

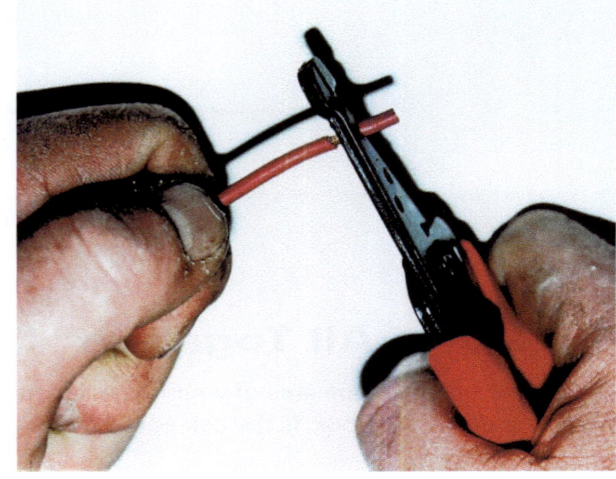

Figure 18 Stripping the insulation.

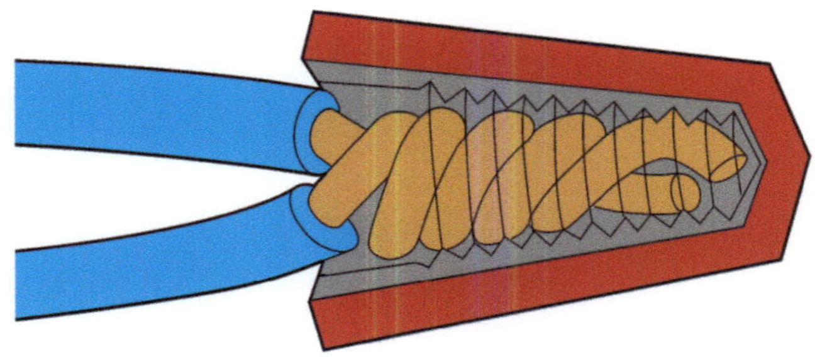

Figure 19 Wires installed in wire nut.

Wire Nuts for Wet Locations

Specially designed wire nuts are made for use in wet locations and/or direct burial applications. These wire nuts have a water repellent, non-hardening sealant inside the body that completely seals out moisture to protect the conductors against moisture, fungus, and corrosion. The sealant remains in a gel state and will not melt or run out of the wire nut body throughout the life of the connection. Unlike other types of wire nuts, this type can be used one time only. The wire nut can be backed off, eliminating the need to cut the wires for future or retrofit applications, but once removed, it must be discarded.

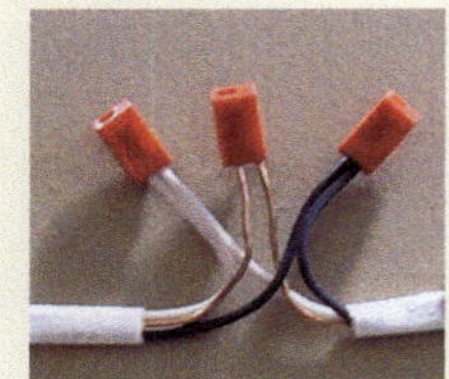

(A) NO. 14/2 NM-B CABLE WITH PUSH-IN WIRE CONNECTORS

(B) NO. 12/2 NM-B CABLE WITH TWIST-ON WIRE NUTS

Putting It All Together

Turn off the power in one area of your home and then remove some of the switch and receptacle plates. Examine the wiring inside each box. Is the box adequately sized for the number of wires and devices?

1.0.0 Section Review

1. A receptacle is usually housed in a(n) ______.

   a. pull box
   b. round box
   c. junction box
   d. device box

2. If a box can hold eight No. 12 wires and contains a cable clamp, a switch, and a looped conductor that is 13" (330 mm) long, the maximum number of additional wires is ______.

   a. two
   b. three
   c. four
   d. five

3. A box installed in a wet location must be ______.

   a. installed upside down
   b. plastic
   c. sealed with caulk
   d. listed for the application

2.0.0 PULL AND JUNCTION BOXES

Objective

Size and install pull and junction boxes.
  a. Size pull and junction boxes.
  b. Install pull and junction boxes.

Performance Task

2. Select the minimum size pull or junction box for the following applications:
    - Conduit entering and exiting for a straight pull.
    - Conduit entering and exiting at an angle.

Trade Terms

Junction box: An enclosure where one or more raceways or cables enter, and in which electrical conductors can be, or are, spliced.

Pull box: A sheet metal box-like enclosure used in conduit runs to facilitate the pulling of cables from point to point in long runs, or to provide for the installation of conduit support bushings needed to support the weight of long riser cables, or to provide for turns in multiple conduit runs.

A **pull box** or **junction box** is provided in an electrical installation to facilitate the installation of conductors, or to provide a junction point for the connection of conductors, or both. In some instances, the location and size of pull boxes are designated on the drawings. In most cases, however, the electricians on the job must determine the proper number, location, and sizes of pull or junction boxes to facilitate conductor installation.

2.1.0 Sizing Pull and Junction Boxes

Pull boxes should be as large as possible. Workers need space within the box for both hands and in the case of the larger wire sizes, workers will need room for their arms to feed the wire. *NEC Sections 314.28(A)* through *(E)* specify that pull and junction boxes must provide adequate space and dimensions for the installation of conductors. For raceways containing conductors of No. 4 or larger, and for cables containing conductors of No. 4 or larger, the minimum dimensions of pull or junction boxes installed in a raceway or cable run shall comply with the following:

- In straight pulls, the length of the box shall not be less than eight times the trade diameter or metric designator (MD) of the largest raceway. (Note that the MD is not a direct conversion between inches and millimeters, but is very close to the trade sizes.)
- Where angle or U pulls are made, the distance between each raceway entry inside the box and the opposite wall of the box shall not be less than six times the trade diameter or MD of the largest raceway in a row. This distance shall be increased for additional entries by the amount of the sum of the trade diameters/MDs of all other raceway entries in the same row on the same wall of the box. Each row shall be calculated individually, and the single row that provides the maximum distance shall be used.
- Also where angle or U pulls are made, the distance between raceway entries enclosing the same conductor shall not be less than six times the trade diameter/MD of the larger raceway.
- When transposing cable size into raceway size, the minimum trade size/MD raceway required for the number and size of conductors in the cable shall be used.

Figure 20 shows a junction box with four runs of conduit. This is a straight pull, and trade size 4" (MD 103) conduit is the largest size in the group. The minimum length required for the box can be determined using the following formula:

$$\frac{\text{Trade size of conduit} \times 8 \; [\text{per } \textit{NEC Section 314.28(A)(1)}]}{} = \text{minimum length of box}$$

When calculating box size, always use the largest conduit size in the group. In the example shown in *Figure 20*, the calculation is as follows:

US Measure:
  Trade size 4" × 8 = 32"

Metric:
  MD 103 × 8 = 824 mm

Therefore, this particular pull box must be at least 32" (824 mm) in length. The width of the box, however, need only be of sufficient size to enable locknuts and bushings to be installed on all the conduits or connectors entering the enclosure.

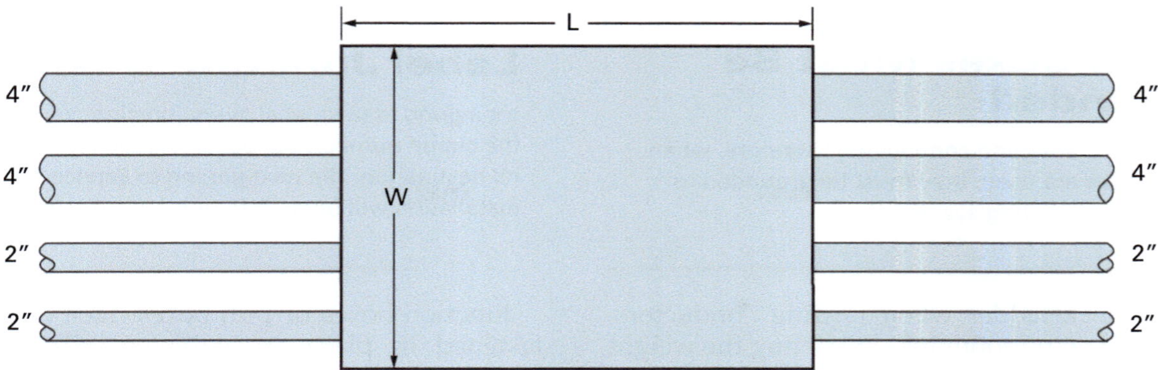

Figure 20 Pull box with straight conduit runs.

Junction or pull boxes in which the conductors are pulled at an angle (*Figure 21*) must have a distance of not less than six times the trade diameter of the largest conduit [*NEC Section 314.28(A)(2)*]. The distance must be increased for additional conduit entries by the amount of the sum of the diameter of all other conduits entering the box on the same side. The distance between raceway entries enclosing the same conductors must not be less than six times the trade diameter of the largest conduit.

Since the 4" (MD 103) conduit is the largest size in this case:

US Measure:
$$L_1 = 6 \times 4" + (3" + 2") = 29$$

Metric:
$$L_1 = 6 \times MD\ 103 + (MD\ 78 + MD\ 53) = 749\ mm$$

Because the same conduit runs are located on the adjacent wall of the box, L_2 is calculated in the same way; therefore, $L_2 = 29"$ (749 mm).

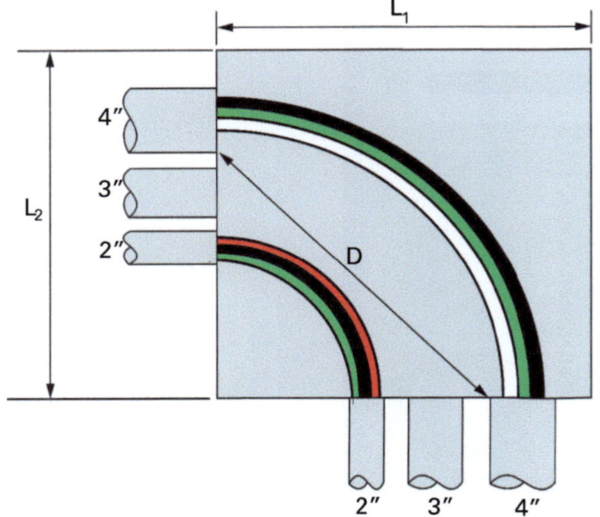

Figure 21 Pull box with conduit runs entering at right angles.

The distance (D) = 6 × 4" or 24" (6 × MD 103 = 618 mm). This is the minimum distance permitted between conduit entries enclosing the same conductor.

The depth of the box need only be of sufficient size to permit locknuts and bushings to be properly installed. In this case, a 6"-deep (152 mm-deep) box would suffice.

2.2.0 Installing Pull and Junction Boxes

Long runs of conductors should not be made in one pull. Pull boxes, installed at convenient intervals, will relieve much of the strain on the conductors. The length of the pull, in many cases, is left to the judgment of the workers or their supervisor, and the condition under which the work is installed.

The installation of pull boxes may seem to cause extra work and expense, but they save a considerable amount of time and hard work when pulling conductors. Properly placed, they eliminate bends and elbows and do away with the necessity of fishing from both ends of a conduit run.

If possible, pull boxes should be installed in a location that allows electricians to work easily and conveniently. For example, in an installation where the conduit comes up a corner of a wall and changes direction at the ceiling, a pull box that is installed too high will force the electrician

Using Pull Boxes

Pull boxes make it easier to install conductors. They may also be installed to avoid having more than 360° worth of bends in a single run. (Remember, if a pull box is used, it is considered the end of the run for the purposes of the *NEC*® 360° rule.)

to stand on a ladder when feeding conductors, and will allow no room for supporting the weight of the wire loop or for the cable-pulling tools.

Unless the contract drawings or project engineer state otherwise, it is just as easy for the pull boxes to be placed at chest height or a similar location that allows workers to stand on the floor with sufficient room for both wire loop and tools.

In some electrical installations, a number of junction boxes must be installed to route the conduit in the shortest, most economical way. The *NEC*® requires all junction boxes to be readily accessible. This means that a person must be able to get to the conductors inside the box without removing plaster, wall covering, or any other part of the building.

Junction boxes or pull boxes must be securely fastened in place on walls or ceilings or adequately suspended. All unused openings, such as knockouts, must be securely covered.

While certain sizes of factory-constructed boxes are available with concentric knockouts, in many instances it will be necessary to have them custom built to meet the job requirements. When it is not possible to accurately anticipate the raceway entrance requirements, it will be necessary to cut the required knockouts on the job.

In the case of large pull boxes and troughs, shop drawings should be prepared prior to the construction of these items with all required knockouts accurately indicated in relation to the conduit run requirements.

Knockout Punch Kit

Often, conduit must enter boxes, cabinets, or panels that do not have precut knockouts. In these cases, a knockout punch can be used to make a hole for the conduit connection.

2.0.0 Section Review

1. A straight pull contains two raceways. One of the raceways has a trade size of 3" and one has a trade size of 2". The length of the box must be _____.

   a. 16"
   b. 24"
   c. 26"
   d. 32"

2. If possible, pull boxes should be installed _____.

   a. high on the wall for security
   b. at a height/location for ease when pulling conductors
   c. behind wall coverings for a neater appearance
   d. with extra knockouts left open for ventilation

SUMMARY

Electricians work with device boxes almost every day on every project. Consequently, you must have a thorough knowledge of the types of boxes available and their applications. *NEC Article 314* covers the installation of boxes.

One of the best ways to learn about boxes and fittings is to study manufacturers' catalogs. You will find a wealth of information in each of these, many of which include detailed instructions on installation techniques. Some of these catalogs also provide simplified *NEC®* explanations of the use of the manufacturer's products.

In this module, you learned that outlet boxes must be sized, installed, and supported to meet current *NEC®* requirements. Because the *NEC®* limits the number of conductors in a given box size, boxes with adequate volume must be selected. The boxes must also be the proper type for the application. Each type of box was discussed, along with references to applicable portions of *NEC Article 314*.

The maximum number of conductors permitted in standard metal boxes is listed in *NEC Table 314.16(A)*. Certain devices and components require deductions in the number of conductors allowed in the box.

In addition to box sizing, there are various other factors to consider when installing boxes. These requirements are discussed in *NEC Sections 314.15* through *314.30*.

1. The maximum weight allowed by the *NEC®* when ceiling fans are mounted directly to an approved, unmarked outlet box is ______.

    a. 25 pounds
    b. 35 pounds
    c. 45 pounds
    d. 55 pounds

2. Square metal boxes are available in sizes of 4" and ______.

    a. $4^{11}/_{16}$"
    b. 5"
    c. $5^{1}/_{4}$"
    d. 6"

3. A box installed under a roofed open porch is considered a ______.

    a. dry location
    b. wet location
    c. damp location
    d. weatherproof location

4. How many conductor(s) must be deducted for each strap-mounted device in a device box?

    a. One
    b. Two
    c. Three
    d. Four

5. The capacity of an outlet box can be increased by using ______.

    a. fixture studs
    b. strap-mounted devices
    c. wire nuts
    d. raised device covers

6. A conductor under 12" (300 mm) long running through a box counts as the conductor equivalent of ______.

    a. 0
    b. 1
    c. 2
    d. 12

7. If there are four equipment grounding conductors in a box, you must deduct ______ conductor(s) from the total allowed.

    a. 1
    b. 2
    c. 3
    d. 4

8. Factors to consider when installing boxes are covered in ______.

    a. *NEC Sections 314.1 through 314.4*
    b. *NEC Sections 314.15 through 314.30*
    c. *NEC Sections 314.40 through 314.44*
    d. *NEC Sections 314.70 through 314.72*

9. If the largest trade diameter of a raceway entering a pull box is 3", and it is a straight pull, the minimum size box allowed is ______.

    a. 20"
    b. 24"
    c. 30"
    d. 36"

10. The depth of a pull box with conduit runs entering at right angles ______.

    a. need only be deep enough to permit the proper installation of locknuts and bushings
    b. must be eight times the diameter of the smallest conduit
    c. must be four times the diameter of the largest conduit
    d. must be six times the diameter of the smallest conduit

Trade Terms Quiz

Fill in the blank with the correct term that you learned from your study of this module.

1. _________ means constructed or protected so that exposure to beating rain will not result in the entrance of water under specified test conditions.

2. A box designed and constructed to withstand an internal explosion without creating an external explosion or fire is called _________.

3. A box constructed so that moisture will not enter the enclosure under specified test conditions is _________.

4. A(n) _________ is an enclosure used to facilitate the installation of cables from point to point in long runs.

5. A box constructed or protected so that exposure to the weather will not interfere with successful operation is referred to as _________.

6. A(n) _________ is a metallic or nonmetallic box installed in an electrical wiring system from which current is taken to supply some apparatus or device.

7. A single-gang outlet box used for surface mounting to enclose receptacles or wall switches on concrete or concrete block construction is called a(n) _________.

8. A(n) _________ is an enclosure where one or more raceways or cables enter, and in which electrical conductors may be spliced.

9. A(n) _________ is a device used to physically connect conduit or cable to an outlet box, cabinet, or other enclosure.

Trade Terms

Connector
Explosion-proof
Handy box

Junction box
Outlet box
Pull box

Raintight
Watertight
Weatherproof

1. All boxes must be sized, installed, and supported to meet the current ___________ requirements.

2. Blank covers are used when boxes serve as ___________.

3. Each conductor originating outside of the box and terminating inside of the box counts as ___________ conductor(s).

4. When sizing outlet boxes, deduct ___________ conductor(s) for each switch.

5. For a straight pull, the minimum length of the box must be at least ___________ times the trade diameter of the largest raceway.

6. The minimum length for a junction box in which the conductors are pulled at an angle and that contains two trade size 3" conduits and one trade size 4" conduit entering each side is ___________.

7. Octagon and round boxes are used mostly for

    ____.

8. When sizing outlet boxes, ___________ conductor(s) must be deducted for each strap-mounted device.

9. An extension ring is used to increase the ___________ of the box.

10. What is the maximum distance between receptacles along the floor line in any wall space?

Gary Edgington
Baker Electric

What made you decide to become an electrician?

I had an early fascination with electricity and used to work on motorized cars, lights, or about anything else that would plug into the wall or run from a battery. My interest continued through high school. After watching an electrician wire a house, I made up my mind to be in the electrical field somewhere.

How did you learn the trade?

After graduating high school in 1972, I called an electrical contractor, Kinsey Electric, on a regular basis until he hired me. Harry was a seasoned electrical contractor, who did residential, commercial, and agricultural wiring. Because it was a one-man operation, I had an advantage in working with the owner and getting some good hands-on training along with acquiring his good work ethics. Harry was a well-respected person in the community, with a reputation of being the person to call if you wanted it done right. He was my first role model. Along with the electrical training, Harry taught me that if you're going to do a job, be the best at it and take pride in what you do.

What factor or factors have contributed most to your success?

After working with Harry for a few years, I went to work at a manufacturing facility as an industrial electrician. I was fortunate to meet two more role models who greatly influenced me in my career. One was a retired Air Force electrician, who taught me the importance of knowledge and the need to be able to find answers to what you don't know. The second was an electrician from Manchester, England. He helped me recognize my abilities and inspired me to acquire more. Above everything, he gave me confidence in myself which has helped me throughout my life.

What kind of jobs did you hold on the way to your current position?

In addition to my job as an electrician helper/ electrician for Kinsey and as an industrial electrician, I became a licensed electrical contractor in 1978. After being in the electrical contracting industry for 22 years, and employing up to fifty people at times, I sold my business and became an electrical consultant. As the owner of the business I have found myself doing everything from electrical work in the field to estimating to project and business management. During this time and since then I have taught electrical apprentice classes and *National Electrical Code*® classes for thirteen years.

What does an electrical consultant do?

As a consultant I have taught electrical classes and code updates, been involved in the electrical design process, done code review on projects, and have been an expert in court cases.

What advice would you give to someone entering the electrical field?

My advice to someone entering the field is to embrace every opportunity to learn something new, to develop good work practices and ethics, and to think safety all of the time.

What advice do you have for trainees?

As a trainee in this field you need to practice at your profession. Be accurate in your work and pursue knowledge. You will find the ability and the confidence you need to succeed in this field.

What would you say was the single greatest factor that contributed to your success?

The single greatest factor for me has been getting close to a professional in the field and trying to follow in their footsteps.

Connector: Device used to physically connect conduit or cable to an outlet box, cabinet, or other enclosure.

Explosion-proof: Designed and constructed to withstand an internal explosion without creating an external explosion or fire.

Handy box: Single-gang outlet box used for surface mounting to enclose receptacles or wall switches on concrete or concrete block construction of industrial and commercial buildings; nongangable; also made for recessed mounting; also known as a utility box.

Junction box: An enclosure where one or more raceways or cables enter, and in which electrical conductors can be, or are, spliced.

Outlet box: A metallic or nonmetallic box installed in an electrical wiring system from which current is taken to supply some apparatus or device.

Pull box: A sheet metal box-like enclosure used in conduit runs to facilitate the pulling of cables from point to point in long runs, or to provide for the installation of conduit support bushings needed to support the weight of long riser cables, or to provide for turns in multiple conduit runs.

Raintight: Constructed or protected so that exposure to a beating rain will not result in the entrance of water under specified test conditions.

Watertight: Constructed so that moisture will not enter the enclosure under specified test conditions.

Weatherproof: Constructed or protected so that exposure to the weather will not interfere with successful operation.

Additional Resources

This module presents thorough resources for task training. The following resource material is suggested for further study:

National Electrical Code®Handbook, Latest Edition. Quincy, MA: National Fire Protection Association.

Section Review Answer Key

Answer	Section Reference	Objective
Section One		
1. d	1.1.3	1a
2. b	1.2.0	1b
3. d	1.3.0	1c
Section Two		
1. b	2.1.0	2a
2. b	2.2.0	2b

2.0.0 SECTION REVIEW

Question 1

Use the formula for determining box size:

$$\frac{\text{Trade size of conduit} \times 8 \text{ [per } \textit{NEC Section 314.28(A)(1)} \text{]}}{= \text{minimum length of box}}$$

Use the largest trade size to determine the minimum length of the box:

$3" \times 8 = 24"$

The length of the box must be **24"**.

NCCER CURRICULA — USER UPDATE

NCCER makes every effort to keep its textbooks up-to-date and free of technical errors. We appreciate your help in this process. If you find an error, a typographical mistake, or an inaccuracy in NCCER's curricula, please fill out this form (or a photocopy), or complete the online form at **www.nccer.org/olf**. Be sure to include the exact module ID number, page number, a detailed description, and your recommended correction. Your input will be brought to the attention of the Authoring Team. Thank you for your assistance.

Instructors – If you have an idea for improving this textbook, or have found that additional materials were necessary to teach this module effectively, please let us know so that we may present your suggestions to the Authoring Team.

NCCER Product Development and Revision

13614 Progress Blvd., Alachua, FL 32615

Email: curriculum@nccer.org
Online: www.nccer.org/olf

❏ Trainee Guide ❏ Lesson Plans ❏ Exam ❏ PowerPoints Other ________________________

Craft / Level: _______________________________________ Copyright Date: _______________

Module ID Number / Title: ____

Section Number(s): ____

Description: ____

____

____

____

Recommended Correction: ____

____

____

____

Your Name: ____

Address: ____

____

Email: ____ Phone: ___________________

Glossary

90° bend: A bend that changes the direction of the conduit by 90°.

Accessible: Able to be reached, as for service or repair.

Ammeter: An instrument for measuring electrical current.

Amperes (A): The basic unit of measurement for electrical current, represented by the letter A

Approved: Meeting the requirements of an appropriate regulatory agency.

Architect's scale: A special ruler with various measurement scales that can be used when drafting or making measurements on architectural drawings. Architect's scales measure distances in inches.

Architectural drawings: Working drawings consisting of site plans, floor plans, elevations, sectional views, details, and other information necessary for the construction of a building.

Atoms: The smallest particles to which an element may be divided and still retain the properties of the element.

Back-to-back bend: Any bend formed by two 90° bends with a straight section of conduit between the bends.

Backfeed: The reverse flow of electrical power in a distribution system caused by power induced into the system from outside sources, such as when using a generator during outages or when using a solar photovoltaic (PV) system. Backfeeds can also be caused by equipment overloads.

Battery: A DC voltage source consisting of two or more cells that convert chemical energy into electrical energy.

Block diagram: A single-line diagram used to show electrical equipment and related connections. See power-riser diagram.

Blueprint: An exact copy or reproduction of an original drawing.

Bonding wire: A wire used to make a continuous grounding path between equipment and ground.

Cable trays: Rigid structures used to support electrical conductors.

Charge: A quantity of electricity that is either positive or negative.

Circuit: A complete path for current flow.

Coil: A number of turns of wire, especially in spiral form, used for electromagnetic effects or for providing electrical resistance.

Concentric bends: 90° bends made in two or more parallel runs of conduit with the radius of each bend increasing from the inside of the run toward the outside.

Conductors: Materials through which it is relatively easy to maintain an electric current.

Conduit: A round raceway, similar to pipe, that houses conductors.

Connector: Device used to physically connect conduit or cable to an outlet box, cabinet, or other enclosure.

Continuity: An electrical term used to describe a complete (unbroken) circuit that is capable of conducting current. Such a circuit is also said to be closed.

Coulomb: A unit of electrical charge equal to 6.25×10^{18} electrons (or 6.25 quintillion electrons). A coulomb is the common unit of quantity used for specifying the size of a given charge.

Current: The movement, or flow, of electrons in a circuit. Current (I) is measured in amperes.

d'Arsonval meter movement: A meter movement that uses a permanent magnet and moving coil arrangement to move a pointer across a scale.

Detail drawing: An enlarged, detailed view taken from an area of a drawing and shown in a separate view.

Developed length: The actual length of the conduit that will be bent.

Dimensions: Sizes or measurements printed on a drawing.

Double-insulated/ungrounded tools: Electrical tools that are constructed so that the case is insulated from electrical energy. The case is made of a nonconductive material.

Electrical drawing: A means of conveying a large amount of exact, detailed information in an abbreviated language. Consists of lines, symbols, dimensions, and notations to accurately convey an engineer's designs to electricians who install the electrical system on a job.

Electrical service: The electrical components that are used to connect the serving utility to the premises wiring system.

Electrons: Negatively charged particles that orbit the nucleus of an atom.

Elevation drawing: An architectural drawing showing height, but not depth; usually the front, rear, and sides of a building or object.

Engineer's scale: A special ruler with various measurement scales that can be used when drafting or making measurements on architectural drawings. Engineer's scales measure distances in decimal units.

Explosion-proof: Designed and constructed to withstand an internal explosion without creating an external explosion or fire.

Exposed location: Not permanently closed in by the structure or finish of a building; able to be installed or removed without damage to the structure.

Fibrillation: Very rapid irregular contractions of the muscle fibers of the heart that result in the heartbeat and pulse going out of rhythm with each other.

Floor plan: A drawing of a building as if a horizontal cut were made through a building at about window level, and the top portion removed. The floor plan is what would appear if the remaining structure were viewed from above.

Frequency: The number of cycles completed each second by a given AC voltage; usually expressed in hertz. One hertz equals one cycle per second.

Gain: Because a conduit bends in a radius and not at right angles, the length of conduit needed for a bend will not equal the total determined length. Gain is the distance saved by the arc of a 90° bend.

Ground fault circuit interrupter (GFCI): A protective device that functions to de-energize a circuit or portion thereof within an established period of time when a current to ground exceeds some predetermined value. This value is less than that required to operate the overcurrent protective device of the supply circuit.

Grounded tool: An electrical tool with a three-prong plug at the end of its power cord or some other means to ensure that stray current travels to ground without passing through the body of the user. The ground plug is bonded to the conductive frame of the tool.

Handy box: Single-gang outlet box used for surface mounting to enclose receptacles or wall switches on concrete or concrete block construction of industrial and commercial buildings; nongangable; also made for recessed mounting; also known as a utility box.

Insulator: A material through which it is difficult to conduct an electric current.

Joule (J): A unit of measurement for doing work, represented by the letter J. One joule is equal to one newton-meter (Nm).

Junction box: An enclosure where one or more raceways or cables enter, and in which electrical conductors can be, or are, spliced.

Kick: A bend in a piece of conduit, usually less than 45°, made to change the direction of the conduit.

Kilo: A prefix used to indicate one thousand (for example, one kilowatt is equal to one thousand watts).

Kirchhoff's current law: The statement that the total amount of current flowing through a parallel circuit is equal to the sum of the amounts of current flowing through each current path.

Kirchhoff's voltage law: The statement that the sum of all the voltage drops in a circuit is equal to the source voltage of the circuit.

Matter: Any substance that has mass and occupies space.

Mega: A prefix used to indicate one million; for example, one megawatt is equal to one million watts.

Neutrons: Electrically neutral particles (neither positive nor negative) that have the same mass as a proton and are found in the nucleus of an atom.

Nucleus: The center of an atom. It contains the protons and neutrons of the atom.

Occupational Safety and Health Administration (OSHA): The federal government agency established to ensure a safe and healthy environment in the workplace.

Offset: An offset is two bends placed in a piece of conduit to change elevation to go over or under obstructions or for proper entry into boxes, cabinets, etc.

Ohm's law: A statement of the relationships among current, voltage, and resistance in an electrical circuit: current (I) equals voltage (E) divided by resistance (R). Generally expressed as a mathematical formula: $I = E/R$.

Ohmmeter: An instrument used for measuring resistance.

Ohms (Ω): The basic unit of measurement for resistance, represented by the symbol Ω.

On-the-job learning (OJL): Job-related learning an apprentice acquires while working under the supervision of journey-level workers. Also called on-the-job training (OJT).

One-line diagram: A drawing that shows, by means of lines and symbols, the path of an electrical circuit or system of circuits along with the various circuit components. Also called a single-line diagram.

Outlet box: A metallic or nonmetallic box installed in an electrical wiring system from which current is taken to supply some apparatus or device.

Parallel circuits: Circuits containing two or more parallel paths through which current can flow.

Plan view: A drawing made as though the viewer were looking straight down (from above) on an object.

Polychlorinated biphenyls (PCBs): Toxic chemicals that may be contained in liquids used to cool certain types of large transformers and capacitors.

Power-riser diagram: A single-line block diagram used to indicate the electric service equipment, service conductors and feeders, and subpanels. Notes are used on power-riser diagrams to identify the equipment; indicate the size of conduit; show the number, size, and type of conductors; and list related materials. A panelboard schedule is usually included with power-riser diagrams to indicate the exact components (panel type and size), along with fuses, circuit breakers, etc., contained in each panelboard.

Power: The rate of doing work, or the rate at which energy is used or dissipated. Electrical power is measured in watts.

Proton: The smallest positively charged particle of an atom. Protons are contained in the nucleus of an atom.

Pull box: A sheet metal box-like enclosure used in conduit runs to facilitate the pulling of cables from point to point in long runs, or to provide for the installation of conduit support bushings needed to support the weight of long riser cables, or to provide for turns in multiple conduit runs.

Raceway system: An enclosure that houses the conductors in an electrical system (such as fittings, boxes, and conduit).

Raceways: Enclosed channels designed expressly for holding wires, cables, or busbars, with additional functions as permitted in the *NEC*®.

Raintight: Constructed or protected so that exposure to a beating rain will not result in the entrance of water under specified test conditions.

Relays: Electromechanical devices consisting of a coil and one or more sets of contacts. Used as a switching device.

Resistance: An electrical property that opposes the flow of current through a circuit. Resistance (R) is measured in ohms.

Resistors: Any devices in a circuit that resist the flow of electrons.

Rise: The length of the bent section of conduit measured from the bottom, centerline, or top of the straight section to the end of the conduit being bent.

Rough-in: The installation of the raceway system, wiring, or cable.

Scale: On a drawing, the size relationship between an object's actual size and the size it is drawn. Scale also refers to the measuring tool used to determine this relationship.

Schedule: A systematic method of presenting equipment lists on a drawing in tabular form.

Schematic diagram: A detailed diagram showing complicated circuits, such as control circuits.

Schematic: A type of drawing in which symbols are used to represent the components in a system.

Sectional view: A cutaway drawing that shows the inside of an object or building.

Segment bend: A large bend formed by multiple short bends or shots.

Series circuit: A circuit that has only one path for current flow.

Series-parallel circuits: Circuits that contain both series and parallel current paths.

Shop drawing: A drawing that is usually developed by manufacturers, fabricators, or contractors to show specific dimensions and other pertinent information concerning a particular piece of equipment and its installation methods.

Site plan: A drawing showing the location of a building or buildings on the building site. Such drawings frequently show topographical lines, electrical and communication lines, water and sewer lines, sidewalks, driveways, and similar information.

Solenoids: Electromagnetic coils used to control a mechanical device such as a valve.

Splice: Connection of two or more conductors.

Stub-up: Another name for the rise in a section of conduit at 90°. Also, a term used for conduit penetrating a slab or the ground.

Tap: Intermediate point on a main circuit where another wire is connected to supply electrical current to another circuit.

Transformers: Devices consisting of one or more coils of wire wrapped around a common core. Transformers are commonly used to step voltage up or down.

Trim-out: The installation and termination of devices and fixtures after rough-in.

Trough: A long, narrow box used to house electrical connections that could be exposed to the environment.

Underwriters Laboratories, Inc. (UL): An agency that evaluates and approves electrical components and equipment.

Valence shell: The outermost ring of electrons that orbit about the nucleus of an atom.

Voltage drop: The change in voltage across a component that is caused by the current flowing through it and the amount of resistance opposing it.

Voltage: The driving force that makes current flow in a circuit. Voltage, often represented by the letter E, is also referred to as *electromotive force (emf)*, *difference of potential*, or *electrical pressure*.

Voltmeter: An instrument for measuring voltage. The resistance of the voltmeter is fixed. When the voltmeter is connected to a circuit, the current passing through the meter will be directly proportional to the voltage at the connection points.

Volts (V): The unit of measurement for voltage, represented by the letter V. One volt is equivalent to the force required to produce a current of one ampere through a resistance of one ohm.

Watertight: Constructed so that moisture will not enter the enclosure under specified test conditions.

Watts (W): The basic unit of measurement for electrical power, represented by the letter W.

Weatherproof: Constructed or protected so that exposure to the weather will not interfere with successful operation.

Wireways: Steel troughs designed to carry electrical wire and cable.

Written specifications: A written description of what is required by the owner, architect, and engineer in the way of materials and workmanship. Together with working drawings, the specifications form the basis of the contract requirements for construction.